EFFECTIVE FIELD THEORIES

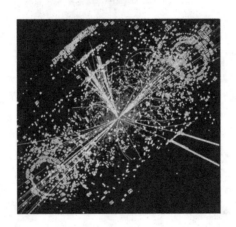

EFFECTIVE FIELD THEORIES

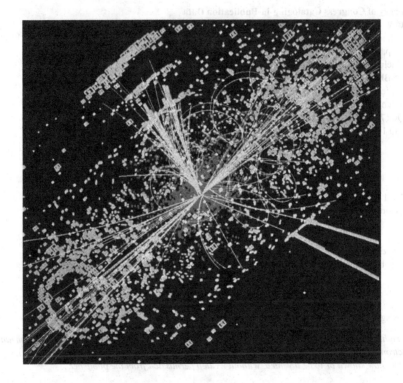

Alexey A. Petrov • Andrew E. Blechman
Wayne State University, USA

Published by

World Scientific Publishing Co. Pte. Ltd.
5 Toh Tuck Link, Singapore 596224
USA office: 27 Warren Street, Suite 401-402, Hackensack, NJ 07601
UK office: 57 Shelton Street, Covent Garden, London WC2H 9HE

Library of Congress Cataloging-in-Publication Data
Petrov, Alexey A., 1971–   author.
  Effective field theories / Alexey A. Petrov, Andrew E. Blechman, Wayne State University, USA.
    pages cm
  Includes bibliographical references and index.
  ISBN 978-9814434928 (hard cover : alk. paper)
  1. Field theory (Physics). 2. Quantum field theory. 3. Gravitational fields.
I. Blechman, Andrew E., author. II. Title.
  QC173.7.P48 2015
  530.14--dc23
                        2015006575

British Library Cataloguing-in-Publication Data
A catalogue record for this book is available from the British Library.

Image credit for book cover: Lucas Taylor. Copyright: CERN,
for the benefit of the CMS Collaboration. Licence: CC-BY-SA-4.0

Copyright © 2016 by World Scientific Publishing Co. Pte. Ltd.

All rights reserved. This book, or parts thereof, may not be reproduced in any form or by any means, electronic or mechanical, including photocopying, recording or any information storage and retrieval system now known or to be invented, without written permission from the publisher.

For photocopying of material in this volume, please pay a copying fee through the Copyright Clearance Center, Inc., 222 Rosewood Drive, Danvers, MA 01923, USA. In this case permission to photocopy is not required from the publisher.

Printed in Singapore

To our families

Preface

It has been the observation of both of the authors that graduate students throughout the world often have a hard time learning how to apply many techniques they pick up in classes to their research. Most universities teach one or two semesters of quantum field theory at the level of, say, [Peskin and Schroeder (1995)], and then students are expected to take off from there. Students usually have to resort to multiple papers and review articles suggested to them by their advisors. Many of these reviews are excellent, but they all have different approaches (and even notations!) and could often lead to more confusion. We were discussing this issue one day, and agreed that it would be much better if there was a book that gave a consistent approach to applying the techniques of Effective Field Theory (EFT). This was the beginning of what would become a long and challenging project.

Our primary motivation for this book is to provide the readers with the basic ideas behind constructing and using effective field theories in elementary particle and nuclear physics, as well as in condensed matter physics. Our starting point was a set of lectures given by one of the authors (AAP) to advanced graduate students at Wayne State University, and a separate set of lectures given by the other author (AEB) to advanced graduate students at the University of Toronto. Each of these lectures focussed on different aspects of how to use EFT techniques in research applications. The first step was to get these notes typed up and synced so that their notations and approaches were consistent; then we began to expand on them. We agreed to divide the types of effective field theory into three forms, based on the approach of constructing the theory. What was originally planned as an 18 month project ended up taking five long years to complete!

Although painful at times, we feel the wait was worth it. The book has been significantly updated and expanded in scope. We tried to make it self-contained, so most of the needed background material can be found in the text, as well as in the Appendices. Still, we strongly recommend that the readers have at least a basic understanding of quantum field theory as given, for example, in the well-known graduate texts such as [Peskin and Schroeder (1995)] or [Ryder (1996)].

We wish to express our appreciation to our colleagues Johannes Heinonen, Tobias Huber, and Gil Paz for carefully reading parts of the manuscript and for providing useful comments that clearly made this book better. We have also benefited greatly from various conversations with Linda Carpenter, Thorsten Feldmann, Sean Gavin, Ben Grinstein, Gordon Kane, Thomas Mannel, George Moschelli, Siew-Phang Ng, Will Shepherd, Brock Tweedie, Mikhail Voloshin, Scott Watson, James Wells, and Jure Zupan.

We also want to thank John F. Donoghue, Adam Falk, Eugene Golowich, Barry Holstein, David E. Kaplan, Mike Luke, Matthias Neubert and Erich Poppitz, who taught us the basics of effective field theory techniques.

AAP would like to thank his wife, Tatiana, and kids Anna and Danil for their help, understanding and support that they provided during the writing of this book. AAP is forever indebted to his mother Ludmila and his father Anatoly for being the prime examples of optimism and kindness. He also would like to thank Jim and Toni Restaneo for their friendship and support.

AEB is eternally grateful to his wife, Carolyn, parents Karl and Rosanne, and in-law relations Richard and Randa Feldman and Caron and Mikey Goldfine, who as a group have been invaluable to keeping his spirits up during the hardest moments of the writing process, and helping to keep things in perspective.

AAP is also grateful to Thomas Mannel and Alexander Khodjamirian at the University of Siegen (Germany), as well as to Andreas Kronfeld and other scientists at the Theory Group of Fermi National Accelerator Laboratory and University of Michigan at Ann Arbor, where parts of this book were written, for their hospitality. AEB also wants to thank Lisa Baker and Jamie Benigna at The Roeper School, who have been beyond understanding and have done much to give him more time to work and complete this project.

A. E. Blechman, A. A. Petrov
Detroit, Michigan, USA
August, 2015

Contents

Preface   vii

1. Introduction   1
   - 1.1 Wherefore EFT?   1
   - 1.2 EFT vs MFT   3
   - 1.3 An example from Newton   4
   - 1.4 A theorem of Weinberg   6
   - 1.5 Organization of the book   7

2. Symmetries   9
   - 2.1 Introduction   9
   - 2.2 Noether's Theorem   9
   - 2.3 Examples of Noether currents   11
   - 2.4 Gauged symmetries and Noether's procedure   13
   - 2.5 Broken symmetries and Goldstone's Theorem   15
     - 2.5.1 Nonrelativistic NG-bosons   18
   - 2.6 The BEHGHK Mechanism of Anderson   20
     - 2.6.1 A little history   21
     - 2.6.2 An example   23
     - 2.6.3 An interlude: superconductivity   25
   - 2.7 CCWZ construction of EFT   27
   - 2.8 Explicit breaking; spurion analysis   30
   - 2.9 Anomalous symmetries   32
     - 2.9.1 Anomalies in the path integral   33
     - 2.9.2 The chiral anomaly and its consequences   35
     - 2.9.3 't Hooft anomaly matching   37

	2.10	Notes for further reading	38
3.	Elementary Techniques		41
	3.1	Canonical (engineering) dimensions	41
		3.1.1 Dimensional analysis	41
		3.1.2 Example: hydrogen atom	44
	3.2	Dimensional transmutation	46
	3.3	Callan-Symanzik Equation	49
	3.4	The renormalization group	53
		3.4.1 Engineering dimensions of $\Gamma^{(n)}$	53
		3.4.2 Physics of the anomalous dimension	54
	3.5	Renormalizability and Effective Field Theory	57
		3.5.1 Dropping renormalizability	57
		3.5.2 Matching	57
	3.6	Subtraction schemes as part of EFT definition	58
	3.7	Decoupling. Appelquist-Carrazone theorem in various schemes	60
	3.8	Notes for further reading	63
4.	Effective Field Theories of Type I		65
	4.1	Introduction	65
	4.2	Real physics: Euler-Heisenberg Lagrangian	70
	4.3	Fermi Theory of Weak Interactions as an effective theory	71
	4.4	Fermi theory to one loop: $\Delta S = 2$ processes in EFT . . .	73
		4.4.1 Setting up the EFT approach	74
		4.4.2 A more detailed calculation	76
	4.5	QCD corrections in EFTs	78
		4.5.1 Matching at one loop in QCD	79
		4.5.2 Renormalization group improvement and EFTs .	83
		4.5.3 Complete basis. Penguin operators	85
	4.6	Chiral perturbation theory	88
		4.6.1 Goldstone bosons and their properties	90
		4.6.2 Sources in chiral perturbation theory	93
		4.6.3 Applications: Gell-Mann-Okubo relation	97
		4.6.4 Power counting. Chiral loops and higher orders in χPT	99
		4.6.5 Naive dimensional analysis	101
		4.6.6 Baryons and chiral perturbation theory	103

	4.7	Notes for further reading	104

5. **Effective Field Theories of Type II. Part A.**   107

- 5.1 Introduction ... 107
- 5.2 Heavy Quark Effective Theory ... 108
  - 5.2.1 Quantum mechanics of heavy particles ... 109
  - 5.2.2 From quantum mechanics to field theory: HQET. Field redefinitions ... 111
  - 5.2.3 Spin symmetry and its consequences ... 115
  - 5.2.4 More symmetry: reparameterization invariance ... 117
  - 5.2.5 HQET Green's functions. Radiative corrections ... 118
  - 5.2.6 External currents and external states ... 122
- 5.3 Different degrees of freedom: heavy mesons ... 130
  - 5.3.1 Heavy meson states. Tensor formalism ... 130
  - 5.3.2 Leading-order Lagrangian ... 133
  - 5.3.3 Subleading Lagrangians ... 134
  - 5.3.4 Calculations with HHχPT ... 135
- 5.4 Light baryons in heavy particle formalism ... 139
  - 5.4.1 Leading-order Lagrangian ... 139
- 5.5 Notes for further reading ... 142

6. **Effective Field Theories of Type-II. Part B.**   145

- 6.1 Introduction. Non-relativistic QED and QCD ... 145
- 6.2 NRQCD Lagrangian at the scale m_Q ... 147
- 6.3 Going down: non-perturbative scales $m_Q v$ and $m_Q v^2$ ... 151
  - 6.3.1 Electromagnetic decays of the η_c ... 155
  - 6.3.2 Inclusive decays of the η_c into light hadrons ... 156
  - 6.3.3 Inclusive decays of the χ_{cJ} into light hadrons ... 158
- 6.4 Going down: perturbative scales $m_Q v$ and $m_Q v^2$. pNRQCD ... 160
  - 6.4.1 Example: heavy quarkonium potential ... 161
- 6.5 Different degrees of freedom: hadronic molecules ... 163
- 6.6 Notes for further reading ... 168

7. **Effective Field Theories of Type-III. Fast Particles in Effective Theories**   171

- 7.1 Infrared divergences ... 171
- 7.2 Soft-Collinear Effective Theory ... 174

		7.2.1	Quantum mechanics of fast particles	174

- 7.2.1 Quantum mechanics of fast particles 174
- 7.2.2 SCET power counting 175
- 7.2.3 SCET action 178
- 7.3 Symmetries of SCET 182
  - 7.3.1 Gauge invariance 182
  - 7.3.2 Reparametrization invariance 184
- 7.4 Examples 185
  - 7.4.1 $B \to X_s \gamma$ 185
  - 7.4.2 $B \to D\pi$ 188
  - 7.4.3 Deep Inelastic Scattering 191
  - 7.4.4 Jet production 197
- 7.5 Notes for further reading 199

8. Standard Model as an Effective Field Theory — 203

- 8.1 Introduction 203
- 8.2 Standard model as the leading term in the EFT expansion — 205
  - 8.2.1 Dimension-5 operators: fermion number violation — 206
  - 8.2.2 Dimension-6 operators: parameterizing new physics 208
  - 8.2.3 Experimental tests and observables 209
- 8.3 BSM particles in EFT 211
  - 8.3.1 Dark matter at colliders: effective operators . . . 212
  - 8.3.2 Mono-Higgs signatures of dark matter at LHC . . 213
- 8.4 Notes for further reading 217

9. Effective Field Theories of Gravity — 219

- 9.1 Introduction 219
- 9.2 Review of general relativity 221
  - 9.2.1 Geodesics and affine connection 221
  - 9.2.2 General relativity and the weak field limit 222
  - 9.2.3 Gravity sources: energy-momentum tensor 226
- 9.3 Building an effective field theory 227
  - 9.3.1 Quantization. Feynman rules 228
  - 9.3.2 Quantum EFT for gravity 231
  - 9.3.3 Newtonian potential 233
  - 9.3.4 Postscript 234
- 9.4 Classical EFT: NRGR 235
  - 9.4.1 Setting up the problem 236

	9.4.2	Gravition modes	238
	9.4.3	Feynman rules	241
	9.4.4	Gravitational radiation	242
	9.4.5	Renormalization	245
9.5		Notes for further reading	247

10. Outlook 249

- 10.1 Supersymmetry 249
- 10.2 Extra dimensions 250
- 10.3 Technicolor and compositeness 251
- 10.4 High-T_c superconductivity 252

Appendix A Review of Group Theory 253

- A.1 General definitions 253
- A.2 Continuous groups 255
- A.3 Representation theory of Lie groups 258
- A.4 Young Tableaux for $SU(N)$ 262
- A.5 Group theory coefficients 266
- A.6 Notes for further reading 267

Appendix B Short Review of QED and QCD 269

B.1		Quantum electrodynamics	269
B.2		Quantum chromodynamics	270
	B.2.1	QCD Lagrangian and Feynman rules	271
	B.2.2	Symmetries of the QCD Lagrangian	277

Appendix C Useful Features of Dimensional Regularization 279

- C.1 Overview of dimensional regularization 279
- C.2 Useful formulas 280
- C.3 Dimensional regularization vs other schemes 282
- C.4 Advanced features: scaleless integrals 283
- C.5 Advanced features: integration by parts 285
- C.6 Advanced features: method of regions 287

Bibliography   291

Index   301

Chapter 1

Introduction

> *What's in a name? That which we call a rose*
> *By any other name would smell as sweet.*
>
> W. Shakespeare

1.1 Wherefore EFT?

This book is about effective field theories (EFTs). What makes a field theory effective? Is it better or worse than a "regular" field theory? We shall argue in this book that the way calculations are set up in EFTs makes them the most natural and convenient tools to address multi scale problems. Problems with separated scales often appear in Nature, and we intuitively know that it is most convenient to only work with degrees of freedom that are relevant for a particular scale – otherwise the problem quickly becomes intractable! You never worry about physics of the atoms when designing bridges, nor try to track each and every molecule of a gas through phase space; you instead define some "macroscopic" variables, and once you know how to relate those variables to the more "fundamental" laws, you can stop thinking about these laws and focus only on the relevant large-scale physics. EFT techniques codify this principle when working with problems in quantum field theory.

It is interesting to note that scale separation is very natural in physics. In quantum mechanics, we are not concerned with the value of the top quark mass when we calculate the energy levels of a hydrogen atom. Of course, given certain precision of an experimental measurement, we might want to be concerned about that.[1] Having this in mind, however, we would still insist that only degrees of freedom relevant to the problem in hand are needed to perform the calculation. In the language of quantum field theory this implies that operators that are responsible for experimental observables only include fields describing light degrees of freedom. By doing so we effectively eliminate all heavy particles with masses well above the scale associated with the problem at hand (say, a hydrogen atom).[2] They have not disappeared completely: quantum theory allows the possibility of these particles to be created and destroyed on very short time scales, and this leads to coupling constants and other parameters of our theory changing with scale – and this change could be affected by the removed degrees of freedom. Also, we might improve the accuracy of our predictions by introducing more operators in our theory. This certainly happens when it is not forbidden, for example, by the symmetries of the system we are studying. In a sense, symmetry becomes the guiding principle for our construction of effective description of physical systems: we do not even need to know what heavy particles we integrated out! We can keep adding more operators, provided we know the way to assess the importance of those new contributions – or, colloquially speaking, *power count* them. The structure and coefficients in front of those operators, when fit to experimental data, might tell us something about the heavy particles that we integrated out. Thus, EFTs become a very convenient way of studying new, undiscovered physics.

As we shall see in this book, this is one way of using techniques of effective field theories. There are many others. In the previous paragraph we discussed removing some (heavy) degrees of freedom completely. Instead of doing that, we can remove only "parts" of the fields. That is, we can remove the known solution, say, for a static or fast-moving particle, only concentrating on the corrections to the known result. This might reveal new symmetries of the theory, simplifying the overall description. The EFT method will allow us to do this as well.

---

[1] One of the most precise measurements known so far is the measurement of anomalous magnetic moment of the muon. That measurement is sensitive to the effects of heavy quarks, but not the top quark.

[2] Thinking in terms of a path integral formulation of a field-theoretic problem, we *integrate out* all heavy degrees of freedom.

1.2 EFT vs MFT

In condensed matter physics, scientists have often employed the technique of "Mean Field Theory" (MFT), and for those that are familiar with it, you might think that we are simply re-casting an old idea in new clothes. It is certainly true that MFT and EFT have a common heritage, and they borrow a lot from each other; but there are a few philosophical differences between the two approaches that should be understood from the outset.

In MFT, the program is to try and calculate a background state of the dynamical degrees of freedom, which is an average, or "mean" field. This quantity is usually called the *order parameter*. You can set up your action (or in condensed matter systems, free energy) as a function of this order parameter, and use the result to make predictions. Like EFT, there are several ways to set up this free energy functional. One approach is to attempt to compute the order parameter directly and then re-express the free energy as a function of this mean field; this is the so-called *Bragg-Williams MFT*. Another thing you can try to do is identify the order-parameter and symmetries using physical arguments, and then write down the most general free energy as a function of this order parameter that is consistent with all the symmetries you identified; this is the so-called *Landau MFT*. There are other approaches to MFT that have been tried and tested over the last century as well. A nice review of these approaches to MFT is given in [Chaikin and Lubensky (1995)].

MFT has its advantages and disadvantages. It is simple and intuitive, and it often does a great job making *qualitative* predictions, such as the general structure of the phase diagram for a system. It also tends to do a good job making quantitative predictions far away from any phase transition or other breakdown of the assumptions that go into constructing it. However, MFT explicitly ignores fluctuations away from the mean field. For a system in equilibrium, this is not generally a problem, although there is a certain limit of accuracy; but when describing a phase transition, for example, it leads to nonsense! Phase transitions are precisely the point at which fluctuations can dominate the system, and as a result, MFT calculations of things like critical exponents are often far from the mark. At this point, physicists need to look elsewhere. There are many clever approaches that have been developed to correct for MFT's missing information. For a wonderful review of these methods, see [Zinn-Justin (2002); Parisi (1988)], for example.

EFT does not share the problems of MFT, at least not on the surface.

One can think of MFT as a "leading-order" EFT result. Many of the methods we alluded to in the above paragraph are actually built into EFT's general make-up. Furthermore, EFT takes many of the ideas from quantum field theory, such as Feynman diagrams, path integrals and renormalization, and builds it in directly to the theory, rather than as an add-on to correct for MFT's deficiencies. In short, EFT is a more general approach to a problem than MFT.

Of course, since both methods are applied to similar problems, they often borrow from each other, and so becoming an expert in EFT will help you to have a better understanding and appreciation for the MFT approach to condensed matter systems. As both of the authors of this book are trained as particle physicists, most of our examples are focussed in that direction. Nevertheless, we hope that seeing how the techniques are generally applied will help readers from many different fields have a better understanding of how to perform effective calculations, and the power you gain by casting your theory in this language.

1.3 An example from Newton

As a simple example that nevertheless shows many features of a real application of effective Lagrangians, let us consider Galileo Galilei's Leaning Tower of Pisa experiment. According to Galileo's student, Vincenzo Viviani, Galileo dropped balls of different mass m from the Leaning Tower of Pisa. We can write a Lagrangian for one of the balls,

$$L = \frac{mv^2}{2} - V(h) = \frac{mv^2}{2} - mgh, \quad (1.1)$$

where g is the free-fall acceleration, h is the ball's height, say, above the ground, and $v = \dot{h}$ is the velocity of the ball,.

We argue that Eq. (1.1) represents an *effective Lagrangian* to the full mechanical description of this problem. We are all taught that the zero level of potential energy $V(h)$ can be chosen arbitrarily, it is only the potential difference that is physical. Since constant terms can be dropped in the definition of a Lagrangian in Eq. (1.1), this fact is represented by a manifest shift symmetry, $h \to h + a$, where a is constant distance. Another way of saying this is that the force acting on the ball, $F = mg$, is independent of the height of the tower.

Now this is not, strictly speaking, correct! If the Leaning Tower of Pisa is moved to the top of Mount Everest, the force and the potential energy would change, simply because free-fall acceleration, $g(R) = GM/R^2$, depends on how high the object is located above the Earth.[3] In fact, one can show that $g(R)$ satisfies the following differential equation,

$$R\frac{\partial}{\partial R}g(R) = \gamma_g g(R), \qquad (1.2)$$

with $\gamma_g = -2$. This equation looks very much like a renormalization group equation.[4]

We already know that Eq. (1.1) is not an exact expression, in fact, it is an approximation. One can try to make it better by observing that the radius of the Earth is much bigger than the height of the Leaning Tower of Pisa, i.e. $h/R \to 0$. A better approximation to a true potential can take a form of a power expansion,

$$V(h) = C_1(R)\, m\left(\frac{h}{R}\right) + C_2(R)\, m\left(\frac{h}{R}\right)^2 + \cdots . \qquad (1.3)$$

Here $C_i(R)$ are unknown coefficients, which can be found if precise experimental data is available. In principle, an exact potential can be guessed if a sufficient number of terms in the expansion of Eq. (1.3) is determined.

In the case at hand, however, we can do better. Indeed, as Sir Isaac Newton tells us, the *full theory* is described by a Newtonian interaction potential between the Earth of mass M and a ball, separated by distance r,

$$V(h) = G\frac{Mm}{r} = G\frac{Mm}{R+h}, \qquad (1.4)$$

which, however, does not have a manifest symmetry $h \to h+a$ that the *effective* Lagrangian of Eq. (1.1) possesses. Expanding $V(R)$ in h/R and *matching* Eqs. (1.3) and (1.4) we can determine the unknown coefficients,

$$C_1(R) = -C_2(R) = \ldots = \frac{GM}{R}, \qquad (1.5)$$

which results in the expansion of the potential,

$$V(h) = \frac{GM}{R} m\left(\frac{h}{R}\right) - \frac{GM}{R} m\left(\frac{h}{R}\right)^2 + \cdots$$
$$= mgh - \frac{mg}{R}h^2 + \cdots, \qquad (1.6)$$

where we dropped the constant term. We indeed found that the leading term in Eq. (1.6) is the potential term of the original Lagrangian of Eq. (1.1). It is interesting that even in such a simple example we performed all the steps needed in a derivation of a classical effective Largangian! It simply shows that using effective Lagrangian techniques is very natural.

---

[3] Indeed, g is not the same at the top and at the bottom of the Leaning Tower of Pisa. In making this argument we neglected the variation of g along the height h.

[4] This curious observation was pointed out to us by T. Huber.

1.4 A theorem of Weinberg

Much more useful for practical calculations is the notion of *quantum* effective Lagrangians. A theoretical basis of any quantum effective field theory can be formulated in terms of a theorem, first given by S. Weinberg [Weinberg (1979b)]

Theorem 1.1. *To any given order in perturbation theory, and for a given set of asymptotic states, the most general possible Lagrangian containing all terms allowed by the assumed symmetries will yield the most general S-matrix elements consistent with analyticity, perturbative unitarity, cluster decomposition and assumed symmetry principles.*

Initially, this theorem was written to conjecture the equivalence of current algebra results and methods of effective Lagrangians in pion physics, where it was shown to work in all cases. Nothing in this theorem says that it is only applicable to pion physics, so it is expected that this theorem should work in any EFT. So far, there are no known counterexamples of this theorem.

Theorem 1.1 is very plausible, but not as trivial as it seems. As we know from our quantum field theory courses, an important part of quantum theory is renormalization, i.e. a correct treatment of the unknown behavior of that theory at ultra-small scales. According to the theorem, we need to write the most general set of operators in a Lagrangian consistent with given symmetries in order to get the most general S-matrix elements. That set would surely contain a very large number of operators; moreover, in general the number of such operators is infinite! How can one make any predictions when there are an infinite number of contributions to take into account?

We will see that the situation is not hopeless: for a given precision of measurements, effective field theories will provide consistent and testable predictions even if they are not renormalizable in a "classical" sense. We shall show that computing quantum loops in effective theories is exactly the same as it is in (conventionally) renormalizable field theories: we would still integrate over *all* values of momentum, even if our EFT is only valid to some momentum scale μ.[5] Also, even in quantum loops, we would only deal with the degrees of freedom given in the effective Lagrangian, i.e the ones with which we started our calculations. While this might appear strange, as the integration over momenta greater that μ certainly misses

---

[5]Since we are using a "natural" system of units where we set $\hbar = c = 1$, we shall use the notions of energy and momentum interchangeably.

some physics, it is actually quite natural – and not different from what is done in conventional QFTs. After all, if structures of the terms generated by loop effects are the same as the structures already present in the original local Lagrangian, those "incorrect" loop contributions would be corrected by shifting the parameters of the Lagrangian, i.e. by renormalization! Thus, in order to execute the EFT program, we will need to find a way to assess the importance of loop-induced contributions in our calculations, and, if needed, introduce proper counterterms to render the result finite. That is, we would have to find a proper *power counting* scheme.

1.5 Organization of the book

As mentioned above, there are many ways to construct an effective field theory, depending on what it is you are trying to describe. We find it convenient to group different EFTs into one of three categories, based on what degrees of freedom they include.[6]

The first kind of EFT we discuss, which we call "Type-I," refers to the famous classic example of an effective field theory, where the list of degrees of freedom only includes those fields that can contribute at the energy and momentum scale of the interaction. For example, beta decay is described by a theory that never makes any mention of the W-boson; atomic and molecular physics does not make use of quarks; standard-model particle physics processes make no mention of any super-heavy particle such as GUT remnants; etc. These EFTs are the most straightforward examples, and we will use them to start the discussion.

The second kind of EFT, which we call "Type-II," refers to problems where some fields no longer participate in *dynamics*, but they are still a part of the Fock space. These objects are taken as (nearly) infinitely heavy, and simply sit there while other lighter degrees of freedom are bouncing off of it in totally elastic collisions. One example of a Type-II theory is the Newtonian example we considered earlier. In that problem, the Earth was taken as infinitely massive, and therefore does not recoil when objects hit it, but it would be wrong to say that we have "integrated out the Earth!" The function of the Earth was to provide the gravitational field (if you wish, it is a massive, static reservoir of gravitons!) that was represented by the potential in the Lagrangian. Other examples of Type-II EFT are when you

---

[6]Similar suggestions for the breakdown into EFT-types were made in a talk by Michael Luke at SCET2007 conference at MIT.

have a bound state between a heavy particle and a light particle, like the Hydrogen atom. In that case, the proton is a source of photons (generating an electrostatic field), but we ignore its recoil as it interacts with the lighter electron. We can then reincorporate that recoil as subleading terms in our effective Lagrangian (just like we included higher-order terms in h/R with the Earth example.

The third kind of EFT, called "Type-III," is the newest and also the most controversial construction. This is an attempt to describe objects that have large energy-momentum transfers, but *only in a given, fixed direction*. This implies that we need to attempt a Type-II construction, but *only on the components of the field that can create the large momentum*. This means that, roughly speaking, we should integrate out the part of the field with momentum in the z-direction, but leave the parts of the field that create particles moving in the x, y-directions in the dynamics. This is a strange theory, to say the least, and it is still not clear whether such a construction is even self-consistent. From the point of view of studying EFT for its own sake, Type-III is definitely "where the action is!" But whatever its philosophical and technical problems might be, its prime example of Soft-Collinear Effective Theory (SCET) has proven to be incredibly useful in helping us understand heavy-to-light particle decays; parton showering; event-shape distributions; IR factorization theorems; the list goes on!

While it is important to have a basic knowledge of quantum field theory techniques, we made an attempt to make this book self-consistent by providing all of the needed background and introductory material. Group theory techniques are reviewed in Appendix A; QED and QCD are briefly discussed in Appendix B; and the ideas and some more advanced uses of Dimensional Regularization, one of the most popular regulators in EFT, are reviewed in Appendix C; so if the readers feel like some brush-up is needed, those resources are there for their convenience.

In the following chapters we shall provide needed background, discuss the "mechanics" of EFT building and then consider several classes of EFTs. We suggest the readers try and solve problems at the end of the chapters. After each chapter we also provided references for further studies of the topics discussed in the chapter.

We employ the Minkowski metric with the mostly-minus sign convention $g_{\mu\nu} = \text{diag}(1, -1, -1, -1)$ throughout this book.

Chapter 2

Symmetries

2.1 Introduction

The most vital part of effective field theory is knowing what symmetries apply to your system. This knowledge can get you very far in describing the nature of the problem, constructing model-independent equations to describe dynamics, and put constraints on matrix elements. This chapter will be a review of some of the more important results that follow from symmetries. We will discuss Noether's and Goldstone's theorem, and the consequences that arise from them. We will discuss various examples, but for simplicity we will stick mostly with scalar fields wherever we can, avoiding fermions until we need them for anomalies.

2.2 Noether's Theorem

The chief reason why symmetries are important is due to a theorem in Lagrangian mechanics known as *Noether's Theorem*:

Theorem 2.1 (Noether). *Every continuous symmetry of the action (and path integral measure) implies a conservation law.*

The caveat about the measure being invariant is important to handle the possibility of quantum anomalies, as we will see at the end of this chapter.

Proof. Consider a Quantum Field Theory (QFT) with fields ϕ^a and action $S[\phi]$ that is presumed to be invariant under a *global* transformation of the form:

$$\phi^a \longrightarrow \phi^a + \epsilon \Delta \phi^a \qquad (2.1)$$

where ϵ is an infinitesimal parameter. Although the action is only supposed to be invariant under these transformations for constant ϵ, let us consider the behavior of the action under the above field transformation when ϵ is allowed to vary with space-time:

$$S[\phi + \epsilon(x)\Delta\phi] = \int d^d x\ \mathcal{L}(\phi + \epsilon(x)\Delta\phi, \partial_\mu\phi + \epsilon(x)\Delta\partial_\mu\phi)$$

$$= \int d^d x\ \mathcal{L}(\phi, \partial_\mu\phi) + \int d^4 x \left\{\epsilon\Delta\phi^a \frac{\partial\mathcal{L}}{\partial\phi^a} + \partial_\mu(\epsilon\Delta\phi^a)\frac{\partial\mathcal{L}}{\partial\partial_\mu\phi^a}\right\}$$

$$= S[\phi] + \int d^d x (\partial_\mu \epsilon)\Delta\phi^a \frac{\partial\mathcal{L}}{\partial\partial_\mu\phi^a}$$

$$+ \int d^d x\ \epsilon \left\{\frac{\partial\mathcal{L}}{\partial\phi^a}\Delta\phi^a + \frac{\partial\mathcal{L}}{\partial\partial_\mu\phi^a}\Delta\partial_\mu\phi^a\right\}$$

(2.2)

The last term is there regardless of whether ϵ is constant or a function of space-time, so if $S[\phi]$ is to be invariant under the global transformation in Eq. (2.1) this term must take the form $\int d^d x \epsilon \partial_\mu \mathcal{J}^\mu$. The quantity $\mathcal{J}^\mu$ will generally be nonzero if $\Delta\phi$ explicitly involves changes in space-time coordinates, such as translations and rotations.

Integrating the second term by parts and dropping a surface term, we have:

$$S[\phi + \epsilon(x)\Delta\phi] = S[\phi] - \int d^d x\ \epsilon(x)\ \partial_\mu \left[\Delta\phi^a \frac{\partial\mathcal{L}}{\partial\partial_\mu\phi^a} - \mathcal{J}^\mu\right] \quad (2.3)$$

If we insert this into the path integral and use the invariance of the measure we get:

$$Z = \int \mathcal{D}\phi\ e^{iS[\phi]} = \int \mathcal{D}(\phi + \epsilon\Delta\phi)\ e^{iS[\phi + \epsilon\Delta\phi]}$$

$$= \int \mathcal{D}\phi\ e^{iS[\phi]}\ e^{-i\int d^d x \epsilon \partial_\mu j^\mu} \quad (2.4)$$

where in the first line we shifted the dummy functional integration variable $\phi(x)$ and in the second step we have defined

$$j^\mu = \Delta\phi^a \frac{\partial\mathcal{L}}{\partial\partial_\mu\phi^a} - \mathcal{J}^\mu\ . \quad (2.5)$$

Since ϕ is a dummy field variable, Z must be independent of ϵ and so we can write:

$$0 = \frac{i}{Z}\frac{\delta Z}{\delta \epsilon}\bigg|_{\epsilon=0} = \frac{\int \mathcal{D}\phi\ (\partial_\mu j^\mu) e^{iS}}{\int \mathcal{D}\phi\ e^{iS}} = \langle \partial_\mu j^\mu \rangle\ . \quad (2.6)$$

This is just the correspondence principle at work: j^μ is the classically conserved Noether current, and it is conserved quantum mechanically in all correlation functions. □

Eq. (2.5) is the Noether current corresponding to the symmetry transformation in Eq. (2.1). Throughout most of this book we will be concerned with internal symmetries, where you can set $\mathcal{J}^\mu = 0$. In that case, the Noether current only involves kinetic (gradient) terms in the Lagrangian. This tells us that Noether currents are *universal* in the sense that they are independent of the potential (non derivative) interactions in the action. Thus, even if a symmetry is violated by potential terms, one can still construct a Noether current: *the symmetry structure of any quantum field theory is dictated by the kinetic terms alone!* Although contact interactions can (and often do) break symmetries, we can still identify a Noether current and treat the symmetry breaking effects as corrections to the theory.

Noether currents are not uniquely defined: we could construct a new current from the old one:

$$j_2^\mu = j_1^\mu + \partial_\nu K^{\mu\nu} \qquad (2.7)$$

where j_1^μ is the current in Eq. (2.5) and $K^{\mu\nu} = -K^{\nu\mu}$ is an arbitrary antisymmetric two-indexed tensor. Both currents are locally conserved, as can be easily checked. What is the meaning of j_2^μ? To answer that, consider the charge that is conserved:

$$Q_1 = \int d^{d-1}\vec{x}\, j_1^0 \qquad (2.8)$$

$$Q_2 = \int d^{d-1}\vec{x}\, j_2^0 = \int d^{d-1}\vec{x}\, (j_1^0 + \partial_\nu K^{0\nu}) = Q_1 + \int d^{d-1}\vec{x}\, \partial_i K^{0i}$$

$$= Q_1 + \int_\infty d\mathbf{S} \cdot \mathbf{K} \qquad (2.9)$$

So the extra term represents a surface charge at infinity. So long as we are considering the case that all charges are confined to the interior of our space-time, this extra term has no physical effects. However, in cosmology or condensed matter systems, where boundary effects might be important, you must remember to take this term into account.

2.3 Examples of Noether currents

Let us consider one of the most important starting actions for an effective field theory of any kind: that of a (complex) scalar field with arbitrary potential that only depends on the modulus of the field:

$$S[\phi] = \int d^d x\, \{(\partial_\mu \phi^*)(\partial^\mu \phi) - \mathcal{V}(|\phi|)\} \qquad (2.10)$$

This action is invariant under a constant rephasing of the scalar field, which in infinitesimal form is

$$\phi \longrightarrow \phi + i\epsilon\phi \tag{2.11}$$

In the notation of the previous section, $\Delta\phi = i\phi$ and so the Noether current is

$$\begin{aligned}j^\mu &= \Delta\phi^a \frac{\partial \mathcal{L}}{\partial \partial_\mu \phi^a} \\ &= (i\phi)(\partial^\mu \phi^*) + (-i\phi^*)(\partial^\mu \phi) \\ &= -i(\phi^* \partial^\mu \phi - \phi \partial^\mu \phi^*)\end{aligned} \tag{2.12}$$

Notice that we have to consider the transformation of every field – in this case, that means *both* ϕ and ϕ^*. Also note that this result is independent of the choice of $\mathcal{V}(\phi)$ as we previously advertised.

The above calculation generalizes to the case of N scalar fields[1] that can transform into each other and still leave the action invariant:

$$S[\phi^a] = \int d^d x \, \frac{1}{2} g_{ab} (\partial_\mu \phi^a)(\partial^\mu \phi^b) \tag{2.13}$$

where g_{ab} is a real, nonsingular $N \times N$ matrix, and we're considering real fields for concreteness, although you can generalize to complex fields if you like. This action will be invariant under the transformation

$$\phi^a \longrightarrow \phi^a + i\epsilon^A [T^A]^a_{\ b} \phi^b \tag{2.14}$$

where ϵ^A is a vector of infinitesimal parameters, and iT^A are a set of real, $N \times N$ matrices that satisfy the rule:

$$g_{ac} g^{db} [T^A]^c_{\ d} = -[T^A]^b_{\ a} \tag{2.15}$$

Here we use the usual Einstein convention that g^{ab} is the inverse matrix of g_{ab}, and repeated upper-lower indices are summed. We will never lower the capital letter indices, however.

Since there is one symmetry transformation for each A, there is a corresponding current. We can group them together from the arguments above to find

$$j^{\mu A} = \phi_a [T^A]^a_{\ b} \partial^\mu \phi^b \tag{2.16}$$

---

[1] We will index the fields by lowercase letters from the front of the alphabet. Upper case letters will be used to label the matrices; note that a set of $N \times N$ matrices are spanned by N^2 matrices, although depending on the details of the action, the precise form of g_{ab}, and any additional terms, not all of these matrices need lead to symmetries.

where $\phi_a \equiv g_{ab}\phi^b$. Usually we will suppress the lowercase indices as understood, but we include them above so you can see how they contract. In an index-free notation:
$$j^{\mu A} = \overline{\phi} T^A \partial^\mu \phi \tag{2.17}$$
Let us consider one more example: the real (free) scalar field
$$S[\theta] = \int d^d x \, \frac{1}{2}(\partial_\mu \theta)(\partial^\mu \theta) \tag{2.18}$$
This action is invariant under the shift symmetry:
$$\theta(x) \longrightarrow \theta(x) + \epsilon v \tag{2.19}$$
where v is a constant with the same units as θ. Then $\Delta\theta = v$ and the Noether current is
$$j^\mu = v \partial^\mu \theta \tag{2.20}$$
The conserved current is just the gradient of the scalar field! We will see how this current appears in applications shortly, as well as the meaning of the number v introduced in the transformation rule.

2.4 Gauged symmetries and Noether's procedure

When a transformation like Eq. (2.1) remains a symmetry of the action even when the parameter ϵ is a function of space-time, then the symmetry is said to be a *local* or *gauged* symmetry. There are a few general points about gauging a symmetry that should be pointed out:

(1) Gauge symmetries imply global symmetries, since we can always take $\epsilon(x)$ to be a constant. Thus everything mentioned previously about Noether's theorem still holds.
(2) The "gauge" part of the symmetry is strictly speaking not a *symmetry* but a *redundancy* coming from the fact that (in $d = 4$) a massless spin-1 field has two degrees of freedom, but is described by a four-vector potential that allows you to maintain Lorentz covariance. "Gauge invariance" is simply the statement that the unphysical degrees of freedom do not contribute to any physical amplitude.
(3) For a global symmetry, Noether's theorem is only valid *on-shell* – the quantum version of the theorem we proved above only had the current being conserved within correlation functions, not as an operator equation. However, when the symmetries are gauged, Noether currents are conserved *as operators!* Thus people sometimes say that gauged symmetries are more "fundamental" than global symmetries.

Recall that when we promoted a global transformation to a local one we generated an extra term that was proportional to the Noether current:

$$S_0[\phi + \epsilon(x)\Delta\phi] = S_0[\phi] + \int d^d x \, (\partial_\mu \epsilon) j^\mu \qquad (2.21)$$

For the action to remain invariant under this local symmetry we must include a term that cancels this; we accomplish this by introducing a four-vector potential A_μ that transforms as

$$A_\mu \longrightarrow A_\mu + \partial_\mu \epsilon \qquad (2.22)$$

and couple it to the Noether current:

$$S_1[\phi, A_\mu] = -\int d^d x \left[A_\mu j^\mu + \frac{1}{4} F_{\mu\nu} F^{\mu\nu} \right] \qquad (2.23)$$

The first term cancels the shift in S_0 upon transforming ϕ, while the second term is the usual kinetic energy operator for A_μ. In this simple case, $F_{\mu\nu}$ and j^μ are already invariant under the combined transformation of Eqs. (2.1) and (2.22) and so the total action $S_0 + S_1$ is the minimum extension of the globally invariant action that is also invariant under the gauge symmetry.

This procedure of correcting the global action step by step by adding terms designed to cancel changes from the modified transformation law is known as the *Noether's procedure* and is a very powerful way of building gauge invariant actions from knowledge of globally invariant ones. This case we have considered is an example of an Abelian gauge theory like Maxwell's electromagnetism. Let us see how to extend this idea to more complicated gauge symmetries.

Consider a collection of vector fields A_μ^a $a = 1, \ldots, N$ which transform in the adjoint representation of some Lie algebra[2]:

$$A_\mu^a \longrightarrow A_\mu^a + g c^{abc} \epsilon^b A_\mu^c \quad \Rightarrow \quad \Delta^b A_\mu^a = g c^{abc} A_\mu^c \qquad (2.24)$$

where c^{abc} are the structure constants from Eq. (A.7) and g is a small constant (the gauge coupling). The kinetic term for a collection of spin-1 vector bosons,

$$-\int d^d x \, \frac{1}{4} f_{\mu\nu}^a f^{\mu\nu a}, \qquad f_{\mu\nu}^a \equiv \partial_\mu A_\nu^a - \partial_\nu A_\mu^a \qquad (2.25)$$

is invariant under the global transformation in Eq. (2.24) and generates a Noether current:

$$j^{\mu a} = (g c^{abc} A_\nu^c) \frac{\partial \mathcal{L}}{\partial \partial_\nu A_\mu^a} = g c^{abc} f^{\mu\nu b} A_\nu^c \qquad (2.26)$$

---

[2] See Appendix A.2 for details on algebra theory.

Notice that this current is *not* invariant under the transformation!

Noether's procedure then says that to make this theory invariant under a local transformation, we must add a new term to the action

$$S_1 = -\int d^d x\, j^{\mu a} A^a_\mu \tag{2.27}$$

and also modify the transformation law to

$$A^a_\mu \longrightarrow A^a_\mu + g c^{abc} \epsilon^b A^c_\mu + \partial_\mu \epsilon^a \tag{2.28}$$

$S_0 + S_1$ is now invariant under the transformation of Eq. (2.28) up to first order in g, but not at order g^2:

$$\delta S_0 + \delta S_1 = -\int d^d x\, g^2 c_{abc} c_{bde} A^a_\mu A^c_\nu A^{\nu e} \partial^\mu \epsilon^d \tag{2.29}$$

To cancel this term, one must add another term to the action

$$S_2 = +\int d^d x\, \frac{g^2}{4} c_{abc} c_{bde} A^a_\mu A^c_\nu A^{\nu e} A^{\mu d} \tag{2.30}$$

Now $S_0 + S_1 + S_2$ is invariant under Eq. (2.28) to order g^2, and in fact to all orders in g, so we may stop here. Our final result is in fact the usual gauge invariant action of Yang-Mills:

$$S = -\int d^d x\, \frac{1}{4} F^a_{\mu\nu} F^{\mu\nu a}, \qquad F^a_{\mu\nu} = f^a_{\mu\nu} - g c^{abc} A^b_\mu A^c_\nu \tag{2.31}$$

$$A^a_\mu \longrightarrow A^a_\mu + D_\mu \epsilon^a \tag{2.32}$$

as it should be.

In this case, Noether's procedure terminated at $\mathcal{O}(g^2)$, but that does not have to be the case. The process can go much further, often never terminating at all, and you must modify both the action and the field transformation rules at each order in the coupling. In practice, this is a very powerful way to derive the action of a theory based on knowing the symmetries of the problem. We will see how this works in some exercises.

2.5 Broken symmetries and Goldstone's Theorem

Consider a complex scalar field with a potential of the form

$$V(\phi) = m^2 |\phi|^2 + \lambda |\phi|^4 \tag{2.33}$$

This potential has a symmetry

$$\phi \longrightarrow e^{i\alpha} \phi \tag{2.34}$$

regardless of the value of (m^2, λ). However, the spectrum of the theory is very sensitive to these parameters. In particular, the ground state of the theory where $\mathcal{V}$ is minimized occurs when

$$\frac{\partial \mathcal{V}}{\partial \phi} = \phi^*(m^2 + 2\lambda |\phi|^2) = 0 \Rightarrow |\phi_0|^2 = 0, \ -\frac{m^2}{2\lambda} \quad (2.35)$$

We will assume that $\lambda > 0$ so that $\mathcal{V}$ is bounded from below. Then if $m^2 > 0$, the only minimum is at $\phi_0 = 0$ which is clearly still invariant under Eq. (2.34). On the other hand, if $m^2 < 0$ the minimum actually occurs at the nonzero value, and so the ground state is *not* invariant under the symmetry. The field forms a background *condensate* in the ground state, and we can no longer treat the other fields in the theory as living in a vacuum. When the action of a QFT has a symmetry while the ground state forms a condensate that breaks that symmetry, we say that the symmetry is *spontaneously broken*.

A word of warning: the name is unfortunate, as the symmetry is *not* really broken! The action is still symmetric, and therefore there is still a conserved Noether current (remember that the current was independent of the potential).

We will define the nonzero value of ϕ_0 in the vacuum state as

$$v^2 \equiv -\frac{m^2}{\lambda} \quad (2.36)$$

and write the field as[3]

$$\phi(x) = \frac{1}{\sqrt{2}} (v + \varphi(x)) \quad (2.37)$$

If we plug this back into our potential we find

$$\mathcal{V}(\varphi) = \left[\frac{1}{2} m^2 v^2 + \frac{1}{4} \lambda v^4\right] + \frac{1}{2}(m^2 + \lambda v^2)\left[v(\varphi + \varphi^*) + |\varphi|^2\right]$$
$$+ \frac{\lambda}{4}\left[v^2(\varphi + \varphi^*)^2 + 2v|\varphi|^2(\varphi + \varphi^*) + |\varphi|^4\right] \quad (2.38)$$

The first term in brackets ($\mathcal{V}(\varphi = 0)$) is a constant and corresponds to the value of the potential in the ground state. It gives us information about the condensate, and represents a cosmological constant, but has no other meaning. This term is called the *Mean Field Theory* result in condensed matter systems, and v is the *order parameter*. The second term vanishes for

---

[3]The factor of $1/\sqrt{2}$ is conventional, so $|\phi_0| = v/\sqrt{2}$, but not every author uses this convention – make sure you know what convention is being used to avoid frustrating mistakes!

the value of v in Eq. (2.36), leaving just the third term. We can understand its meaning by decomposing our field even further into real and imaginary components:

$$\varphi(x) = \phi_1(x) + i\phi_2(x) \qquad (2.39)$$

Then our potential is

$$\mathcal{V}(\phi_1, \phi_2) = V_0(v) + \lambda v^2 \phi_1^2 + \lambda v \phi_1 (\phi_1^2 + \phi_2^2) + \frac{\lambda}{4}(\phi_1^2 + \phi_2^2)^2 \qquad (2.40)$$

So upon spontaneous symmetry breaking our theory has three points of interest:

(1) A condensate described by a number v.
(2) A (real) massive scalar field (ϕ_1) with mass-squared equal to $2\lambda v^2$.
(3) A (real) *massless* scalar field (ϕ_2).

The presence of a massless mode is not a coincidence of the problem, nor is it only true classically or at leading order in perturbation theory. It turns out that whenever a symmetry is spontaneously broken, we expect to see a massless particle appear. This is a consequence of *Goldstone's Theorem* which we turn to now.

Theorem 2.2 (Goldstone). *For every spontaneously broken symmetry, there corresponds a massless particle whose quantum numbers are the same as that symmetry's Noether charge.*

The massless modes of the theorem are called **(Nambu-)Goldstone (NG) Fields**.

Proof. Consider a QFT with N fields whose action (and path integral measure) are invariant under a field transformation:

$$\phi^i \longrightarrow \phi^i + i\epsilon T^i_j \phi^j \qquad (2.41)$$

where T is a finite, imaginary matrix. Rather than study the usual action, we will consider the effective action that generates the 1PI correlation functions:

$$\Gamma[\phi] = -\log Z[J] - \int d^d x \, J(x)\phi(x) \qquad (2.42)$$

This will allow us to include quantum effects without any more work. In terms of the effective action, invariance under the transformation in Eq. (2.41) implies

$$\left(\frac{\delta \Gamma}{\delta \phi^n}\right) T^n_m \phi^m = 0 \qquad (2.43)$$

Let us take the functional derivative of this equation and evaluate it at $\phi(x) = \phi_0$, the minimum of the full effective potential

$$\left(\frac{\delta^2 \Gamma}{\delta\phi^l \delta\phi^n}\right)_{\phi=\phi_0} T_m^n \phi^m + \left(\frac{\delta \Gamma}{\delta\phi^n}\right)_{\phi=\phi_0} T_l^n = 0 \qquad (2.44)$$

But since ϕ_0 is the vacuum solution, defined as the minimum of the effective action, the first derivative vanishes, and the second derivative is precisely the two-point 1PI correlation function evaluated with all external momenta equal to zero. This is the definition of the mass matrix M^2 of the theory, and so we have

$$\left[M^2\right]_{ln} T_m^n \phi_0^m = 0 \qquad (2.45)$$

Thus the vectors $T_m^n \phi_0^m$ are eigenvectors of the mass matrix with eigenvalue zero. If $\phi_0 = 0$, then there is nothing to be said; but if $\phi_0 \neq 0$, there must be a zero-eigenvalue of M^2 – that is, a massless mode. Furthermore, if $1 \leq k \leq N$ is the number of nonzero ϕ_0^m, there are k such zero-eigenvectors, and this is also the precise number of broken generators. Thus there is one massless mode for each broken generator, as we wished to show.

We will only sketch the proof of the statement that the quantum numbers of these massless modes must be the same as those of the broken charges. One can use the Källèn-Lehmann spectral techniques to show that

$$\langle B | j^0(x) | 0 \rangle \neq 0 \qquad (2.46)$$

where B is one of the NG modes; see [Weinberg (1996)] for a nice explanation of this calculation. But this matrix element will certainly vanish if the B state has different parity, helicity or internal quantum numbers as the current. Therefore the NG fields must have the same quantum numbers as the broken generators. □

In most cases of interest, the space-time properties of the generators of symmetry groups are simple – they are space-time scalars. Therefore, the NG fields themselves must be spin-0. The famous exception to this is supersymmetry, where the current is a vector-spinor (so the SUSY charges are spinors) and the corresponding NG field is a spinor, affectionately known as a *goldstino*.

2.5.1 Nonrelativistic NG-bosons

Although Goldstone's theorem does not require Lorentz invariance to hold, the form of the theorem and the proof we sketched out is most appropriate for relativistic systems. It turns out that when the system is nonrelativistic, it is still true that each broken generator gives you a "flat direction" in the potential, but the corresponding *gapless excitation*,[4] which we can interpret as a massless (quasi-)particle, can be a bit more tricky to identify. In particular, it is no longer true that each broken generator corresponds to its own NG-boson.

The flaw in our proof comes from how we interpreted Eq. (2.45). We pointed out that if $\phi_0^m \neq 0$, then $T_m^n \phi_0^m$ is an eigenvector of the mass matrix with zero-eigenvalue. But it is also possible that a linear combination of the charges annihilates ϕ_0, which would mean that some of the purported massless modes are actually null-vectors! This happens if the symmetry-breaking field ϕ corresponds to one of the charges. This cannot occur in a Lorentz-invariant theory, but it is quite possible in condensed matter systems.

Perhaps the most famous example of this is the Heisenberg Ferromagnet. The model is a collection of spins on a lattice with Hamiltonian:

$$H = -J \sum_{\langle ij \rangle} \vec{S}_i \cdot \vec{S}_j \; ; \qquad (2.47)$$

The notation $\langle ij \rangle$ refers to sums over neighboring spins. The model has a global $O(3)$ symmetry that rotates all the spins in the same way. J is a constant, taken to be positive for the ferromagnet. The ground state (at zero temperature) is the state where all the spins are aligned in one direction (the z direction, say). This state spontaneously breaks the $O(3)$ to an $O(2)$ symmetry, leaving only rotations about the z axis. The other two rotation generators represent broken symmetries, and therefore you might have expected to find two NG-bosons: gapless states that come from the broken generators acting on the spontaneous magnetization. This state is often called the *magnon* in the condensed matter literature. But there is something very strange: there is only one magnon state, while the Goldstone theorem seems to suggest that there should be two, since two symmetry generators are broken. The resolution to this apparent paradox is that the spontaneous magnetization (ϕ_0) is directly related to the symmetry generators, and so there is a null-vector. That leaves only one eigenvector left as a Goldstone boson.

This point was clarified further in a powerful theorem by [Nielsen and Chadha (1976)]. Their result states that as long as certain conditions are

---

[4]By this we mean a quantum excited state that can have infinitely long wavelength (or low momentum); that is, it obeys a dispersion relation such that $E(\vec{p} \to 0) = 0$.

satisfied by the theory, such as translational invariance and proper asymptotic behavior of the correlation functions, spontaneous symmetry breaking of N generators will always be accompanied by a collection of gapless modes that come in two types:

$$\text{Type}-1 \text{ modes}: \ E(\vec{p}) \propto |\vec{p}|^{2l+1},$$
$$\text{Type}-2 \text{ modes}: \ E(\vec{p}) \propto |\vec{p}|^{2l}.$$

Say that there are n Type-1 modes, and m Type-2 modes; then the Nielsen-Chadha theorem says that $n + 2m = N$. In other words, if the dispersion relation of the NG mode is even in (quasi-)momentum, then it counts twice! Indeed, if you work out the dispersion relation for a Heisenberg Ferromagnet, you will find that $E \to |\vec{p}|^2/2M$, where M is a nonuniversal constant that depends on the details of the model. As the magnon has a quadratic dispersion relation, we see that there is only one magnon in the spectrum, since it counts both of the broken generators.

On the other hand, the Heisenberg antiferromagnet (same as Eq. (2.47), but with $J < 0$) undergoes spontaneous symmetry breaking, where ϕ_0 is the *staggered magnetization*. This is not directly related to the generators and therefore we expect both generators to represent two independent NG modes. Sure enough, there are two gapless modes in the low-energy spectrum with a *linear* dispersion: $E \to \pm C|\vec{p}|$, and so there are two NG bosons, according to the Nielsen-Chadha theorem.

There are never any Type-2 modes in a relativistic theory; indeed the Ferromagnetic magnon dispersion relation violates our understanding of a relativistic dispersion relation, since it seems to represent a particle of effective mass M that is also gapless, violating Einstein's formula, $E = Mc^2$! Clearly, we can only have this sort of thing happen when dealing with nonrelativistic effective field theory.

We will see another example of nontrivial NG counting in the problems.

2.6 The BEHGHK Mechanism of Anderson

In 1964, a series of papers were published in the *Physical Review* in which an exception to Goldstone's theorem was noted.[5] These papers realized that if a continuous symmetry was *gauged*, then there need not be massless modes in the spectrum. To be more precise, the NG fields that you get are purely gauge artifacts and can be removed by a suitable choice of gauge.

---

[5] "BEHGHK" (pronounced "Beck") comes from a talk by Ben Kilminster.

2.6.1 *A little history*

This seems like a very good place to insert a bit of background on where the "Higgs Mechanism" came from. For those less interested in the history, feel free to skip this section.

After Yang and Mills published their theory of non-Abelian gauge invariance, it was hoped (and that hope was ultimately realized) that one can use this idea to explain the forces of nuclear and particle physics. The problem was that these theories required the existence of massless spin-1 particles, and there were no candidates (remember that gluons were not discovered until much later). When Nambu and Goldstone published their results on spontaneous symmetry breaking, it was found that in addition to the vector bosons, you also have a collection of spin-0 bosons that must be massless. The situation has gone from bad to worse!

It was Philip Anderson (motivated by a result from Julian Schwinger) who first seemed to realize that the situation was not as bad as it seemed. Schwinger had suggested that the requirement of massless gauge bosons was only a requirement in the weak coupling limit. When you have strongly interacting forces, he showed that you need not necessarily have massless gauge fields! Anderson was an expert in condensed matter physics, and he presented a physical example of Schwinger's ideas by studying the behavior of a non-relativistic free-electron gas, where it was known that transverse EM waves do not propagate below the plasma frequency, while above this frequency there are three propagating modes (one longitudinal, two transverse). He showed how this result followed from Schwinger's ideas, and by making the connection to superconductivity, he realized that there need be no massless modes in the system. He then made a very interesting point at the conclusion of his paper.

> It is noteworthy that in most of these cases, upon closer examination, the Goldstone bosons do indeed become tangled up with the Yang-Mills gauge bosons and, thus, do not in any true sense really have zero mass.... We conclude, then, that the Goldstone zero-mass difficulty is not a serious one, because we can probably cancel it off against an equal Yang-Mills zero-mass problem. [Anderson (1963)]

As an amusing afterthought, Anderson points out that when one takes gravity into account, that the breakdown of translational and rotational invariance leads to the presence of three phonons, which, when combined with the two graviton helicities, precisely gives you the right number of degrees of freedom for a massive, spin-2 field. Anderson was way ahead of his time!

The following year, three papers came out, all of which were published in *Physical Review*, Volume 13. All of these papers show, to various degrees, that Goldstone's theorem need not apply to gauge theories, and that when a gauge theory is spontaneously broken, there are no massless particles in the spectrum.

(1) The first paper was by Robert Brout and Francois Englert [Englert and Brout (1964)]. They showed that when a gauge theory is spontaneously broken, the vector boson acquires a mass in perturbation theory. They compute this mass, but admit: "We have not yet constructed a proof in arbitrary order..." although they point out that there should be no problems generalizing the result.

(2) The second paper was by Peter Higgs [Higgs (1964)]. He pointed out that, as Brout and Engert discovered, Goldstone's theorem need not hold when the symmetry is gauged, and furthermore, that when you send the gauge coupling to zero, the longitudinal modes of the gauge field manifest as the required NG fields. So it was Higgs who first makes the connection of "gauge fields eating the NG bosons". In addition, he proposes a model for how this might work; we will use this model in our example below. Higgs's approach is to treat the theory classically – he derives the field equations from a gauge invariant action, and shows that the result is a Proca Lagrangian with an extra (massive) scalar boson, forever known as the "Higgs boson".

(3) The third paper was by Gerald Guralnik, Carl Hagen and Tom Kibble [Guralnik *et al.* (1964)];. Their paper took a somewhat different approach. They pointed out that when a gauged symmetry is spontaneously broken, the corresponding (global) charge *is no longer conserved* despite the existence of a local conservation law. The problem comes from the gauge-fixing conditions: for example, in the radiation gauge, the theory is not manifestly covariant and you can get non-trivial surface terms at infinity. In a Lorentz-covariant gauge (such as Lorenz gauge), the authors point out that you *do* have massless modes as Goldstone's theorem requires, but that these are "gauge [artifacts] rather than physical particles." Indeed, this can be seen explicitly in terms of the later discovered R_ξ gauge with $\xi = 0$ where ghosts and NG bosons are realized as massless particles [Peskin and Schroeder (1995)]. In a physical gauge, however, these extra modes do not appear in the

spectrum, since Goldstone's theorem requires a global conservation law that is explicitly broken by gauge fixing terms!

We now know that the BEHGHK mechanism does seem to play a major role in the standard model of particle physics, with the residual boson mass at around 125 GeV. After this discovery was confirmed at the Large Hadron Collider, Englert and Higgs shared the 2013 Nobel Prize in Physics.

2.6.2 *An example*

We will start by studying the model of a complex, electrically charged scalar field. This is known as the *Landau-Ginzburg model* by condensed matter physicists; the *Coleman-Weinberg model* by particle physicists; and *scalar QED* among students:

$$
\begin{aligned}
S[\phi, A_\mu] &= \int d^d x \left\{ |D_\mu \phi|^2 - m^2 |\phi|^2 - \lambda |\phi|^4 - \frac{1}{4} F_{\mu\nu} F^{\mu\nu} \right\} \\
&= \int d^d x \Big\{ |\partial_\mu \phi|^2 - m^2 |\phi|^2 - \lambda |\phi|^4 \\
&\qquad + e A_\mu j^\mu + e^2 |\phi|^2 A_\mu A^\mu - \frac{1}{4} F_{\mu\nu} F^{\mu\nu} \Big\}
\end{aligned}
\tag{2.48}
$$

where $D_\mu \phi = \partial_\mu \phi + i e A_\mu \phi$, and j^μ is the current defined in Eq. (2.12). As we have seen in Section 2.5, when $m^2 < 0$ the phase symmetry is spontaneously broken and the field picks up a vacuum expectation value $v/\sqrt{2} = \sqrt{-m^2/2\lambda}$. We should therefore expect to see a massless field appear upon going to a physical basis for the fields. In this section we will decompose ϕ in terms of real fields with a different parametrization:

$$
\phi(x) = \frac{1}{\sqrt{2}} \left(v + \varphi(x) \right) e^{i\theta(x)/v}
\tag{2.49}
$$

rather than splitting the field into real and imaginary parts. It is a straightforward exercise to relate our fields (φ, θ) to the previously chosen basis (ϕ_1, ϕ_2), and we leave it to you.

There is a good reason why we chose to decompose ϕ in this way: since the potential only depends on the modulus of ϕ, the dependence on θ is restricted to very few terms:

$$\begin{aligned}S[\varphi,\theta,A_\mu] &= \int d^dx \Big\{ \frac{1}{2}(\partial_\mu\varphi)^2 + \frac{1}{2}(\partial_\mu\theta)^2 - \mathcal{V}(v+\varphi) \\ &\quad + eA_\mu \tilde{j}^\mu + \frac{g^2}{2}(v+\varphi)^2 A_\mu A^\mu - \frac{1}{4}F_{\mu\nu}F^{\mu\nu} \Big\} \\ &= \int d^dx \Big\{ \frac{1}{2}(\partial_\mu\theta)^2 + \frac{1}{2}g^2v^2 A_\mu A^\mu + gA_\mu \tilde{j}^\mu \Big\} + \cdots \end{aligned}$$
(2.50)

where

$$\tilde{j}^\mu = v\partial^\mu\theta \tag{2.51}$$

is the current in terms of our physical fields, and the ellipses in Eq. (2.50) refer to θ-independent terms. Notice right away that φ does not appear in the current – this represents a *neutral* particle. Also notice that the phase field $\theta(x)$ is the massless NG field that we expect, but in this choice of parametrization it makes no appearance in the potential. This is why this choice of parametrization of real fields is so useful. You should also recognize this current as our Noether current that follows from a *shift symmetry* in the field from Eq. (2.20). This is precisely correct, since under the symmetry:

$$\phi \longrightarrow e^{i\alpha}\phi \quad \Rightarrow \quad \theta \longrightarrow \theta + \alpha v \tag{2.52}$$

while φ does not transform at all, and therefore does not appear in the current. However, the situation is much more interesting now that our symmetry is gauged, since now α can be a function of space-time. Therefore, why not choose $\alpha(x) = -\theta(x)/v$ everywhere? In that case, the field is completely cancelled by the gauge transformation and should not appear in the action at all!

If we look at Eq. (2.50) and use Eq. (2.51) we find

$$\begin{aligned}S &= \int d^dx \frac{1}{2}\Big\{ (\partial_\mu\theta)^2 + 2gvA_\mu\partial^\mu\theta + g^2v^2 A_\mu A^\mu \Big\} + \cdots \\ &= \int d^dx \frac{1}{2}g^2v^2 \Big[A_\mu + \partial_\mu\Big(\frac{\theta}{gv}\Big) \Big]^2 + \cdots \\ &= \int d^dx \frac{1}{2}g^2v^2 A'_\mu A'^\mu + \cdots \end{aligned}$$
(2.53)

where A'_μ is precisely the gauge field suitably transformed after the transformation in Eq. (2.52). So θ really has completely vanished from our action after a gauge transformation, and there is no NG field! The final result is a massive vector-boson, and a massive, neutral (real) scalar field.

It might bother you that a field has suddenly vanished from our theory – what about unitarity? How can we start with a theory with two degree of freedom (θ, φ) and lose one? The answer comes from keeping track of the degrees of freedom in the vector boson. In d space-time dimensions, a massless vector boson has $d-2$ degrees of freedom, while a massive vector boson has $d-1$ degrees of freedom. So where did that extra longitudinal degree of freedom come from? It is nothing more than θ! This has led to the idea that, "The massless vector boson *ate* the NG boson and became massive!"

2.6.3 *An interlude: superconductivity*

One of the amazing things about local spontaneous symmetry breaking is that it describes most of the phenomena of superconductivity, without the need to resort to a model! This was first noted in a paper by Steven Weinberg dedicated to Nambu [Weinberg (1986)], and is explained with even more detail in [Weinberg (1996)]. Although it is a bit of an aside from the main thrust of the book, we feel remiss if we do not give an explanation to how superconductivity follows from an EFT of a spontaneously broken U(1) gauge symmetry, since it is such a prime example of how far you can get without knowing "the man behind the curtain!"

The only assumptions we will make are that a U(1) gauge symmetry (E&M) is spontaneously broken by a charge-2 field. It could be a fundamental field (like a Higgs boson) or a composite object (like a Cooper pair of electrons) – those kinds of details are irrelevant for our purposes. We will parametrize this field as

$$\phi(x) = \frac{1}{\sqrt{2}} (v+h) e^{2ie\theta} \tag{2.54}$$

where θ is the (dimensionless) NG boson, and h is a massive excitation. Now the broken gauge invariance acts on θ through a shift symmetry as in Eq. (2.52), so a gauge transformation $\alpha(x)$ shifts $\theta \to \theta + \alpha$. But since this is a charge-2 object, there is a residual $\mathbb{Z}_2$ symmetry, so that the fields

$$\theta \cong \theta + \frac{\pi}{e} \tag{2.55}$$

must represent the same physics.

Specifying the action $S[\theta, A_\mu, h]$ would require us to pick a model for our theory, but whatever we choose, we know that it must be a function of $A_\mu - \partial_\mu \theta$ in order for the dynamics to remain gauge invariant: remember, only the ground state breaks gauge symmetry, not the action. Therefore, any charge densities and currents in our superconducting material are given by

$$J^0 = -\frac{\delta S}{\delta A^0} = -\frac{\delta S}{\delta \dot\theta}, \qquad (2.56)$$

$$\vec{J} = \frac{\delta S}{\delta \vec{A}}. \qquad (2.57)$$

This is enough to describe several phenomena relating to superconductivity. It is particularly nice that we have this kind of universality, since there is still a lot of debate as to the nature of high-temperature superconductors.

(1) **Meissner effect:** In the broken phase, the vector potential that describes the photon is described by a Proca Lagrangian:

$$\mathcal{L} = -\frac{1}{4}F_{\mu\nu}F^{\mu\nu} + \frac{1}{2}m^2 A_\mu A^\mu, \qquad (2.58)$$

with $m_A = 2ev$. The resulting field equation for the photon implies that there will be a penetration depth $\lambda \sim m^{-1}$, and the magnetic field inside $|\vec{B}| \sim e^{-\lambda r}$ as you penetrate into the superconductor.

(2) **Flux quantization:** The absence of magnetic fields deep inside the superconductor does not imply that the vector potential vanishes, but it does imply that $A_\mu = \partial_\mu \theta$, whose spacial components are $\vec{A} = \vec{\nabla}\theta$. Doing a line integral of this expression around a closed loop inside the superconductor gives:

$$\oint \vec{A} \cdot d\vec{x} = \oint \vec{\nabla}\theta \cdot d\vec{x}$$

$$\int (\vec{\nabla} \times \vec{A}) \cdot d\vec{S} = \Delta\theta.$$

The left-hand side of this equation is just the magnetic flux through the area bound by the integration loop, while the right-hand side is the change in the Goldstone field going all the way around the loop. This change can be any integer times π/e, since changes in θ by this amount do not affect any physical observable, according to Eq. (2.55). We have therefore proved the famous flux-quantization result:

$$\Phi_B = n\frac{\pi}{e}. \qquad (2.59)$$

(3) **Infinite conductivity:** The fact that magnetic flux is quantized is enough to show that the conductivity is infinite deep inside the superconductor since the flux (and therefore the current, by Ampère's Law) cannot decay continuously, but only drop in discrete jumps, which is impossible without large changes in energy.

Another way to see how infinite conductivity arises is by using Eq. (2.56), which tells us that the charge density and $-\theta$ are canonically conjugate variables, so the Hamiltonian of the superconductor is a natural function of J^0 and θ. We therefore have

$$\dot{\theta} = -\frac{\partial \mathcal{H}}{\partial J^0} = -V, \qquad (2.60)$$

where V is the energy per unit charge, which is simply the voltage. This tells us that a superconductor carrying a steady current with time-independent fields must have a vanishing potential difference. This is equivalent to the statement that the conductivity is infinite (zero voltage drop for a fixed current).

(4) **AC Josephson Junction:** Consider a gap between two superconducting slabs (labeled 1 and 2). The NG field is nonvanishing in either slab, but $\theta_1 \neq \theta_2$ in general. We can conclude that the dynamics in the gap will be determined by the difference between θ_1 and θ_2: $S_{\text{gap}} = \mathcal{A}F[\Delta\theta]$, where $\mathcal{A}$ is the area of the gap. Now thanks to Eq. (2.55), we know that both $\theta_{1,2}$ can change by $n\pi/e$ independently with no consequence to any observable, so it must be that F is periodic with period π/e.

Now imagine there is a nonzero vector potential in the gap. By gauge invariance, we must replace $\Delta\theta \to \int (\vec{\nabla}\theta - \vec{A}) \cdot d\vec{x}$, integrated across the gap. Then from Eq. (2.57), we have

$$\vec{J}_{\text{gap}} = \frac{\delta S_{\text{gap}}}{\delta \vec{A}} = \frac{\delta S_{\text{gap}}}{\delta \Delta\theta} \cdot \frac{\delta \Delta\theta}{\delta \vec{A}} = -\hat{n}F'[\Delta\theta], \qquad (2.61)$$

where $\hat{n}$ is a unit vector pointing across the gap. So we find that a current flows across the gap. Furthermore, if we additionally impose a fixed potential difference (ΔV) between the two superconducting slabs, Eq. (2.60) gives us $\Delta\theta = -t\Delta V$. Since S_{gap}, and therefore $\vec{J}_{\text{gap}}$, is periodic, we have a an alternating current oscillating with frequency

$$\nu = \frac{e\Delta V}{\pi}. \qquad (2.62)$$

This is the same expression Josephson derived using a specific model, but it is clear that the frequency is model independent.

2.7 CCWZ construction of EFT

In 1969, Callan, Coleman, Wess and Zumino (CCWZ) showed that physics truly is invariant to how you choose to parametrize your fields [Coleman et al. (1969); Callan et al. (1969)]. In particular, this means that we can use the phase parametrization for NG fields, removing them completely from the effective potential and placing them in the current with derivative interactions. There is no physics missed by doing this. The original two CCWZ papers are deep and difficult – and required reading for all serious students wishing to master the techniques of effective field theory! We will only summarize their results and show how they can be applied.

Consider the situation where the action (and path integral measure) is invariant under a Lie group of transformations G, and suppose that G is spontaneously broken to a Lie subgroup $H \subset G$. Let T^a be a set of generators of H and X^a be a set of generators that, together with the T^a, generate G. CCWZ showed that you can always choose to parametrize your fields as

$$\phi^a(x) = \Sigma^a_b(x)\varphi^b(x) \qquad (2.63)$$

where $\Sigma(x) \equiv e^{i\pi \cdot X}$ with $\pi^a(x)$ a collection of fields, one for each broken generator X^a, and φ transforms linearly under H:

$$h : \varphi^a \longrightarrow [D(h)]^a_b \varphi^b \qquad (2.64)$$

At this point, the choice of parametrization (that is, the precise definition of the $\pi^a(x)$) is arbitrary, but it is very convenient to choose φ to satisfy the condition

$$\varphi^\dagger_n [T^a]^n_m \langle \phi^m \rangle = 0 \qquad (2.65)$$

This is just an orthogonality condition of the $\varphi(x)$ fields, requiring that they have no components in the broken symmetry directions. With this choice, $\varphi(x)$ does not contain *any* of the NG fields, which are therefore parametrized completely in $\Sigma(x)$ (or equivalently, $\pi(x)$).

Notice that this condition still does not define Σ uniquely since any transformation $h \in H$ leaves $\langle \phi^m \rangle$ invariant; so plugging into Eq. (2.65) we find:

$$0 = \varphi^\dagger_n [T^a]^n_m \langle \phi^m \rangle = \varphi^\dagger_n [T^a]^n_m [D(h)]^m_l \langle \phi^l \rangle \qquad (2.66)$$

Therefore, we can choose to let $\phi(x) = \Sigma(x) D(h) \varphi(x)$ for any $h \in H$ and still satisfy Eq. (2.66). So Σ is only defined up to (right) multiplication by elements in H; it is called a *right coset of H*.[6]

Let us study how the fields φ and Σ transform.[7] Starting with H transformations (and suppressing indices):

$$h : \phi = \Sigma \varphi \longrightarrow D(h) \Sigma \varphi = (D(h) \Sigma D(h^{-1})) D(h) \varphi = \Sigma' \varphi' \qquad (2.67)$$

---

[6]See Appendix A.1 for a review of cosets in group theory.
[7]See Appendix A.3 for a review of group actions on the fields.

where we have used Eq. (2.64). So this tells us that $\Sigma(x)$ should transform as

$$\Sigma(x) \longrightarrow D(h)\Sigma D(h^{-1}) \tag{2.68}$$

Assuming that ϕ is in the fundamental representation of H, this means that Σ is in the adjoint representation.

For more general transformations from the larger group G, we use a fact from group theory that for any $g \in G$ we can write

$$g = e^{i\alpha \cdot X} e^{i\beta \cdot T} \equiv \tilde{g}h \tag{2.69}$$

Then we have:

$$g : \phi = \Sigma\varphi \longrightarrow \tilde{g}h\Sigma h^{-1}h\varphi = \tilde{g}\Sigma'\varphi' = \Sigma''\varphi' \tag{2.70}$$

where $\Sigma'' = e^{i\alpha \cdot X} e^{i\pi' \cdot X} \equiv e^{i\pi'' \cdot X} e^{i\alpha' \cdot T}$. Now this has a component in H, but recall that Σ was only defined up to right multiplication by elements of H; thus physics should not care if we use Σ'' or if we chose to use Σ' (without the α' factor).

So we find that the theory in terms of Σ and φ is explicitly invariant under H transformations, and under general G transformations, the theory remains invariant *up to field redefinitions which do not affect S matrix elements*. This is the fundamental result of CCWZ, who proved it for general symmetry groups.

If we plug our expression for ϕ into the original action, we find

$$\mathcal{V}(|\phi|) \longrightarrow \mathcal{V}(|\varphi|), \quad \text{since } \Sigma^\dagger \Sigma = 1 \tag{2.71}$$

$$\partial_\mu \phi \longrightarrow \Sigma(x)\left[\partial_\mu \varphi + (\Sigma^\dagger \partial_\mu \Sigma)\varphi\right] \tag{2.72}$$

When plugged into the kinetic terms, the overall factor of Σ will cancel, and so we find the NG fields with this choice of parametrization will only appear in the combination $\Sigma^\dagger \partial_\mu \Sigma$. This is a Lie algebra element, and therefore we can expand it in terms of the generators:

$$\Sigma^\dagger \partial_\mu \Sigma = iT^a \mathbb{V}_\mu^a[\pi] + iX^a \mathbb{A}_\mu^a[\pi] \tag{2.73}$$

where $\mathbb{V}_\mu$, $\mathbb{A}_\mu$ are functionals of π and depend linearly on $\partial_\mu \pi$. Now under the unbroken subgroup H, our fields transform as

$$\varphi \longrightarrow D(h)\varphi \tag{2.74}$$

$$\mathbb{A}_\mu \longrightarrow D(h)^{-1}\mathbb{A}_\mu D(h) \tag{2.75}$$

$$\mathbb{V}_\mu \longrightarrow D(h)^{-1}\mathbb{V}_\mu D(h) + D(h)^{-1}i\partial_\mu D(h) \tag{2.76}$$

Notice in particular that $\mathbb{V}_\mu$ transforms precisely as a *gauge field* under H transformations – this suggests we define a covariant derivative of sorts:

$$\mathbb{D}_\mu \varphi \equiv \partial_\mu \varphi + i\mathbb{V}_\mu \varphi \qquad (2.77)$$

and a field strength

$$[\mathbb{D}_\mu, \mathbb{D}_\nu]\varphi = i\mathbb{F}_{\mu\nu}\varphi \qquad (2.78)$$

with

$$\mathbb{F}^a_{\mu\nu} = \partial_\mu \mathbb{V}^a_\nu - \partial_\nu \mathbb{V}^a_\mu + c^{abc}\, \mathbb{V}^b_\mu \mathbb{V}^c_\nu \qquad (2.79)$$

that also transforms under the adjoint representation of H like Eq. (2.75); c^{abc} are the structure constants of H.

We now know how to construct the most general action for a theory that is invariant under a symmetry group G that has spontaneously broken to H. We need only construct the most general functional of φ, $\mathbb{D}_\mu \varphi$, $\mathbb{F}_{\mu\nu}$, and $\mathbb{A}_\mu$, that is invariant under the H transformations in Eqs. (2.74-2.76), as well as Lorentz invariance if the theory is relativistic. In particular, such a theory is automatically invariant under G transformations, although such transformations are nonlinearly realized. In practice, this can be used to describe how Nambu-Goldstone bosons couple to other fields, such as pions coupling to nucleons, for example.

2.8 Explicit breaking; spurion analysis

Even if a symmetry is not exactly realized in the action, we can still apply many of the results of this chapter, so long as the symmetry breaking effects are small. Consider a QFT whose action $S_0[\phi]$ (and path integral measure) are invariant under a symmetry transformation

$$\phi \longrightarrow D(g)\phi \qquad (2.80)$$

where $D(g)$ is some representation of a Lie group. Now imagine that we add a term to the action that breaks this symmetry[8]

$$\Delta S = \int d^d x\, \lambda_a \mathcal{O}^a[\phi] \qquad (2.81)$$

where $\mathcal{O}^a[\phi]$ is some operator functional of ϕ that is not invariant but transforms under some representation of the Lie group, and λ_a are a set

---

[8] It is well known from the theory of symmetries that a non-invariant term can always be written as an operator that transforms in some nontrivial (not necessarily reducible) representation of the group. This is what we do here.

of small[9] coupling constants. Then under the symmetry transformation in Eq. (2.80) we find

$$S_0[\phi] + \Delta S \longrightarrow S_0[\phi'] + \int d^d x\, \lambda_a\, [\tilde{D}(g)]^a_b \mathcal{O}^b[\phi'] \qquad (2.82)$$

where $\tilde{D}(g)$ is the representation of g under which $\mathcal{O}$ transforms. This is not invariant, but it would be if λ behaved like a field that transformed as

$$\lambda \longrightarrow \tilde{D}(g)^{-1}\lambda \qquad (2.83)$$

Now λ is a constant, and does not know anything about transforming, but if we replace it by a *field* $\hat{\lambda}$ and write a new action:

$$S_0[\phi] + \int d^d x\, \hat{\lambda}_a \mathcal{O}^a[\phi] \qquad (2.84)$$

This action *is* invariant under the transformation in Eq. (2.80). All we have to do is treat $\hat{\lambda}$ like any other field, and then set $\langle \hat{\lambda}_a \rangle = \lambda_a$ at the end of the calculation. This "spurious" new field is known as a *spurion*, and this form of argument is called a spurion analysis. In the end, we want to develop a power series expansion in λ, so this kind of analysis will only make sense if λ_a is small in the sense described earlier. Then we are allowed to do perturbation theory in λ.

Let us see how this works in an example. Consider a complex scalar field with an additional holomorphic mass term:

$$S_0[\phi] = \int d^d x\, \{(\partial_\mu \phi^*)(\partial^\mu \phi) - m^2|\phi|^2 - \lambda|\phi|^4\} \qquad (2.85)$$

$$\Delta S = \int d^d x\, \{\mu^2 \phi^2 + \text{c.c}\} \qquad (2.86)$$

S_0 is preserved by the phase symmetry $\phi \longrightarrow e^{i\alpha}\phi$, while ΔS breaks the symmetry explicitly. The symmetry would be restored if we replaced the troublesome parameter[10] μ^2 with a spurion that transforms as:

$$\hat{\mu}^2 \longrightarrow e^{-2i\alpha}\hat{\mu}^2 \qquad (2.87)$$

With this new spurion field, the most general operators that are invariant under the phase symmetry must take the form

$$\hat{\mu}^{2a} \phi^b \phi^{*c} \longrightarrow e^{i(-2a+b-c)\alpha} \hat{\mu}^{2a} \phi^b \phi^{*c} \qquad (2.88)$$

---

[9] We must be careful here about what we mean by "small" - if the couplings are dimensionless, they should be much less than any other dimensionless coupling constants in the theory; if they carry mass dimension, they should be much less than any other scale in the problem, so that ratios of the form λ/M are small for any scales M in the unbroken theory.

[10] Notice that the parameter is μ^2, *not* μ – to get μ we would have to get $\sqrt{\mu^2}$, which implies a dangerous loss of analyticity.

which implies

$$a = \frac{b-c}{2} \tag{2.89}$$

We take $b > c$ to avoid $\hat{\mu}^2$ being put in the denominator (remember – it is a field!); the case $b < c$ is taken care of by complex conjugating. Similarly, we cannot have $\sqrt{\hat{\mu}^2}$ as we mentioned in the previous footnote, which means that (b, c) must be both even or odd. Here are a few facts about the matrix elements of operators in this theory:

(1) Operators that do not involve μ^2 at all must be of the form $(\phi^*\phi)^n$ for some integer n. This is the only way to make $a = 0$ in Eq. (2.89).
(2) For $a = 1$, we need $b = c + 2$. Keeping in mind that m^2 is the only other scale in the problem, this means that

$$\hat{\mu}^2(\phi^*\phi^3) \times f(\phi^*\phi) \tag{2.90}$$

with f an arbitrary function, is the only invariant operator with a single power of $\hat{\mu}^2$. This tells us right away about the matrix elements of such operators; for example

$$\langle 0|\phi^*\phi^3|0\rangle \sim \mu^2 \tag{2.91}$$

(3) Similarly to the previous point, $\langle 0|\phi^*\phi^5|0\rangle \sim \mu^4$ and so forth, but $\langle 0|\phi^*\phi^4|0\rangle = 0$, because it would have to have a fractional a, which cannot be allowed by our spurion analysis.

So we have learned the behavior of many matrix elements, including which ones vanish when the symmetry is restored ($\mu^2 \to 0$), and which ones vanish even when the symmetry is broken; and all without doing a single Feynman integral!

2.9 Anomalous symmetries

Everything that we have been doing in this chapter has assumed that if we have a symmetry at the classical level, then it is there at the quantum level as well. If it is broken spontaneously we know what to do, and if it is broken explicitly we can still get far using the technique of spurion analysis. But what if the symmetry is not broken by the vacuum or an explicit operator, but by the quantum effects themselves? When this happens, we say that the symmetry is *anomalous*. In this section, we will consider how this can happen, and what it does to our effective theory.

2.9.1 Anomalies in the path integral

At the quantum level, a typical (perturbative) QFT is defined by an action (dynamics), a vacuum, *and* a regulator. This last item is vital – QFT perturbation theory *requires* a regulator to extract finite resuts, except for very special circumstances. Furthermore, one regulator need not (and generally will not) give you the same answer as another regulator. While it is true that *physical* results must of course be regulator independent, the perturbative QFT itself is not. With this in mind, we can make a definition:

A symmetry is *anomalous* if it is realized in the action either linearly or nonlinearly, but is violated by the regulator – for *all* regulators.

The last point about the symmetry breaking happening for every regulator is very important, as can be shown by a famous counterexample. Consider QED with momentum cutoff. The momentum cutoff breaks the gauge symmetry, since it removes some of the momentum modes from the theory so gauge transformations cannot be completely cancelled. Sure enough, at one loop you have a quadratically divergent photon mass appearing in the photon propagator

$$\Pi^{\mu\nu} \sim g^{\mu\nu}\Lambda^2 + \cdots \tag{2.92}$$

However, if instead of using momentum cutoff, you used Pauli-Villars or dimensional regularization, which *does* maintain the gauge symmetry, this no longer happens:

$$\Pi^{\mu\nu} = (g^{\mu\nu}q^2 - q^\mu q^\nu)\Pi(q^2) , \tag{2.93}$$

where $\Pi(q^2)$ can only be logarithmically divergent at worst. This expression explicitly satisfies the Ward identity ($q_\mu \Pi^{\mu\nu}(q) = 0$), so even though there is a particular regulator that breaks the gauge symmetry, this is not anomalous.

Effective field theory approaches rely most often on the path integral formalism, so we will sketch a proof of how anomalies appear within path integrals, first derived in [Fujikawa (1979)]. Recall that if you have a set of fields ϕ^a, not necessarily scalars, and an action on these fields $S[\phi]$, the path integral is given by

$$Z[J] \equiv \int \mathcal{D}\phi \, e^{iS[\phi(x)] + i\int d^d x J(x)\phi(x)} , \tag{2.94}$$

and in principle, all the physics, both classical and quantum, is contained in Z.

Recall from our earlier treatment of Noether's theorem that the action changes under a local version of the global transformation in Eq. (2.1) as:

$$\delta S[\phi] = \int d^d x \, \epsilon \partial_\mu j^\mu \qquad (2.95)$$

where j^μ is the Noether current. Thus the path integral becomes:

$$Z \to \int \mathcal{D}\phi \, e^{iS[\phi] + \int d^d x \, \epsilon(\partial_\mu j^\mu)} \qquad (2.96)$$

Since this is supposed to be a symmetry, Z cannot depend on ϵ, so

$$\frac{1}{Z}\frac{\delta Z}{\delta \epsilon} = \langle \partial_\mu j^\mu \rangle = 0 \, . \qquad (2.97)$$

and we recover Noether's theorem through the correspondence principle in correlation functions.

However, there is a big flaw in this argument: we have been very cavalier about the behavior of the measure during all of these changes of coordinates. Indeed, the measure generally transforms with a Jacobian factor:

$$\mathcal{D}\phi' = \mathcal{D}\phi \, \mathcal{J}[\phi] \, , \qquad (2.98)$$

where

$$\mathcal{J}[\phi] \equiv \det\left(\frac{\delta \phi'}{\delta \phi}\right) = e^{\mathrm{Tr} \log\left(\frac{\delta \phi'}{\delta \phi}\right)} = e^{\mathrm{Tr} \log(1 + i\epsilon C)} = e^{i\mathrm{Tr}(\epsilon C)} \, . \qquad (2.99)$$

We use the fact that ϵ is infinitesimal and define

$$C = -i\frac{\delta(\Delta\phi)}{\delta\phi} \qquad (2.100)$$

So in fact, instead of Eq. (2.96) we really have

$$Z' = \int \mathcal{D}\phi \, e^{iS[\phi] + \int d^d x \, \epsilon(\partial_\mu j^\mu + \mathrm{tr}(C))} \qquad (2.101)$$

and Noether's theorem becomes

$$\frac{1}{Z}\frac{\delta Z}{\delta \epsilon} = \langle \partial_\mu j^\mu + \mathrm{tr}(C) \rangle = 0 \, . \qquad (2.102)$$

If $\mathrm{tr}\, C = 0$ then nothing has changed. But if $\mathrm{tr}\, C \neq 0$ then the conservation law is altered. Classically, $\mathrm{tr}\, C$ will vanish, but once you go to the quantum theory, it might not. In particular: upon renormalization the regulator might cause it to develop a finite piece.

2.9.2 The chiral anomaly and its consequences

The classic example of an anomaly is the *chiral anomaly*. Recall that massless fermions have two (classically) conserved currents:

$$j^\mu = \overline{\Psi}\gamma^\mu\Psi , \tag{2.103}$$

$$j_5^\mu = \overline{\Psi}\gamma^\mu\gamma^5\Psi. \tag{2.104}$$

j^μ is the fermion number current that couples to gauge bosons. If it is anomalous, then the gauge symmetry is broken by quantum effects; we will discuss this possibility at the end of this section. j_5^μ is the "axial number" current; it is well known that it is anomalous [Peskin and Schroeder (1995)], and the anomaly equation is

$$\partial_\mu j_5^\mu = -\frac{g^2 T(R)}{16\pi^2}\epsilon^{\mu\nu\rho\sigma}F^a_{\mu\nu}F^a_{\rho\sigma} , \tag{2.105}$$

for an $SU(N)$ theory coupled to (massless) fermions in the representation R, where $T(R)$ is the Dynkin index for R.[11]

In terms of left and right handed currents for fermions that form $SU(N)$ multiplets, we have

$$j_L^{\mu a} = \overline{\Psi}\gamma^\mu P_L T^a \Psi , \tag{2.106}$$

$$j_R^{\mu a} = \overline{\Psi}\gamma^\mu P_R T^a \Psi , \tag{2.107}$$

where $P_{L,R} = \frac{1}{2}(1 \pm \gamma^5)$, so that $j^\mu = j_R^\mu + j_L^\mu$ and $j_5^\mu = j_R^\mu - j_L^\mu$. Then in general, any chiral theory will have an anomaly in both j_L^μ and j_R^μ such that the anomaly cancels in the sum and adds in the difference:[12]

$$\partial_\mu j_L^{\mu a} = +\frac{g^2 \mathcal{A}^{abc}}{32\pi^2}\epsilon^{\mu\nu\rho\sigma}F^b_{\mu\nu}F^c_{\rho\sigma} , \tag{2.108}$$

$$\partial_\mu j_R^{\mu a} = -\frac{g^2 \mathcal{A}^{abc}}{32\pi^2}\epsilon^{\mu\nu\rho\sigma}F^b_{\mu\nu}F^c_{\rho\sigma} , \tag{2.109}$$

where $\mathcal{A}$ is a new group theory index object that we must try to work out. To compute it, consider the correlation function of three L currents $C(x,y,z) = \langle j_L^{\mu a}(x) j_L^{\nu b}(y) j_L^{\rho c}(z)\rangle$. To leading order, this correlator is given by the diagrams in Figure 2.1.

Let us compute the group theory structure. Since they involve closed loops, there is a trace over generators. We have for fermions in representation R

$$C \sim \text{Tr}[T_R^a T_R^b T_R^c] + \text{Tr}[T_R^a T_R^c T_R^b] = \text{Tr}\{T_R^a [T_R^b, T_R^c]_+\} \tag{2.110}$$

---

[11] See Appendix A.5 for explanations of group factors.

[12] A *chiral* theory is any theory of fermions that treat left and right fermions differently. This is the opposite of a *vectorlike* theory such as QED, which treats left and right fermion fields the same way.

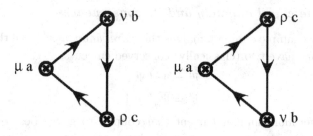

Fig. 2.1 Triangle diagrams that are generated by the three point correlation function of chiral currents.

But this is nothing more than the expression for the anomaly coefficient. This is where it gets its name. In addition, remember that if $\{T_1^a\}$ generate the group G_1 and $\{T_2^a\}$ generate the group G_2, then the traces factorize $\text{Tr}[T_1 \otimes T_2] = \text{Tr}[T_1] \cdot \text{Tr}[T_2]$.

Thus to compute anomalies, one must first compute the anomaly coefficients of the theory, for which there are tables and rules discussed in the group theory appendix. In particular, there is the very important rule that $A(R) = -A(\overline{R})$. This means that a theory of *Dirac* fermions is always anomaly free! Also, if you have fermions that transform in a real (or pseudoreal) representation, either vectorlike or something like the Adjoint representation, then there is no anomaly. Finally, only $SU(N)$ groups with $N \geq 3$ or $U(1)$ groups[13] can have anomalies, since they're the only groups with complex representations.

An amusing sidenote: in Fig. 2.1 we can attach gravitons to two of the vertices (replacing the chiral currents with energy-momentum tensors), and then the anomaly would be proportional to $\text{Tr}(T^a)$. This can be nonvanishing only for $U(1)$, and it leads to the important result that unless all the charges of the fields sum to zero, the theory has a "gravitational anomaly."

If the symmetries are global, anomalies can lead to interesting physics. For example, in the chiral Lagrangian, it is the dominant contribution to the decay of the π^0 to two photons. However, if we gauge the chiral symmetry, it is *very* important that there are no anomalies, otherwise there would be the usual story of unitarity breakdown. This is a very important constraint on model building. The standard model, for example, is anomaly free because of its funny "flavor structure".

---

[13]The generator for a $U(1)$ is just the charge matrix, so for anomalies to vanish, the charges, or products of charges, must sum to zero.

2.9.3 't Hooft anomaly matching

There is a very powerful use of anomalies discovered by 't Hooft when trying to understand the behavior of strongly coupled theories. As we will see, when a theory becomes strongly coupled, there is often a paradigm shift in the relevant degrees of freedom. For example, at short distances the strong nuclear force is described by quarks and gluons, while at longer distances, the proper degrees of freedom are the hadrons. In plasma physics, at low temperatures the plasma can be thought of as a gas of electrons, while at higher pressures and densities, the real degrees of freedom are collective modes of the plasma ("plasmons" and other "excitons"). Each time the theory undergoes a strong coupling, you must restate your effective theory to describe the new relevant degrees of freedom. If you do not, your theory's approximations break down.

So the question we ask is, what happens when a strongly coupled theory has an anomalous symmetry? 't Hooft showed that in certain cases, you can use this knowledge to predict the spectrum of the theory after it has become strongly coupled. To be concrete, imagine a theory that is weakly coupled (so perturbation theory works) as long as you are above or below a certain energy scale Λ. On the other side of this scale, your theory is strongly coupled and you cannot use perturbation theory to describe it. What do you do?

If the weakly-coupled theory has an anomalous symmetry, then you have an advantage! 't Hooft showed ['t Hooft (1980)] that, regardless of the strength of interactions, anomalies must be preserved on *both* sides of Λ. This allows you to identify (at least partially) the fermionic sector of your effective theory.

Let us imagine a theory that has N left-handed fermions that transform in the fundamental representation of $SU(N)$. According to our rules for calculating the anomaly coefficients, this theory will have an anomaly coefficient of $\mathcal{A} = 1$ coming from the correlation function of three currents. Now imagine at the scale Λ, the theory becomes strongly coupled, and so the fermions will form (quasi) bound states and no longer be the correct degrees of freedom to use. But we know that even when the theory is strongly coupled, there must be an $SU(N)$ anomaly with the same anomaly coefficient. This might occur many ways: there might be a fermion that transforms in the N representation; or their might be three fermions that transform in the two-index symmetric, two-index antisymmetric, and anti-fundamental representation. This list goes on!

As you can see, 't Hooft anomaly matching does not give you the answer right away, but it does tell you some things: if there is an anomaly, then there must be some fermionic modes in the strongly coupled sector to reproduce the anomaly; and if their is no anomalies, then you are very restricted in the types of bound states that you might generate in the strongly coupled limit. While it may not sound like much, remember that we are talking about strong coupling, where virtually no approximation will work, so if we can say *anything* concrete, we should be impressed!

2.10 Notes for further reading

This chapter was a basic review of many concepts that are taught in a typical quantum field theory course. A great reference to review this material and fill in any gaps is [Peskin and Schroeder (1995)] or [Itzykson and Zuber (1980)]. The best all-around explanation of the CCWZ results we know of is [Weinberg (1996)], although it is challenging. The bible of the symmetry principles applied to effective field theory is Sidney Coleman's collection of Erice Lectures [Coleman (1985)].

Problems for Chapter 2

(1) **Forms of the Noether current**

 (a) Repeat the derivation of the current in Eq. (2.12) in terms of real and imaginary components of ϕ. What does the transformation law look like in this basis?

 (b) What if we had a single real scalar field. What is the current in that case?

(2) **Nielsen-Chadha Theorem**

 Consider a theory describing a complex scalar field $\Phi = \begin{pmatrix} \phi_1 \\ \phi_2 \end{pmatrix}$ that transforms in the fundamental representation of $U(2)$, similar to the Higgs boson of the standard model of particle physics. Now let us imagine that this scalar field is in a particle bath, which can be accounted for with the introduction of a chemical potential μ. We can incorporate μ into our effective Lagrangian by making the replacement $\partial_0 \to \partial_0 - i\mu$; this will explicitly break the Lorentz invariance of the theory. The final result is

$$\mathcal{L} = \partial_\mu \Phi^\dagger \partial^\mu \Phi - (m^2 - \mu^2)\Phi^\dagger \Phi - \lambda(\Phi^\dagger \Phi)^2 + i\mu \left(\Phi^\dagger (\partial_0 \Phi) - (\partial_0 \Phi^\dagger)\Phi\right). \tag{2.111}$$

Notice that the chemical potential appears in two places: it couples to the number density $j_0(x)$, as a chemical potential should, but it also shifts the mass term of the scalar field.

(a) When $\mu > m$, the $U(2)$ symmetry is broken to $U(1)$ even with a positive m^2. This leaves three broken generators, and so we naively expect to have three Goldstone bosons π^a. Find an expression for the vacuum expectation value v for the field.

(b) Parametrize the broken phase using the CCWZ parametrization

$$\Phi = \frac{1}{\sqrt{2}} e^{-i\sigma^a \pi^a / v} \begin{pmatrix} 0 \\ v + h \end{pmatrix}, \tag{2.112}$$

where v is the vev you computed in Part 1, and σ^a are the Pauli matrices. The π^a fields are the presumed NG fields, and h is the remaining field, akin to the "Higgs boson". Write down the quadratic part of the Lagrangian in terms of these four fields.

(c) Show that for $\mu > m \neq 0$, there are only two NG bosons (gapless modes) and two "massive" modes. One of the NG modes has a linear dispersion, while the other has a quadratic dispersion. Thus the Nielsen-Chadha sum rule is still maintained.

(3) **CP Violation**

Show that even though there is a holomorphic mass in Eq. (2.86) (which would imply the presence of a phase), there is no CP violation in this theory.

(4) **CCWZ Construction**

Suppose you have a theory with a global $SU(N)$ symmetry that is spontaneously broken to an $SU(N-1)$ symmetry by a scalar field that transforms in the fundamental representation. Suppose also that there is another heavy scalar field that does not get a vev, but that also transforms in the fundamental of $SU(N)$. Using the CCWZ formalism, write down an action that describes the couplings between this heavy field and the Nambu-Goldstone bosons.

Chapter 3

Elementary Techniques

Now that we have reviewed the important topics from field theory related to symmetry management, we would like to start working out some Effective Field Theories. The best way to see how the subject works is to consider several examples, and point out the methods that can be used to solve problems. But before we dive into the field, we want to present a collection of results that apply for nearly all EFTs. In this chapter we will discuss these general, all-purpose results, most notably the concept of the *Renormalization Group*, the ideas that lead up to it, and its many consequences. It is primarily because of this result that "effective field theory" works at all!

We will begin with the most important first step in solving any physics problem: dimensional analysis. From there, we will find ourselves going deep into the ideas behind renormalization that are central to field theory, both quantum and thermal. Once we understand how to apply these core ideas, we will be in a much better position to study real world examples.

3.1 Canonical (engineering) dimensions

3.1.1 *Dimensional analysis*

Let us begin by considering the simplest theory: that of a free relativistic scalar field in d dimensions:

$$S = \frac{1}{2} \int d^d x \left(\partial_\mu \phi \partial^\mu \phi - m^2 \phi^2 \right). \tag{3.1}$$

In natural units ($\hbar = c = 1$), the action is dimensionless. Coordinates have dimension $[x] = -1$, therefore we require the field ϕ to have dimensions $[\phi] = (d-2)/2$. Similarly, we see that the parameter m^2 must be dimension

+2, as we would expect for a mass term. These numbers are called the *canonical* (or *engineering*) *dimension* of the field, and computing it is simply a matter of dimensional analysis. Other engineering dimensions are given for different fields and parameters in Table 3.1. Note that the coupling constant of a composite operator of dimension d_e would itself be of mass dimension $d - d_e$.

Table 3.1   Engineering dimensions for various fields and operators.

Operator	Canonical dimension, d_e
Boson ϕ in d-dimensions	$(d-2)/2$
Fermion ψ in d-dimensions	$(d-1)/2$
Derivative, ∂_μ	$+1$
Coordinate x^μ	-1
Generic operator, $\partial^l \phi^n (\bar\psi \psi)^m$	$(n/2 + m)d - (n+m) + l$

It is amazing what dimensional analysis will get us. The goal of quantum field theory is to compute various correlation functions

$$G^{(n)}(x_1, \cdots, x_n) = \langle \phi(x_1) \cdots \phi(x_n) \rangle_S, \qquad (3.2)$$

where the subscript is to remind you that the expectation value is to be taken with respect to the action S:

$$\langle \mathcal{O} \rangle_S = \frac{\int \mathcal{D}\phi \, \mathcal{O} \, e^{iS[\phi]}}{\int \mathcal{D}\phi \, e^{iS[\phi]}}. \qquad (3.3)$$

One way to gain information about these correlation functions is to use a scaling argument relating $G^{(n)}$ evaluated at different points in spacetime. To see how this works, consider the following change of coordinates and field:

$$x = sx' \qquad \phi(x) = s^{\frac{2-d}{2}} \phi'(x'). \qquad (3.4)$$

If we replace x and ϕ in the action in Eq (3.1) we will get a new result:

$$S' = \frac{1}{2} \int d^d x' (s^d) \left[(\partial'_\mu \phi')^2 \left(\frac{s^{(2-d)/2}}{s} \right)^2 - m^2 \phi'^2 \left(s^{(2-d)/2} \right)^2 \right] \qquad (3.5)$$

$$= \frac{1}{2} \int d^d x \left(\partial'_\mu \phi' \partial'^\mu \phi' - m^2 s^2 \phi'^2 \right). \qquad (3.6)$$

For a massless theory ($m^2 \equiv 0$) we therefore have $S = S'$, and the action is said to be *scale invariant*. Correlation functions computed with S' are related to $G^{(n)}$ by the relation

$$\langle \phi(sx_1) \cdots \phi(sx_n) \rangle_S = s^{\frac{n(2-d)}{2}} \langle \phi'(x_1) \cdots \phi'(x_n) \rangle_{S'}, \qquad (3.7)$$

where we also must redefine $m' = sm$.

Often, we are interested in the long distance (low momentum) behavior of the correlation functions, known as the infrared (IR) limit. From Eq. (3.7) we see that we can study this limit if we start with x_i near each other and take $s \to \infty$. Then the left hand side corresponds to the long distance limit of the correlation function, while the expectation value on the right hand side is a short-distance expectation value, but evaluated in a theory with a rescaled action. This connection between short and long distance physics is quite generic in quantum field theory.

Notice that in the rescaled action S', the mass term grows with s. Thus, as we take the long distance limit, we find that the mass term becomes more and more important when evaluating the expectation values. This is our first example of a *relevant operator*. In contrast, the kinetic term is independent of s in the rescaled action and so has equal importance at any scale. Such operators are called *marginal operators*, and they correspond to operators whose engineering dimension is equal to the spacetime dimension. Finally, if we were to include operators of higher engineering dimension than spacetime dimension, these operators would have coefficients that got *smaller* as $s \to \infty$; such operators are called *irrelevant operators*, as they play a weaker and weaker role in long distance physics.

Since each operator that appears in the Lagrange density must have mass dimension equal to the spacetime dimension (so the action remains dimensionless), it follows that the coefficients of an operator of engineering dimension d_e must come with a coupling constant of mass dimension $d - d_e$. The above classification then applies to the coefficients of operators: any coupling constant with positive dimension corresponds to a relevant operator; dimensionless couplings to marginal operators; and negative dimension to irrelevant operators.

It is absolutely vital that *any* quantum field theory only has a finite number of relevant and marginal operators, while having an infinite number of irrelevant operators. Irrelevant operators do not concern us: their effects manifest only at the shortest distances (highest energies) and therefore don't have a strong effect on observations. They can therefore be treated generically in perturbation theory and do not have a strong effect on observations.[1] On the other hand, relevant and marginal operators can never be ignored – they must be taken into account in the leading order theory, and determine things like universality classes, vacuum states, etc.

---

[1] Except, of course, when they do!

The fact that there are only a finite (and small) number of these operators is fortunate for us, since it means that there are effectively only a finite number of "theories" one can write down – the many irrelevant operators then correspond to calculable corrections to this small list of theories.

3.1.2 Example: hydrogen atom

Let us show that understanding dimensional power-counting alone will allow us to estimate decay probabilities. We will analyze a famous problem from atomic physics: calculating the decay width of the $2p$ state of Hydrogen, i.e.: a transition probability for $H(2p) \to H(1s) + \gamma$.

To describe the transition, we need to introduce two fields that annihilate the $2p$ and $1s$ states of Hydrogen. These are thought of as completely independent particles in this picture. The $1s$ state is a $J^P = 0^+$ state and therefore is annihilated by a real scalar field ϕ (it should be real since it will appear linearly in any operator that describes this transition). The $2p$ state has three polarizations[2], and so can be described by a massive vector field χ_μ. The photon is described in the usual way by a massless vector A_μ.

The next step is to identify the symmetries. There is electromagnetic gauge invariance, and since χ_μ and ϕ describe neutral states, we can only use the field strength $F_{\mu\nu}$. There is also Lorentz invariance which will force us to contract every index. There is also the discrete PT symmetry that QED respects. Then the leading operator we can write down has the form

$$\partial_\mu \cdot \phi \cdot F_{\nu\alpha} \cdot \chi_\beta \cdot X^{\mu\nu\alpha\beta} ,$$

where we have not yet said how the derivative can act, and the indices are contracted every possible way:

$$X^{\mu\nu\alpha\beta} \equiv a g^{\mu\nu} g^{\alpha\beta} + b g^{\mu\alpha} g^{\nu\beta} + c g^{\mu\beta} g^{\nu\alpha} + d \epsilon^{\mu\nu\alpha\beta} . \qquad (3.8)$$

Finally, PT symmetry forces $d = 0$ but the remaining contractions are in principle allowed.

Finally, we can use the tree level equations of motion and integrations by parts to eliminate some of the above possibilities. The relevant equations

---

[2] Note that we are completely ignoring the spin of the electron and proton. This means that we are neglecting the fine structure information, which should be acceptable as long as we do not push our approximations too far.

are
$$\partial_\mu F^{\mu\nu} = 0, \tag{3.9}$$
$$\partial_\mu \chi^\mu = 0. \tag{3.10}$$

Also recall that $F^{\mu\nu}$ is antisymmetric. In the end, we find only one operator survives. Using the result that in a relativistic QFT, bose fields have engineering dimension $d_e = 1$, as does the derivative operator, the relevant interaction is

$$\mathcal{L}_I = \frac{eg}{\Lambda} \phi F^{\mu\nu} G_{\mu\nu}, \tag{3.11}$$

where $G_{\mu\nu} \equiv \partial_\mu \chi_\nu - \partial_\nu \chi_\mu$. Notice that we could have derived this even earlier if we pretended that the vector field χ_μ had a spurious gauge symmetry. This symmetry is broken by the mass, but this only affects higher dimension operators. You should convince yourself this is true using a spurion argument!

Since we are emitting a single photon, we have included an explicit factor of the electron charge e. g is then an $\mathcal{O}(1)$ number, but what about Λ? The rule is that we should chose the lowest scale where we expect additional physics will have a strong effect on our predictions. This is the scale at which we expect our EFT to break down. In the hydrogen system, there are three obvious choices for such a scale: $\{m_e, a_0^{-1} \sim m_e \alpha, \mathbb{R} \sim m_e \alpha^2\}$. $\mathbb{R}$ is the Rydberg constant: at these energies we can start to excite the Hydrogen atom. However, we *are* exciting the atom, so our theory is still valid at this scale. The next higher scale is the Bohr radius, at which point fine structure of the hydrogen atom is relevant. Since we do not include these effects, this is the correct cutoff to use.

That was the hard part! The rest is straightforward QFT plug-and-chug. We want to calculate the matrix element for the transition

$$\begin{aligned}\mathcal{M} &= \frac{ge}{\Lambda} \langle H(1s); \gamma | \phi F^{\mu\nu} G_{\mu\nu} | H(2p) \rangle \\ &= \frac{ge}{\Lambda} (k^\mu \varepsilon^{*\nu}(k) - k^\nu \varepsilon^{*\mu}(k))(q_\mu \eta_\nu(q) - q_\nu \eta_\mu(q)),\end{aligned} \tag{3.12}$$

where k, ε are the momentum and polarization vector of the photon, and q, η are the momentum and polarization vectors of the $H(2p)$ state. It makes sense to do everything in the rest frame of the decaying $H(2p)$ state. In that case, $q = (M_{2p}, \mathbf{0})$, and $k = \omega(1, \mathbf{n})$ for some unit vector $\mathbf{n}$, with $\omega = \frac{3}{8} m \alpha^2$, the energy gap between the 2p and 1s state in natural units; note that this is much less than $\Lambda = m\alpha$, which suggests our power-counting scheme should work. Multiplying out the amplitude gives

$$\mathcal{M}_0 = \frac{2eg}{\Lambda} M_{2p} [\omega (\varepsilon \cdot \eta) - \varepsilon^0 (k \cdot \eta)]. \tag{3.13}$$

To calculate the full rate, we need to average over initial polarizations of the $H(2p)$ state and sum over final polarizations of the photon. The average gives us a factor of $1/3$, and the spin sums are

$$\sum_{\text{pols}} \varepsilon^\mu \varepsilon^{*\nu} = -g^{\mu\nu}, \tag{3.14}$$

$$\sum_{\text{pols}} \eta^\mu \eta^{*\nu} = -\left(g^{\mu\nu} - \frac{q^\mu q^\nu}{M_{2p}^2}\right). \tag{3.15}$$

Using these sums and identifying $\alpha = \frac{e^2}{4\pi}$ gives

$$\langle |\mathcal{M}|^2 \rangle = \frac{32\pi}{3} \alpha g^2 M_{2p}^2 \left(\frac{\omega}{\Lambda}\right)^2. \tag{3.16}$$

Finally, we have the formula for the rate of the process

$$\Gamma(H(2p) \to H(1s) + \gamma) = \frac{1}{2M_{2p}} \int d\Pi_2 \langle |\mathcal{M}|^2 \rangle$$

$$= \left(\frac{1}{2M_{2p}}\right)\left(\frac{1}{4\pi}\frac{|\mathbf{k}|}{E_{CM}}\right)\left(\frac{32\pi}{3} \alpha g^2 M_{2p}^2 \left(\frac{\omega}{\Lambda}\right)^2\right)$$

$$= \left(\frac{4}{3}g^2\right) \cdot \alpha a_0^2 \omega^3, \tag{3.17}$$

using $\Lambda = a_0^{-1}$, $|\mathbf{k}| = \omega$ and $E_{CM} = M_{2p}$. The correct answer from your favorite quantum mechanics textbook is

$$\Gamma(H(2p) \to H(1s) + \gamma) = \frac{2^{17}}{3^{11}} \cdot \alpha a_0^2 \omega^3. \tag{3.18}$$

This is in perfect agreement with our result if $g = 0.74$, which is certainly $\mathcal{O}(1)$!

3.2 Dimensional transmutation

In this section, we will restrict ourselves to massless scalar field theory in four dimensions, setting all irrelevant operators to zero. The action then has two terms[3]

$$S = \frac{1}{2}\int d^4x \left(\partial_\mu \phi \partial^\mu \phi - \frac{\lambda}{4!}\phi^4\right). \tag{3.19}$$

where $[\lambda] = 0$, corresponding to a marginal operator.

Beyond the leading order in perturbation theory, renormalization is required to obtain finite correlation functions. This means that the numerical

---
[3] We can forbid the ϕ^3 operator with a symmetry $\phi \to -\phi$.

value of λ is sensitive to how we chose to set our renormalization conditions. For example, suppose we chose to renormalize with dimensional regularization (DimReg) and minimal subtraction (MS), which we will see later is a very convenient scheme for EFT calculations. This scheme introduces a mass scale into the problem, which we call μ, and allows us to define a coupling at that scale $\lambda(\mu)$. This new scale is a new parameter – what is its meaning? Physical results cannot depend on the value we chose for μ, so we can ask: if we fix a value for $\lambda(\mu_1)$, what can be said for the value of $\lambda(\mu_2)$?

If λ is always small for all values of μ between μ_1 and μ_2, then we can safely use perturbation theory. Since λ is dimensionless and there are no other scales in the problem, the only way μ can appear explicitly is through logarithms:

$$\lambda(\mu_2) = \lambda(\mu_1) + B\lambda^2(\mu_1)\log\left(\frac{\mu_2}{\mu_1}\right) + \mathcal{O}(\lambda^3(\mu_1)), \qquad (3.20)$$

where B is some number that can be computed. Let us pass from μ_1 to μ_2 in small steps. We will define

$$\Delta \equiv \Delta \log(\mu) = \frac{1}{N}\log\left(\frac{\mu_2}{\mu_1}\right), \qquad (3.21)$$

with N some sufficiently large number. Then for Δ small enough we have

$$\lambda(\mu_1 e^\Delta) = \lambda(\mu_1) + B\lambda^2(\mu_1)\Delta + \cdots$$
$$\lambda(\mu_1 e^{2\Delta}) = \lambda(\mu_1 e^\Delta) + B\lambda^2(\mu_1 e^\Delta)\Delta + \cdots$$
$$\vdots$$
$$\lambda(\mu_2) = \lambda(\mu_1) + \sum_{j=0}^{N-1} B\lambda^2(\mu_1 e^{j\Delta})\Delta + \cdots$$

or, after taking the limit $N \to \infty$:

$$\lambda(\mu_2) = \lambda(\mu_1) + \int_{\mu_1}^{\mu_2} \frac{d\mu}{\mu} B\lambda^2(\mu) + \mathcal{O}(\lambda^3(\mu_1)). \qquad (3.22)$$

Taking the logarithmic derivative of both sides with respect to μ_2 then gives you

$$\mu\frac{d\lambda}{d\mu} = \beta_\lambda(\lambda), \qquad (3.23)$$

where

$$\beta_\lambda(\lambda) = B\lambda^2 + \mathcal{O}(\lambda^3). \qquad (3.24)$$

This is the celebrated *beta function* of ϕ^4 theory. Notice that since the logarithmic derivative of a dimensionless coupling is dimensionless, the beta function can only depend on the coupling constant itself. This is a special property of DimReg and MS, and would not hold if we used another renormalization scheme, as we will see later.

Note that the behavior of the coupling constant as we move from one renormalization scale to another is dependent on the sign of B. If $B > 0$, then the coupling gets stronger as we go to larger renormalization scales, while if $B < 0$ the opposite occurs. We will see how to interpret this below. In the case of ϕ^4 theory, this number is positive.

The good news is that Eq. (3.23) is straightforward to solve. The solution to this order in perturbation theory is given by

$$\lambda(\mu) = \frac{\lambda(\mu_0)}{1 + B\lambda(\mu_0)\log(\mu_0/\mu)},$$

where $\lambda(\mu_0)$ is our initial condition. We can write this in a clever way if we define a new scale Λ where this coupling blows up[4]:

$$1 + B\lambda(\mu_0)\log(\mu_0/\Lambda) = 0 \implies \Lambda = \mu_0 e^{1/B\lambda(\mu_0)} \qquad (3.25)$$

Λ is called the *Landau Pole* of the theory (not to be confused with a momentum cutoff regulator), and in terms of this parameter we have

$$\lambda(\mu) = \frac{1}{B\log(\Lambda/\mu)}. \qquad (3.26)$$

Notice that this is independent of our initial condition $\lambda(\mu_0)$.

So we see that even though we started with a theory with one dimensionless parameter (λ), after renormalization we can interpret it as a theory with one dimensionful parameter (Λ). This phenomenon is known as *dimensional transmutation*, and is quite generic in field theory (both quantum and thermal). It tells us that even though you may start with a theory that is scale invariant (as we did), quantum effects will ultimately make us sensitive to a scale. This will always happen unless $\beta = 0$. Such theories are called *conformal field theories*, and are a subject of their own. It is important to recognize that all the information contained in λ is also contained in Λ: specify one, and you can calculate the other to whatever accuracy your calculation allows. So it is not that quantum corrections generate new information above and beyond what is specified in your action, but that this information can appear in very interesting ways.

---

[4]Of course it doesn't *really* blow up – remember we are working in perturbation theory and our equations are no longer correct once λ gets too large. Therefore, it is better to say that this is the scale where perturbation theory fails.

At this point it is not entirely clear what we have done. Everything up to this point has been in terms of renormalization scales. But such scales do not carry any information – they are a theorist's ploy! The question we should be asking ourselves is: how does this give us insight into the energy dependence of correlation functions? To do that, we need to introduce the "Renormalization Group" methods of Callan and Symanzik.

3.3 Callan-Symanzik Equation

We know from quantum field theory that one particle irreducible (1PI) functions must be renormalized:

$$\Gamma_0^{(n)}(p_1, \cdots, p_n) = Z^{-n/2}(M)\Gamma^{(n)}(p_1, \cdots, p_n; M) , \qquad (3.27)$$

where Z is the field renormalization constant, and M is some scale introduced after renormalization; either the cutoff, renormalization scale, or the parameter of dimensional regularization. As the left hand side does not depend on this scale, it must be that dependence on M cancels between the renormalization constant and the renormalized Green's function. Writing this out explicitly as

$$M\frac{d}{dM}\Gamma_0^{(n)} = 0$$

and noting that there is implicit dependence on M in the masses and coupling constants after renormalization, we have[5]

$$\left[M\frac{\partial}{\partial M} + \beta(\lambda)\frac{\partial}{\partial \lambda} - n\gamma(\lambda) + m^2\gamma_{m^2}(\lambda)\frac{\partial}{\partial m^2}\right]\Gamma^{(n)} = 0 . \qquad (3.28)$$

where we have introduced the *anomalous dimension*

$$\gamma(\lambda) = \frac{\partial}{\partial M}\log\sqrt{Z} , \qquad (3.29)$$

as well as the anomalous mass dimension

$$m^2\gamma_{m^2}(\lambda) = M\frac{\partial m^2}{\partial M} \qquad (3.30)$$

where m^2 is the renormalized scalar mass-squared. A similar equation holds for the n point correlation functions

$$G_0^{(n)}(p_1, \cdots, p_n) = Z^{n/2}(M)G^{(n)}(p_1, \cdots, p_n; M) , \qquad (3.31)$$

---

[5] As we are working with scalar fields for the moment, the mass parameter is m^2, not m, which is why it appears this way. For fermion fields, the parameter would be linear in mass.

Note that the bare correlation functions are proportional to Z rather than inversely proportional like the 1PI functions, leading to a sign change in the γ term:

$$\left[M\frac{\partial}{\partial M} + \beta(\lambda)\frac{\partial}{\partial \lambda} + n\gamma(\lambda) + m\gamma_m(\lambda)\frac{\partial}{\partial m}\right] G^{(n)} = 0 \,. \tag{3.32}$$

Eqs. (3.28) and (3.32) are the *Callan-Symanzik Equations* for the 1PI and correlation functions, respectively. This equation gives us the tool needed to compute the beta function and anomalous dimensions of a theory. We just compute various correlation functions, demand that they satisfy Eq. (3.32), and then read off the coefficient functions. We will see how this works with a few more examples from ϕ^4 theory.

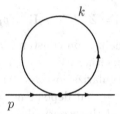

Fig. 3.1   One loop correction to the two-point function in scalar field theory.

Consider first the two-point function, which includes one-loop corrections from diagrams like Fig. 3.1:

$$G^{(2)}(p) = \frac{1}{p^2 - m^2} + \frac{1}{p^2 - m^2}\left[I + (p^2\delta Z - \delta m^2)\right]\frac{1}{p^2 - m^2} \tag{3.33}$$

where

$$I = \frac{\lambda}{2}\int \frac{d^4k}{(2\pi)^4}\frac{1}{k^2 - m^2} \to \frac{\lambda\mu^{2\epsilon}}{2(4\pi)^{d/2}}\frac{\Gamma(1 - d/2)}{(m^2)^{1-d/2}} \quad \text{(DimReg)}$$

$$\to \frac{\lambda}{32\pi^2}\Lambda^2 + \cdots \quad \text{(cutoff)}$$

Notice that this result is independent of p^2, regardless of renormalization scheme, so $\delta Z = 0$ to this order in perturbation theory. Thus $Z = 1 + \delta Z = 1 + \mathcal{O}(\lambda^2)$, and from Eq. (3.29), this means $\gamma = \mathcal{O}(\lambda^2)$. The remaining divergence is cancelled by the mass counterterm.

Now let us consider the four-point function:

$$G^{(4)}(p_1, p_2, p_3, p_4) = \left(\prod_{i=1}^{4}\frac{i}{p_i^2 - m^2}\right)$$
$$\times \left[-i\lambda + (-i\lambda)^2[iV(s) + iV(t) + iV(u)] - i\delta\lambda\right], \tag{3.34}$$

Elementary Techniques

with

$$V(Z) = -\frac{\mu^{2\epsilon}\Gamma(2-d/2)}{2(4\pi)^{d/2}} \int_0^1 \frac{dx}{[m^2 - x(1-x)Z]^{2-d/2}} \,. \quad (3.35)$$

We will need to specify a renormalization condition to nail down the coupling counterterm:

$$G^{(4)} = -i\lambda \quad \text{for} \quad s = t = u = -M^2 \,, \quad (3.36)$$

where M is the subtraction scale[6]. This condition gives us in the massless limit,

$$\delta\lambda = (-i\lambda)^2 \cdot 3V(-M^2) \to \frac{3\lambda^2 \mu^{2\epsilon}}{2(4\pi)^{d/2}} \int_0^1 \frac{\Gamma(2-d/2)\, dx}{[x(1-x)M^2]^{2-d/2}}$$

$$= \frac{3\lambda^2}{32\pi^2}\left[\frac{1}{\epsilon} + \log\left(\frac{\mu^2}{M^2}\right) + \cdots\right]$$

where we work in DimReg and only keep terms that diverge as $\epsilon \to 0$ or that depend on M. Since the only place M will appear is in the counterterm, we have[7]

$$M\frac{\partial G^{(4)}}{\partial M} = M\frac{\partial}{\partial M}(-i\delta\lambda) = \frac{3i\lambda^2}{16\pi^2} \cdot \prod_{i=1}^{4} \frac{i}{p_i^2} \,. \quad (3.37)$$

The next term in the Callan-Symanzik equation is the beta function term:

$$\beta(\lambda)\frac{\partial G^{(4)}}{\partial \lambda} = \beta(\lambda)(-i + \mathcal{O}(\lambda)) \prod_{i=1}^{4} \frac{i}{p_i^2} \,. \quad (3.38)$$

Finally, $\gamma G^{(4)} = \mathcal{O}(\lambda^3)$ and can therefore be ignored at this order in perturbation theory. Putting it together, the Callan-Symanzik equation reads:

$$\left[\frac{3i\lambda^2}{16\pi^2} + (-i)\beta(\lambda)\right]\prod_{i=1}^{4} \frac{i}{p_i^2} = 0$$

or

$$\beta(\lambda) = \frac{3\lambda^2}{16\pi^2} + \mathcal{O}(\lambda^3) \,. \quad (3.39)$$

Let us pause to take note of a few points from this calculation:

---

[6] If $m \neq 0$ then we can set $M = 0$, but since we will be considering the massless limit, we keen $M \neq 0$ to avoid any infrared singularities in the counterterms.

[7] Notice that when $m \neq 0$, we can get terms that go like m/M, which could change our results. We will discuss these terms in Section 3.7.

(1) $\gamma = 0$ at leading order in ϕ^4 theory, but this is very special to this theory. In a generic scalar field theory, we have

$$G^{(2)}(p) = \text{(tree level)} + \text{(one loop)} + \text{(counterterm)}$$

$$= \frac{i}{p^2} + \frac{i}{p^2}\left(p^2 A\left[\frac{1}{\epsilon} + \log\left(\frac{\mu^2}{-p^2}\right)\right] + \text{finite}\right)\frac{i}{p^2}$$

$$+ \frac{i}{p^2}\left(ip^2 \delta Z\right)\frac{i}{p^2} . \tag{3.40}$$

The M dependence comes from the couterterm, so the Callan-Symanzik equation reads

$$\frac{i}{p^2}\left(-M\frac{\partial(\delta Z)}{\partial M} + 2\gamma\right) = 0 ,$$

or

$$\gamma = \frac{1}{2}M\frac{\partial(\delta Z)}{\partial M} . \tag{3.41}$$

Since $\delta Z = -A\log(M^2)$ plus M-independent terms, this means $\gamma = -A$; that is, (minus) the coefficient of the pole.

(2) It is possible to show (and we will do so shortly) that

$$\Gamma^{(n)}(tp_i, \lambda, \mu) = t^{4-n} \exp\left[-n\int_0^t \frac{dt'}{t'}\gamma(\bar\lambda(t'))\right] \tilde\Gamma^{(n)}(p_i, \bar\lambda(t), \mu) , \tag{3.42}$$

where $\bar\lambda$ is the running coupling defined by

$$t\frac{d\bar\lambda}{dt} = \beta(\bar\lambda) . \tag{3.43}$$

Thus we find that the behavior of 1PI Green's functions depends on the beta function in a nontrivial way.

It is an important case when the beta function has a zero, since at that point the running coupling stops running! This leads us to make the following definition: λ_0 is called a *fixed point* if $\beta(\lambda_0) = 0$. If $\beta'(\lambda_0) < 0$ it is a *UV stable* fixed point; and if $\beta'(\lambda_0) > 0$ it is a *IR stable* fixed point. A *domain* is the set of λ between two fixed points. If $\beta(\lambda) < 0$ for small values of λ the theory is said to be *asymptotically free*.

Notice that a vanishing coupling is always a fixed point, called the *Gaussian Fixed Point*, because the action is quadratic at this fixed point. It is also sometimes called the *trivial* fixed point, since there are no interactions.

Condensed matter physicists generally work in position space rather than energy-momentum space, so they usually classify running in terms of changing *length* scales rather than energy scales. As a result, their terminology is reversed ($\beta'(\lambda_0) < 0$ is an IR stable fixed point, etc.) since the slope is change of length scale rather than energy scale.

3.4 The renormalization group

The Callan-Symanzik Equation is the central equation of the renormalization group (RG). After removing divergences from a correlation function, you are often left with a residual scale dependence. Changing the value of that scale cannot have any effect on the correlation function, so any change must be cancelled by a corresponding change in the coupling constants. This fact allows you to absorb higher-order contributions into the changing couplings. The process is sometimes called "resummation of logarithms". In this section, we will introduce the tools you need to accomplish this.

3.4.1 *Engineering dimensions of* $\Gamma^{(n)}$

We begin by working out the dimensions of the correlation functions. Recall that $[\phi] = d/2 - 1$ and

$$G^{(n)}(x_1, \cdots, x_n) = \langle 0|T\phi(x_1)\cdots\phi(x_n)|0\rangle .$$

This means that $[G^{(n)}(x)] = n(d/2 - 1)$, or just n in $d = 4$. As a check, recall the propagator:

$$G^{(2)}(x,y) = \int \frac{d^dp}{(2\pi)^d} \frac{e^{ip\cdot(x-y)}}{p^2 - m^2 + i\varepsilon} \implies [G^{(2)}] = d - 2 .$$

If we take the Fourier transform[8] of this we obtain[9]

$$G^{(n)}(p_1, \cdots, p_n)\, \delta^{(d)}(\sum p_i) = \int \prod_{i=1}^{n}(d^d x_i\, e^{-ip_i \cdot x_i})\, G^{(n)}(x_1, \cdots, x_n) . \tag{3.44}$$

The delta function on the left hand side of this expression has $-d$ dimensions, while the measure of the transform has dimension $-nd$. Plugging this in with the dimensions of $G^{(n)}$ then tells us that $[G^{(n)}(p)] = d - n(1+d/2)$, or $4 - 3n$ in $d = 4$. Again, this is consistent with the 2-point function: $G(p) = 1/(p^2 - m^2)$.

What about 1PI functions? Recall the Legendre transformation:

$$W[J] = -\Gamma[\phi] - \int d^dx\, J(x)\phi(x) , \tag{3.45}$$

---

[8] To avoid ugly notations, we will denote both the function and its Fourier transform by the same symbol, including its arguments of either position or momentum if the context is unclear which it is.

[9] Notice that due to translation invariance, $G^{(n)}(x)$ is only a function of $n - 1$ coordinates, and the Fourier transform of the last coordinate becomes a delta function. We explicitly factor out this delta function in the definition of the transformed correlation function, as we see in this equation.

where $W[J] \equiv \log Z[J]$ is the generator of n point correlation functions (corresponding to the Helmholtz Free Energy), and $\Gamma[\phi]$ is the effective action, (corresponding to the Gibbs Free Energy). We can write $\Gamma[\phi]$ as

$$\Gamma[\phi] = \sum_{n=0}^{\infty} \left(\prod_{i=0}^{n} \int d^d x_i \, \phi(x_i) \right) \Gamma^{(n)}(x_1, \cdots, x_n) \,. \tag{3.46}$$

Since the effective action is dimensionless, and the product in parentheses has dimension $-nd + n(d/2 - 1)$, we have $[\Gamma^{(n)}(x_1, \cdots, x_n)] = n(d/2 + 1)$, or $3n$ in $d = 4$. Taking the Fourier transform as above, we get $[\Gamma^{(n)}(p_1, \cdots, p_n)] = d + n(1 - d/2)$, or $4 - n$ in $d = 4$.

3.4.2 Physics of the anomalous dimension

Now let us see how these considerations help us to understand how the n point correlation functions scale with momenta. Suppose we have calculated $\Gamma^{(n)}(p)$ at some fixed momenta. Consider again the rescaling $p \to tp$ ($x \to sx$ in terms of the position space rescaling, with $s \equiv 1/t$). We will also rescale $m \to tm$, and $\mu \to t\mu$. Then using dimensional analysis along with the results from above, we have

$$\Gamma^{(n)}(tp_i, \lambda, m, \mu) = t^D \Gamma^{(n)}(p_i, \lambda, m/t, \mu/t) \,,$$

where $D = d + n(1 - d/2)$. Taking the derivative of this equation with respect to t gives[10]

$$\left[t \frac{\partial}{\partial t} + m \frac{\partial}{\partial m} + \mu \frac{\partial}{\partial \mu} - D \right] \Gamma^{(n)} = 0 \,, \tag{3.47}$$

and plugging in $\mu \partial \Gamma / \partial \mu$ in from Eq. (3.28) gives us[11]

$$\left[-t \frac{\partial}{\partial t} + \beta(\lambda) \frac{\partial}{\partial \lambda} - n\gamma(\lambda) + m(\gamma_m(\lambda) - 1) \frac{\partial}{\partial m} + D \right] \Gamma^{(n)}(tp_i, \lambda, m, \mu) = 0 \,. \tag{3.48}$$

---

[10] As mentioned in the the footnote above Eq. (3.28), the mass parameter of scalar fields is m^2, but to avoid cluttered notation, and to be more in line with the fermion masses where this is often used, we will switch to calling the mass parameter m. This should not cause confusion, but you should remember this distinction if you ever need to compute the running of a scalar mass.

[11] The Callan-Symanzik Equation took the derivative with respect to the subtraction scale M; but we can always choose to identify our DimReg scale $\mu = M$.

Elementary Techniques

Now let us solve this equation. From our previously gained intuition, we know that a change in t is accompanied by changes in λ and m, so let us try the following ansatz:

$$\Gamma^{(n)}(tp_i, \lambda, m, \mu) = f(t)\tilde{\Gamma}^{(n)}(p_i, \bar{\lambda}(t), \bar{m}(t), \mu) , \qquad (3.49)$$

where $\bar{\lambda}$, $\bar{m}$ are some functions of t. If we take the logarithmic derivative with respect to t:

$$t\frac{d}{dt}\Gamma^{(n)}(tp_i, \lambda, m, \mu) = t\frac{df}{dt}\tilde{\Gamma}^{(n)} + tf(t)\left(\frac{d\bar{m}}{dt}\frac{\partial\tilde{\Gamma}^{(n)}}{\partial m} + \frac{d\bar{\lambda}}{dt}\frac{\partial\tilde{\Gamma}^{(n)}}{\partial \lambda}\right)$$

$$= \left(\frac{t}{f(t)}\frac{df}{dt} + t\frac{d\bar{m}}{dt}\frac{\partial}{\partial m} + t\frac{d\bar{\lambda}}{dt}\frac{\partial}{\partial \lambda}\right)\Gamma^{(n)}(tp_i, \lambda, m, \mu) .$$

Comparing this to Eq (3.28) we find that if we chose the functions $f(t)$, $\bar{m}(t)$, $\bar{\lambda}(t)$ to satisfy

$$t\frac{d\bar{\lambda}}{dt} = \beta(\bar{\lambda}) , \qquad (3.50)$$

$$t\frac{d\bar{m}}{dt} = \bar{m}(\gamma_m(\bar{\lambda}) - 1) , \qquad (3.51)$$

and

$$\frac{t}{f(t)}\frac{df}{dt} = D - n\gamma(\bar{\lambda}) \quad \Rightarrow \quad f(t) = t^D \exp\left[-n\int_0^t \frac{dt'}{t'}\gamma(\bar{\lambda}(t'))\right] , \qquad (3.52)$$

with the boundary conditions

$$\bar{\lambda}(0) = \lambda; \qquad \bar{m}(0) = m , \qquad (3.53)$$

then this constitutes a solution to the Callan-Symanzik equation.

At this point, let us stop and make a few key observations:

(1) If $\gamma \equiv 0$, then our previous work using the engineering dimension comes back. However, if $\gamma \neq 0$ then engineering dimensional analysis fails, and we have more work do to. This is why γ is called the "anomalous dimension".

If γ is a constant, then everything we did before is correct, *except* we must replace our engineering dimension by a new quantity called the *scaling dimension*: $d_s = d_e + \gamma$. However, since γ is a function of couplings, which are in turn a function t, it is not generally true that the anomalous dimension is constant. The exception to this is when you are at a fixed point, in which case $\gamma(\lambda_*) = \gamma_*$, and

$$f(t) = t^{D-n\gamma_*} .$$

Where did this extra dimension come from? Our understanding of the momentum dependence of the correlation functions only used dimensional analysis, so how could a field (or correlation function) acquire a new dimension? Recall that we know how correlation functions behave with energy from dimensional analysis, which followed from identifying all the relevant scales. If there are no scales in the theory (such as the massless ϕ^4 theory), then the analysis is very easy. Adding masses and dimensionful couplings complicates matters, but can still be handled in a straightforward extension.

However, in a quantum field theory, there is always another scale through dimensional transmutation described in Section 3.2. Therefore, if ordinary dimensional analysis would give you a correlation function that behaves like $G(p) \sim \frac{1}{p^s}$, after computing quantum corrections, we can get $G(p) \sim \frac{\Lambda^\eta}{p^{s+\eta}}$, where Λ is the Landau Pole and η is an additional power[12]. Notice that this comes about only after quantum corrections are taken into account, justifying the name, "anomalous." The quantum generation of a new scale is a manifestation of the scale anomaly of quantum field theory.

When we are away from a fixed point, the anomalous dimension is not a constant, and this analysis must be generalized, but the basic idea is the same.

(2) Notice that when $d_e = d$ we have a marginal operator, but after including the anomalous dimension, the scaling dimension is no longer zero! Thus quantum corrections can make a marginal operator relevant or irrelevant by generating an anomalous dimension.

(3) At this point, we can see how the renormalization group helps us to "extend" the validity of perturbation theory. If a correlation function computed in perturbation theory takes the form $A_0 + A_1 \lambda \log(\mu^2/p^2) + \cdots$, then when $\lambda \log(\mu^2/p^2) \gg 1$ perturbation theory fails. This will always happen at some value of the momentum; or in terms of the above notation, at some value of t. However, we found that this correlation function is related to the correlation function evaluated with momentum in the region where this logarithm is not too large, as long as we replace the coupling constants with the running coupling constants evaluated at the appropriate point. *Thus the Callan-Symanzik Equation resums danngerous logarithms!*

---

[12] We call it η here rather than γ to emphasize the relationship to the critical exponent from the theory of phase transitions.

3.5 Renormalizability and Effective Field Theory

3.5.1 *Dropping renormalizability*

As we learned in quantum field theory, a Lagrangian describes a *renormalizable* theory if it is composed of operators such that $[\mathcal{O}] \leq d$, the spacetime dimension. If this holds, any divergences that appear can be absorbed into a finite number of counterterms. Otherwise, we require an infinite number of counterterms to absorb the divergences, and our theory loses its predictive power. This is why theoretical physicists put so much store in renormalizability for a theory to be of any use.

However, there is a way around this problem. Organize the Lagrangian according to the dimensions of the operators

$$\mathcal{L}_{\text{eff}} = \mathcal{L}_d + \mathcal{L}_{d+1} + \mathcal{L}_{d+2} + \cdots$$

where $\mathcal{L}_d$ is a renormalizable Lagrangian, and $\mathcal{L}_a$ contains operators of engineering dimension a for $a > d$. These additional pieces to the Lagrangian violate the renormalizability condition and we require all of these operators to have a complete theory. However, the matrix elements of these operators will go like $\langle \mathcal{L}_{d+n} \rangle \sim (p/\Lambda)^n$ for some mass scale Λ (this may or may not equal the Landau Pole from Section 3.2). Thus, even though we require an infinite number of counterterms to get finite results, we need only consider a finite subset of operators to get an answer accurate to a given order in an energy-momentum expansion. In other words, if loop effects from $d + n$ dimensional matrix elements contribute at energy order $d + n + 1$, we have *approximate* predictive power at a given accuracy. This should be sufficient for any physicist's needs!

3.5.2 *Matching*

The procedure for writing down an effective theory valid for energies below Λ comes in three steps:

(1) Write the most general Lagrangian consistent with the symmetries of the problem containing operators with dimension less than or equal to $d + n$ for some non-negative n.
(2) Supplement it with a power counting scheme.
(3) Compute several physical quantities to fix counterterms and coefficients based on external data.

This procedure will generate an effective theory accurate to $\mathcal{O}((E/\Lambda)^n)$, which can be used for computations. In some cases, this program can be modified slightly if one knows how the theory looks in a certain limit. In that case, the unknown coefficients can be computed rather than fixed by renormalization conditions by the procedure of **matching**, which we will discuss now.

Let us suppose we know the theory at very large scales ($\mu \to \infty$), which we call the "full theory;" for example, imagine two scalar fields, one heavy σ (mass M) and one light ϕ (mass m). Let us write this theory as

$$\mathcal{L}_{\text{tot}} = \mathcal{L}(\phi) + \mathcal{L}_H(\phi, \sigma) , \qquad (3.54)$$

where $\mathcal{L}(\phi)$ are all the operators that do not depend on the heavy field, and $\mathcal{L}_H$ are all the other operators. For a renormalization scale $\mu \gg M$ physics is described by this Lagrangian. When we are interested in correlation functions evaluated at lower energies we can use the Callan-Symanzik equation to derive the result, a process known as "RG Running."

But as we run the theory down to the scale $\mu \sim M$, the σ field stops being a propagating degree of freedom; that is, there is no longer enough energy to produce a real σ state. Therefore we can remove it from our theory and set $\mathcal{L}_H \to 0$. However, even though the σ field is no longer a real degree of freedom, it can still appear as a virtual particle in Feynman diagrams, and this will generate new operators built out of ϕ. Below the scale $\mu = M$, our theory is therefore described by the Lagrangian

$$\mathcal{L}_{\text{eff}} = \mathcal{L}(\phi) + \Delta\mathcal{L}(\phi) ,$$

where $\Delta\mathcal{L}$ contains all the newly generated operators, which are in general nonrenormalizable. These operators are generally constrained by whatever symmetries existed in the full theory.

Once we have this new Lagrangian, we can connect the "full theory" to the "effective theory" by calculating correlation functions in both theories, evaluated at the scale $\mu = M$, and demand that they give the same answer to the order in which we are working. This fixes the coefficients of the effective action at the scale $\mu = M$, and we can then run the theory down to lower scales as before using the Callan-Symanzik equation.

3.6 Subtraction schemes as part of EFT definition

Consider, as an example, the $u_L - u_L - Z^0$ vertex in particle physics:

$$\mathcal{L} = \frac{g}{\cos\theta_W} \left(\frac{1}{2} - \frac{2}{3} \sin^2\theta_W \right) \bar{u}_L \gamma^\mu u_L Z^0_\mu , \qquad (3.55)$$

At tree level, the operator has a coefficient

$$C_{\text{tree}} = \frac{g}{\cos\theta_W}\left(\frac{1}{2} - \frac{2}{3}\sin^2\theta_W\right) \sim \mathcal{O}(1) \text{ in } \frac{1}{M_W}\,.$$

At one loop, we also have dimension 6 operators suppressed by $G_F \sim 1/M_W^2$, such as those described later in Eq (4.28). These operators describe a $u \to q$ transition, and if the intermediate q interacts with the Z^0, they can contribute to the effective vertex,

$$\bar{u}_L C_{1\,\text{loop},6}^\mu u_L \sim \frac{4G_F}{\sqrt{2}} V_{ui} V_{ui}^* \int \frac{d^4k}{(2\pi)^4} \bar{u}\gamma_\mu P_L \frac{i}{\slashed{k} - m_i}$$

$$\times \frac{g}{\cos\theta_W}\left(-\frac{1}{2} + \frac{1}{3}\sin^2\theta_W\right) \gamma_\alpha P_L \frac{i}{\slashed{k}+\slashed{q}-m_i}\gamma^\mu P_L u$$

$$\sim \int \frac{d^4k}{(2\pi)^4} \frac{1}{k^2} \frac{1}{M_W^2} \times \{\cdots\} \qquad (3.56)$$

This integral is quadratically divergent. We can regulate this with a momentum cutoff Λ; what value should we chose[13] for it? Since our theory already has a cutoff at M_W, it would make sense to identify this as our cutoff for the integral. However, that means that

$$C_{1\,\text{loop},6} \sim \frac{1}{M_W^2}\Lambda^2 + \cdots \sim \mathcal{O}(1) \text{ in } \frac{1}{M_W}\,. \qquad (3.57)$$

Thus, the one loop effects are *not* suppressed by M_W!

We can keep going, and consider one loop diagrams with an insertion of $\mathcal{L}_8$, which contain operators with derivatives in it, bringing in powers of momentum into the numerator of the loop integral:

$$C_{1\,\text{loop},8} \sim \frac{1}{M_W^4} \int^\Lambda d^4k \frac{k^2}{k^2} \sim \frac{1}{M_W^4}\Lambda^4 \sim \mathcal{O}(1) \text{ in } \frac{1}{M_W}\,. \qquad (3.58)$$

This is a disaster! It means that our energy expansion is not robust: we need to calculate diagrams with insertions of arbitrarily large dimensional operators to compute something to fixed order in $1/M_W$. This is precisely what the founding fathers of quantum field theory feared: nonrenormalizable theories have no predictive power.

The solution to this problem may be the more important aspect of effective field theory: we must introduce a *mass independent subtraction scheme*, where the renormalization conditions do not depend on a mass scale. This is quite different from what we discussed above, but it solves the problem right away.

---

[13] The more correct question would be: at what subtraction scale should be renormalize? However, both questions lead to the same result, as we will see.

The most famous example of a mass independent scheme is DimReg and Minimal-Subtraction (MS), where rather than fixing the correlation functions at a specified scale, we simply set the counterterms to cancel the pole. In this scheme, the integrals above become

$$C_{1\text{ loop},6} \sim \frac{1}{M_W^2} \int d^d k \frac{1}{k^2 - m^2} \sim \frac{m^2}{M_W^2} \log \mu \,,$$

$$C_{1\text{ loop},8} \sim \frac{1}{M_W^4} \int d^d k \frac{k^2}{k^2 - m^2} \sim \left(\frac{m^2}{M_W^2}\right)^2 \log \mu \,,$$

where we have reintroduced a generic quark mass m. We see some things right away:

(1) As long as $m \ll M_W$, this expansion is robust to quantum corrections. This is true at all orders.
(2) Power counting rules from the Lagrangian follow automatically in this scheme, with "subtraction" of divergences rather than renormalization. The DimReg scale μ now plays the role of the subtraction scale M from our previous work.
(3) This effective field theory with this subtraction scheme behaves exactly as the ordinary renormalizable quantum field theory, despite the fact that it is *not* renormalizable. In particular, dimensional analysis works!
(4) The major difference between this and other schemes is that *the β function is independent of scale!*

3.7 Decoupling. Appelquist-Carrazone theorem in various schemes

What are the effects of heavy particles on low energy Lagrangians in general? The answer is given by the *Decoupling Theorem* of [Appelquist and Carazzone (1975)]:

Theorem 3.1. *Heavy fields of mass m decouple at low energy, generating operators suppressed by factors of m^{-1}, except for their contribution to renormalization effects.*

We will not prove this theorem, but we will motivate it. Heavy fields in loops contribute propagators to the integrands that go like $(p^2 - m^2)^{-1}$, where p is some combination of loop and external momenta. When the external momenta are small, and the integral is convergent, then this contributes a factor of $1/m^2$; if the integral is divergent, then the finite parts go like $1/m^2$, while the divergent parts effect the counterterms, as the theorem dictates.

From the decoupling theorem, it is expected that heavy particles should decouple from low energy processes, not contributing to them in the limit $E/m \to 0$ up to renormalization effects. This must also hold in our effective field theory using our chosen renormalization scheme.

To examine this issue further, let us consider the contributions of heavy charged fermions to the running of the QED coupling constant. We need to compute the QED beta function. This can be done, as before, by considering various correlation functions in QED, and demanding that they satisfy the Callan-Symanzik equation.

We follow a procedure similar to what we used in Eq (3.40). Let us consider the 3-point function of two charged fermions and a photon:

$$G^{(1,1,1)}(p_1, p_2, q) = \langle \bar{\psi}(p_1)\psi(p_2)A^\mu(q) \rangle$$
$$= \text{(tree level)} + \text{(1PI loop contributions)}$$
$$+ \text{(vertex counterterm)} + \text{(external leg corrections)}$$

Suppose that, despite the troubles we alluded to, we use a *mass dependent* subtraction scheme, renormalizing everything by subtracting at a scale $p^2 = -M^2$. Applying the Callan-Symazik equation to the vertex function above gives

$$M\frac{d}{dM}G^{(1,1,1)} = \left[\frac{1}{\not{p}_1}\gamma_\nu \frac{1}{\not{p}_2}\frac{g^{\mu\nu} - q^\mu q^\nu/q^2}{q^2}\right]$$
$$\times \left\{ M\frac{\partial}{\partial M}\left[\delta e - e\sum_{i=1}^n \delta Z_i\right] + \beta(e) \right.$$
$$\left. + e\sum_{i=1}^n \frac{1}{2}M\frac{\partial}{\partial M}\delta Z_i \right\} = 0 , \qquad (3.59)$$

where the last term is our expression for the anomalous dimension from Eq (3.41). This then implies

$$\beta(e) = M\frac{\partial}{\partial M}\left(-\delta e + \frac{1}{2}e\sum_{i=1}^n \delta Z_i\right) . \qquad (3.60)$$

In QED, the Ward identity forces $\delta e = e\delta Z_1 = e\delta Z_2$, and they cancel in the above expression. Z_3 is the photon renormalization constant computed from the bubble graph

$$\Pi_0(p^2) = \frac{e^2}{2\pi^2}\left[\frac{1}{\epsilon} - \gamma - \int_0^1 dx\, x(1-x)\log\left(\frac{m^2 - p^2 x(1-x)}{4\pi\mu^2}\right)\right] - i\delta Z_3 .$$
$$(3.61)$$

Our momentum subtraction scheme gives us the finite result:

$$\Pi(p^2; M) = -\frac{e^2}{2\pi^2} \int_0^1 dx\, x(1-x) \log\left(\frac{m^2 - p^2 x(1-x)}{m^2 + M^2 x(1-x)}\right), \quad (3.62)$$

and the beta function is

$$\beta(e) = \frac{e}{2} M \frac{\partial Z_3}{\partial M} = +\frac{e}{2} \frac{e^2}{2\pi^2} M \frac{\partial}{\partial M} \int_0^1 dx\, x(1-x) \log\left(\frac{m^2 + M^2 x(1-x)}{4\pi\mu^2}\right)$$

$$= \frac{e^3}{2\pi^2} \int_0^1 dx\, x(1-x) \left(\frac{M^2 x(1-x)}{m^2 + M^2 x(1-x)}\right). \quad (3.63)$$

As this is a mass dependent scheme, it should not surprise us that the beta function has M dependence. From this formula, we see two limiting cases:

(1) $m \ll M$: In this case, we have

$$\beta(e) = \frac{e^3}{2\pi^2} \int_0^1 dx\, x(1-x) + \cdots = \frac{e^3}{12\pi^2} + \mathcal{O}\left(\frac{m^2}{M^2}\right). \quad (3.64)$$

This is the usual expression for the QED beta function.

(2) $m \gg M$: In this case, we have

$$\beta(e) = \frac{e^3}{2\pi^2} \int_0^1 dx\, x(1-x) \frac{M^2 x(1-x)}{m^2} + \cdots = \frac{e^3}{60\pi^2} \frac{M^2}{m^2} + \mathcal{O}\left(\frac{M^4}{m^4}\right), \quad (3.65)$$

which vanishes as m becomes large. In addition, Eq (3.62) also vanishes in this limit, as long as you are dealing with energies where $p^2 \ll m^2$. Thus the heavy fermion does in fact decouple from the photon at low energy, both in the beta function and the finite parts, just as the Appelquist-Carrazone Theorem demands.

So the decoupling theorem holds for the beta function in a mass dependent scheme. What about for a mass independent scheme? If we use minimal subtraction, our expression for the beta function becomes

$$\beta(e) = \frac{e}{2} \mu \frac{\partial \Pi_{MS}(p^2)}{\partial \mu} = -\frac{e^3}{4\pi^2} \mu \frac{\partial}{\partial \mu} \int_0^1 dx\, x(1-x) \log\left(\frac{m^2 - p^2 x(1-x)}{4\pi\mu^2}\right)$$

$$= +\frac{e^3}{2\pi^2} \int_0^1 dx\, x(1-x) = \frac{e^3}{12\pi^2}, \quad (3.66)$$

independent of mass, as we expect. However, since this is mass independent, it cannot obey the decoupling theorem! Indeed, the result we get here is identical to the contribution from a *massless* fermion considered above. This would suggest that (up to charges) the top quark and the electron would contribute the same amount to the running of the QED coupling, no matter what scale you do your experiment.

The situation is even worse than it seems. Imagine we are doing an experiment at very small p^2. In that case, the finite part of the polarization becomes

$$\Pi_{MS}(p^2) \simeq \frac{e^2}{2\pi^2} \int_0^1 dx\, x(1-x) \log\left(\frac{m^2}{\mu^2}\right) . \qquad (3.67)$$

and we see that there is a large logarithm, depending on our choice[14] of μ. Thus we see that this mass independent scheme has generated a large logarithm that has the potential to spoil perturbation theory.

This is ironic: before we showed that a mass dependent scheme violates the power counting when computing higher order amplitudes, and that we must have a mass independent scheme to avoid this problem. Now we are finding that the same problem happens in mass independent schemes, albeit in a different way. However, there is a beautiful solution that solves both of the above problems at once. The trick is to continue using a mass independent scheme, and force the decoupling theorem to hold explicitly: when working at energies below the mass of a particle, *we must remove the particle from the dynamics by hand*. This is precisely what we mean when we talk about *integrating out* a particle. Then, when faced with large logarithms as above, we can resum them using the Callan-Symanzik equation. This is perhaps the most striking difference between "ordinary" quantum field theory and "effective field theory."

3.8 Notes for further reading

This chapter reviewed the main ideas behind renormalization and the renormalization group. One of the best textbooks out there that explains the physics behind these concepts is [Ramond (1981)]. [Weinberg (1996)] also explains many of the details, but can be challenging. Coleman's lectures in [Coleman (1985)] are an absolute *must read*. Also the original paper on the "Coleman-Weinberg potential" is a classic, easy to follow and full of great explanations of dimensional transmutation [Coleman and Weinberg (1973)]. Finally, a great text to get some ideas of how the ideas of the text are applied in particle physics is [Georgi (1984)].

---

[14]In this simple case, there is the obvious choice $\mu = m$, but in general, if there are many fermion masses with widely different values, as is the case in nature, then there is no unique choice of μ that minimizes the logarithm.

Problems for Chapter 3

(1) **Blue sky in EFT**
Use EFT methods to prove that the sky is blue during the day by deriving Rayleigh's k^4 scattering law. [Kaplan (1995)]
   (a) First identify the fields describing the degrees of freedom (photons and atoms) and the relevant energy scales. What are the dimensions of the fields?
   (b) Write the allowed operators that appear at leading order (there are two). What is the dimensions of these operators?
   (c) Use this to show that the cross section must go like k^4, where k is the wave vector of the photon.

(2) **More hydrogen atom**
Show that the $H(2s) \to H(1s) + \gamma$ transition is forbidden by proving that there are no effective operators allowed by the symmetries. At what order in the power counting does the $H(2s)$ state decay?

(3) **Yukawa running**
Consider a massless fermion and a massive scalar with a Yukawa coupling between them:
$$\mathcal{L} = \bar{\psi} i \slashed{\partial} \psi + \frac{1}{2}((\partial_\mu \phi)^2 - m^2 \phi^2) - g\phi\bar{\psi}\psi - \frac{\lambda}{4!}\phi^4$$
   (a) Compute the two-point correlation functions, as well as the four-point correlation function for scalars, accurate to one loop. Use DimReg and MS.
   (b) Compute the anomalous dimensions of each field, as well as the beta functions for g and λ.
   (c) Express the one-loop running Yukawa coupling in terms of the Landau Poles.

Chapter 4

Effective Field Theories of Type I

4.1 Introduction

Applications of effective field theories to physics started even before they became powerful computational tools. Calculations of transition amplitudes using only degrees of freedom that appear below a certain energy scale[1] started with Enrico Fermi's model of weak interactions. Since then it was proven that the "infrared models" should be considered more seriously than simple phenomenological fits to the data. Extensive studies of renormalization properties of those models showed that one can produce very accurate predictions for decays and cross sections, provided that measurements of a few other decays and cross-sections are available to fix some parameters that cannot be determined by the symmetries. In this book we shall call the models that only deal with low energy (infrared) degrees of freedom as EFTs of Type I.

As an example, let us first consider a somewhat formal derivation of one such theory, which emerges as a limit of a well-known Yukawa model of a (light) Dirac fermion field ψ interacting with heavy real scalar field ϕ. The Lagrangian for such a model is

$$\mathcal{L}_Y = \overline{\psi} i \slashed{\partial} \psi - m \overline{\psi} \psi + \frac{1}{2} \left(\partial_\mu \phi \right)^2 - \frac{M^2}{2} \phi^2 - g \phi \overline{\psi} \psi, \qquad (4.1)$$

where g is a coupling between the scalar and fermion fields. We shall assume that the fermion mass m is much smaller than the scalar mass M, i.e. $m \ll M$.

All Green's functions for this model can be obtained from the generating functional (statistical sum) of the model, which can be written in terms of

---
[1] We shall call those "infrared degrees of freedom" in this book.

the functional integral,

$$Z_Y = N \int [d\psi][d\phi] \exp\left(i \int d^4x \, \mathcal{L}_Y\right), \quad (4.2)$$

where N is a normalization factor. We can define the action $S = \int d^4x \, \mathcal{L}_Y$ in the usual way.

Let us now imagine that a brave experimentalist is performing a test of this model at a particle accelerator. However, the accelerator that the experimentalist is using can only probe energies E far below M. That is to say, there is a definite energy scale, $\mu \sim E \ll M$ that is set up by the experimental conditions. In this situation we can only deal with initial and final states involving the fermions ψ, while ϕ cannot be produced at all, although it might still contribute to the fermion-initiated processes through off-shell propagation at tree level and in quantum loops.

Thus, as long as the accelerator we are using is the same, we only deal with the infrared degree of freedom of the model given by Eq. (4.1), which in this case is the fermion ψ. In fact, we might not even know that the particles described by ϕ even exist. What would happen if we *integrate ϕ out* of the model, i.e perform functional integral with respect to ϕ in Eq. (4.2)?

In the problem at hand this procedure can be executed straightforwardly, as the integral over ϕ is Gaussian. Let us call $J(x) = -g\overline{\psi}\psi(x)$ in Eq. (4.1) and introduce a new field $\phi'(x)$, which is just a shift of the original field $\phi(x)$

$$\phi'(x) = \phi(x) - i \int d^4y \, D_F(x-y)J(y), \quad (4.3)$$

where $D_F(x-y)$ is a position space Green's function (propagator) of the ϕ-field, which can be written as

$$D_F(x-y) = \int \frac{d^4k}{(2\pi)^4} \frac{i}{k^2 - M^2 + i\epsilon} e^{-ik\cdot(x-y)}. \quad (4.4)$$

This substitution leads to the following structure of the action in Eq. (4.2)

$$S = S_\psi - \int d^4x \left(\frac{1}{2}\phi'(x)\left[-\partial^2 - M^2\right]\phi'(x)\right)$$
$$- \int d^4x d^4y \, J(x)(-iD_F(x-y))J(y), \quad (4.5)$$

where S_ψ is the fermionic part of the action that contains the kinetic and the mass terms. Since $\phi'(x)$ is no longer connected to $J(x)$, we can integrate over it, which would turn it into a determinant

$$Z_Y = N \int [d\psi] \sqrt{\det(\partial^2 - M^2)^{-1}}$$
$$\times \exp\left[S_\psi - i \int d^4x d^4y \, J(x)(-iD_F(x-y))J(y)\right]. \quad (4.6)$$

Now, all Green's functions that we are interested in always involve fermion fields, so the determinant is irrelevant for us. Also, in the limit $M \to \infty$ the non-local part of Eq. (4.6) can be expanded in k^2/M^2. The leading contribution to Eq. (4.6) then is just a delta-function,

$$D_F(x-y) \approx -\frac{i}{M^2} \delta^{(4)}(x-y), \quad (4.7)$$

which makes the expression under the action integral in Eq. (4.6) local. Remembering that $J(x) = -g\overline{\psi}\psi(x)$ we get for the *effective* action,

$$S_Y^{eff} = \int d^4x \left[\overline{\psi} i\partial\!\!\!/\psi - m\overline{\psi}\psi + \frac{g^2}{M^2}\left(\overline{\psi}\psi\right)^2\right] = \int d^4x \, \mathcal{L}_{eff}, \quad (4.8)$$

which gives for the effective Lagrangian

$$\mathcal{L}_{eff} = \overline{\psi} i\partial\!\!\!/\psi - m\overline{\psi}\psi + \frac{g^2}{M^2}\left(\overline{\psi}\psi\right)^2. \quad (4.9)$$

This Lagrangian correctly describes physics of low-energy modes, i.e. scattering of fermions with energies $E \ll M$.[2] There are corrections to Eq. (4.9) that are given by higher dimensional operators involving ψ fields and their derivatives, such as $(\overline{\psi}\partial^2\psi)(\overline{\psi}\psi)$, etc. An important thing is that at any given order in $1/M$ there is a *finite* number of effective operators.

As far as the terminology is concerned, Eq. (4.1) is called "an ultraviolet (UV) completion" of the EFT described by Eq. (4.9), which is known in this case. In reality we might not have such luxury. Such is the case of the standard model – we simply don't know what lies beyond! We shall discuss this situation later in the book. Or we might even be in a worse situation: while the UV completion of the model is known, it is written in terms of different degrees of freedom from the infrared modes. This is the case with chiral perturbation theory, discussed later in the chapter. In both cases, symmetries of the full theory, both known and conjectured, play a crucial role.

In deriving Eq. (4.9) we have used a somewhat formal procedure of "integrating out" heavy fields in the path integral. In practical calculations this procedure is not really needed. In fact, we shall not use it in this book

---

[2] In $1 + 1$ dimensions this model has a name, it is called the Gross-Nevue model.

from now on. According to the Theorem 1 introduced in chapter 1, any Lagrangian which satisfies all symmetries of the full theory should yield the same Green's functions as the full theory. What this means is that if we know how to power-count subleading operators, we can simply write out the most general set of them consistent with the symmetries of the full theory. Each operator would be multiplied by an unknown coefficient, which can be determined by calculating the same Green's functions in the full and the effective field theory and matching the results. This procedure is actually called "matching" in the lingo used by EFT practitioners.

Observe that in our example the relevant power counting scheme for the EFT is a simple counting of operator dimension. In this power counting scheme, the only effective operator that generates non-trivial scattering is $\left(\bar{\psi}\psi\right)^2$, which is an operator of dimension 6. According to our discussion in chapter 3, this operator should be suppressed by two powers of M, so the leading order effective Lagrangian looks like

$$\mathcal{L}_{eff} = \frac{C_1}{M^2}\left(\bar{\psi}\psi\right)^2. \qquad (4.10)$$

Performing a tree-level matching of the $2 \to 2$ scattering amplitude in models given by the Lagrangians of Eq. (4.10) and Eq. (4.1) identifies $C_1 = g^2$, as given in Eq. (4.9). This is the procedure that we shall employ most of the time in this book.

We just derived an effective low energy theory of fermions. But this is not the end of the story. So far we have proven that we can trade a contribution of heavy degrees of freedom for an infinite set of local operators, which are suppressed by powers of inverse masses of those heavy particles. It immediately implies that *at tree level* we have a consistent power counting of those operators. However, if we are to use this theory as a full-fledged QFT, we need to show that contributions of quantum corrections, given by *loop* diagrams, are also suppressed by powers of the heavy scale. In other words, contributions of loops with vertices given by irrelevant operators would not generate marginal or relevant operators.

Let us check to see if our EFT given by Eq. (4.9) is indeed consistent. We can estimate a contribution of the graph for the one loop self energy $\Sigma_2(k)$, given in Fig. 4.1, to the shift of the mass parameter δm of the fermion field

$$\delta m = \Sigma_2(\slashed{k} = m), \qquad (4.11)$$

which can be shown to be

$$\delta m \simeq \frac{2img^2}{M^2}\int \frac{d^4k}{(2\pi)^4}\frac{1}{k^2 - m^2 + i\epsilon}. \qquad (4.12)$$

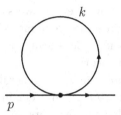

Fig. 4.1 One loop correction to the fermion mass in EFT given by Eq. (4.9).

This integral is quadratically divergent, so we need to regularize it. It might be natural to introduce a cut off to deal with this integral – moreover, a natural value for the cut off is readily available. Our effective theory was obtained by integrating out the scalar field of mass M, which means that the approximation of the propagator as a finite set of local operators suppressed by powers of M breaks down at the scale M. Cutting off the quadratic UV divergence of Eq. (4.12) by the scale M leads, however, to a curious result

$$\delta m \sim \frac{g^2}{M^2} \times M^2 \sim g^2 m. \qquad (4.13)$$

What we found is that a quantum correction generated by an irrelevant operator is *not* suppressed by any power of M! The same thing would happen for any other operator that is suppressed by higher powers of M and contains more derivatives. In fact, the entire expansion in $1/M^n$ breaks down if quantum corrections are included with a simple cut off! This is rather disappointing, albeit not surprising: we saw the same thing happening when we discussed the general problem of mass-dependent subtraction schemes in Section 3.6.

We can trace the problem to the use of a cut off as a regulator of our integrals. Clearly, a high mass scale introduced with regularization spoils the explicit power counting. A cure can be found in using mass-independent regularization, such as dimensional regularization (see Appendix C for a short review). Quadratic divergences are absent in dimensional regularization, so the calculation of the mass renormalization would give

$$\delta m \sim \frac{m^2}{16\pi^2 M^2} \left(2g^2 m \mu^{2\epsilon}\right) \left[\frac{1}{\bar{\epsilon}} + \log\frac{m^2}{\mu^2} - 1\right], \qquad (4.14)$$

where $\bar{\epsilon}$ is defined in Appendix C. It is interesting to see that now the one-loop contribution is properly suppressed by $m^2/(16\pi^2 M^2)$, which is indeed small, as $m \ll M$. The arbitrary scale introduced in dimensional regularization only appears in the logarithm, so it does not spoil power counting, which is now given by inverse powers of M read off directly from the Lagrangian! We can require the use of mass-independent regularization scheme for all EFT calculations: in a sense, we introduced regularization as part of our effective field theory.

This is an important and completely general result. It implies one important consequence. Since, to a given order in $1/M$, our effective theory contains only a finite number of operators and quantum corrections generated by those operators are of the same (or higher) orders in $1/M$, to a *given accuracy* of calculations we would only need a finite number of counter terms to reabsorb UV divergencies generated by quantum loop. What it means is that, order-by-order in $1/M$, our effective field theory behaves like a renormalizable quantum field theory.

4.2 Real physics: Euler-Heisenberg Lagrangian

Photons interact directly with particles carrying electric charge. Since the photon itself is electrically neutral this means that photons will not interact directly. This is just the usual statement of gauge invariance of electromagnetic interactions.

The lightest charged particle we know of is the electron. Consider photons with $E < m_e$. Can these photons interact with each other? The answer is yes, by exchanging virtual electrons.

The two photon vertex is the usual vertex of QED. It requires renormalization as usual and is done in many standard textbooks. In the limit of low energy photons, the answer is

$$\Pi_{\mu\nu} = \frac{\alpha}{15\pi}(q_\mu q_\nu - g_{\mu\nu}q^2)\frac{q^2}{m_e^2} + \mathcal{O}(\alpha^2, (q^2/m_e^2)^2) \ . \tag{4.15}$$

Amusingly, this is precisely the same answer we would have gotten if we used the Lagrangian

$$\mathcal{L}_{\text{LE}} = -\frac{1}{4}F_{\mu\nu}F^{\mu\nu} + \frac{\alpha}{60\pi m_e^2}F_{\mu\nu}\Box F^{\mu\nu} \tag{4.16}$$

The three photon vertex vanishes by Furry's Theorem, which says that the n point correlation function of photons vanishes for n odd; this is most easily seen to follow from charge conjugation invariance.

The four photon vertex comes from a "box diagram" of electrons. The point to realize is that this is the same result we would obtain from a *point like vertex* of four photons coming together. This point like interaction contains momenta and is thus a derivative interaction. In fact, one can show that the QFT described by the Lagrangian

$$\mathcal{L}_{\text{EH}} = -\frac{1}{4}F_{\mu\nu}F^{\mu\nu} + \frac{\alpha^2}{m_e^4}\Big[C_1(F_{\mu\nu}F^{\mu\nu})^2 + C_2(F_{\mu\nu}\tilde{F}^{\mu\nu})^2\Big] \qquad (4.17)$$

will give precisely the same result as a full QED calculation (to this order in E_γ/m_e) provided we set

$$C_1 = \frac{1}{90}, \qquad C_2 = \frac{7}{90}. \qquad (4.18)$$

This is an example of the matching procedure we described above. Notice that $\mathcal{L}_{\text{EH}}$ is the most general Lagrangian one can write down using only photon fields and derivatives, while maintaining gauge invariance, the values of C_i being arbitrary until we specify QED as the "full theory". This effective Lagrangian goes under the name of "Euler-Heisenberg Lagrangian" and does a very good job describing low energy light scattering.

Without doing any more work, we can get a good understanding of $\gamma\gamma$ scattering at low energies. The amplitude from this theory gives

$$\mathcal{M}(\gamma\gamma \to \gamma\gamma) \sim \frac{\alpha^2}{m_e^4} \times E_\gamma^4. \qquad (4.19)$$

where the dependence on E_γ followed from dimensional analysis and the fact that there are no scales left in the theory (m_e is already taken into account). Then again from dimensional analysis, we have:

$$\sigma_{\gamma\gamma} = |\mathcal{M}|^2 \times \text{(phase space)} \times \text{(flux factor)} \sim \left(\frac{\alpha^2 E_\gamma^4}{m_e^4}\right)^2 \times \frac{1}{E_\gamma^2} \qquad (4.20)$$

and low energy (sub X-ray) photon-photon scattering is proportional to E_γ^6, with a coefficient of order α^4/m_e^8, facts confirmed by laser light scattering experiments.

Of course, if we kept higher order terms in E_γ/m_e in the full theory, we can reproduce them with higher order terms the effective theory, such as

$$\frac{1}{m_e^4}F_\mu\Box^2 F^{\mu\nu}, \qquad \frac{1}{m_e^6}F_{\mu\nu}\Box F^{\mu\nu}F_{\alpha\beta}F^{\alpha\beta}.$$

4.3 Fermi Theory of Weak Interactions as an effective theory

For our second example, we turn to one of the most famous examples of an effective field theory in particle physics: Fermi's theory of beta decay (and its generalizations to other particles). We know now that the quarks of the standard model of particle physics have weak force interactions of the form

$$\mathcal{L}_W = \frac{g}{\sqrt{2}} V_{ij} \bar{q}_i \gamma^\mu P_L q_j W_\mu^\pm + \cdots , \quad (4.21)$$

where $P_L = \frac{1}{2}(1 - \gamma^5)$ is the left-handed projection operator and V_{ij} are the CKM matrix elements. Consider a $\Delta S = 1$ transition with a strange quark and three first generation quarks

$$\mathcal{M}^{\Delta S=1} = \left(\frac{ig}{\sqrt{2}}\right)^2 V_{us} V_{ud}^* (\bar{u}\gamma^\mu P_L s)(\bar{d}\gamma^\nu P_L u) \times \frac{-ig_{\mu\nu}}{p^2 - M_W^2} \quad (4.22)$$

where we work in Feynman gauge, neglecting the contributions from gauge artifact fields (they will not affect our arguments here, and are small for light quarks anyway). This matrix element describes $K \to \pi$ transitions, for example. It is a *nonlocal* interaction, as can be seen by the momentum in the denominator. However, in the limit that $p^2 \ll M_W^2$, our matrix element simplifies with the help of the Taylor expansion of the propagator:

$$\frac{1}{p^2 - M_W^2} = -\frac{1}{M_W^2}\left(1 + \frac{p^2}{M_W^2} + \cdots\right).$$

Our matrix element becomes

$$\mathcal{M}^{\Delta S=1} = \frac{i}{M_W^2}\left(\frac{ig}{\sqrt{2}}\right)^2 V_{us} V_{ud}^* (\bar{u}\gamma^\mu P_L s)(\bar{d}\gamma_\mu P_L u) + \mathcal{O}\left(\frac{1}{M_W^4}\right). \quad (4.23)$$

This is the same matrix element one would get if we used the effective Lagrangian

$$\mathcal{L}_{\text{eff}} = -\frac{4G_F}{\sqrt{2}} V_{us} V_{ud}^* (\bar{u}\gamma^\mu P_L s)(\bar{d}\gamma_\mu P_L u) , \quad (4.24)$$

if we apply the matching condition

$$\frac{G_F}{\sqrt{2}} = \frac{g^2}{8M_W^2}. \quad (4.25)$$

G_F is called the Fermi constant[3], and both (the tree level use of) the Lagrangian in Equation (4.21) and the effective theory in Equation (4.24) give the same answer up to $\mathcal{O}(1/M_W^2)$ corrections.

As before, we can keep going, computing contributions from higher powers of $1/M_W$. For example, if we include the next term in the Taylor expansion of the propagator,

$$\Delta \mathcal{M}^{\Delta S=1} = \frac{ip^2}{M_W^4}\left(\frac{ig}{\sqrt{2}}\right)^2 V_{us} V_{ud}^* (\bar{u}\gamma^\mu P_L s)(\bar{d}\gamma_\mu P_L u). \quad (4.26)$$

---

[3] The $\sqrt{2}$ in the definition appears for historical reasons.

The presence of a p^2 tells us that there must be a derivative in our effective Lagrangian. The most general effective Lagrangian we can write down that would generate this term is

$$\mathcal{L}_8 = -\frac{4G_F}{\sqrt{2}} V_{us} V_{ud}^* \Big[C_1 (\bar{u}\gamma^\mu P_L D^2 s)(\bar{d}\gamma_\mu P_L u) \\ + C_2 (\bar{u}\gamma^\mu P_L s)(\bar{d}\gamma_\mu P_L D^2 u) \qquad (4.27) \\ + C_3 (\bar{u}\gamma^\mu P_L D^\nu s)(\bar{d}\gamma_\mu P_L D_\nu u) \Big].$$

where D_μ is the usual (QCD+QED) gauge covariant derivative. By comparing the matrix elements of these three operators with the full theory result, we can match the C_i. Notice that the Fermi constant we matched earlier has been explicitly factored from these coefficients – this is standard practice for Fermi theory calculations.

In addition, one can generalize the effective theory in Eq. (4.24) to include all quarks (and leptons). For example, if we are interested in $u \to u$ quark transitions, we can write

$$\mathcal{L}_{u \to u} = -\frac{4G_F}{\sqrt{2}} \sum_{i,j=d,s,b} (V_{ui} V_{ui}^*)(\bar{u}\gamma^\mu P_L q_i)(\bar{q}\gamma^\mu P_L u_i) . \qquad (4.28)$$

Furthermore, if we also include Z boson exchange, we can have operators like

$$\mathcal{L}_{u \to u}^{(Z)} = -\frac{4G_F}{\cos^2 \theta_W} \sum_{A,B=L,R} (a_u^A a_u^B)(\bar{u}\gamma^\mu P_A u)(\bar{u}\gamma^\mu P_B u) , \qquad (4.29)$$

where $a_u^A = \frac{1}{2}\delta_{AL} - \frac{2}{3}\sin^2 \theta_W$.

4.4 Fermi theory to one loop: $\Delta S = 2$ processes in EFT

Now we will consider the low energy effective theory of the weak nuclear force. We begin by considering $K\bar{K}$ mixing. This occurs in the standard model of particle physics at one loop order. A full calculation can be done employing the usual techniques of quantum field theory without including QCD corrections; we will discuss how they come into play later.

The relevant amplitude comes from box diagrams:

$$\mathcal{A} = \frac{g^4}{128\pi^2 M_W^4} \sum_{ij} \xi_i \xi_j \; E(x_i, x_j) \times \langle K | (\bar{d}\gamma^\mu P_L s)(\bar{d}\gamma_\mu P_L s) | \bar{K} \rangle , \qquad (4.30)$$

where $x_i \equiv m_i^2 / M_W^2$, $\xi_i \equiv V_{is} V_{id}^*$, and these satisfy the vital relation

$$\sum_{i=u,c,t} \xi_i = 0 ; \qquad (4.31)$$

This follows from the unitary nature of the CKM Matrix. $\bar{E}(x_i, x_j)$ is a complicated function whose form does not concern us for the moment, but in the limit $x_u = 0$, $x_{c,t} \ll 1$ we have

$$\mathcal{A} \simeq -\frac{G_F^2}{4\pi^2} \langle K|(\bar{d}\gamma^\mu P_L s)(\bar{d}\gamma_\mu P_L s)|\bar{K}\rangle$$
$$\times \left[\xi_c^2 m_c^2 + \xi_t^2 m_t^2 + 2\xi_c \xi_t m_c^2 \log\left(\frac{m_t^2}{m_c^2}\right)\right]. \quad (4.32)$$

While this limit is not realistic (especially for the top quark mass), it is surprisingly accurate. We would like to reproduce these results with effective field theory techniques, and then see how much further our methods can take us.

4.4.1 Setting up the EFT approach

We will break our theory into three pieces, depending on the scale:

(1) $\mu > M_W$,
(2) $m_c < \mu < M_W$,
(3) $\mu < m_c$.

As we cross each threshold, a particle disappears from our theory, as explained in the last chapter. Our final result after integrating out the c quark is an effective Lagrangian:

$$\mathcal{L}_{\text{eff}} = \mathcal{L}_{\Delta S=1} + C^{(2)}(\mu)(\bar{d}\gamma^\mu P_L s)(\bar{d}\gamma_\mu P_L s). \quad (4.33)$$

Below the scale M_W, the tree level $\Delta S = 1$ part of the effective Lagrangian remains in the full and effective theory; but loop contributions to these operators might need be altered. As for the $\Delta S = 2$ terms, there is work to do, since the full theory result is manifestly convergent, while the effective theory looks to have a quadratically divergent diagram; therefore a new counterterm must be introduced.

However, it turns out that this is not the case, and the effective theory loop diagram is secretly convergent! To see this, let us write the amplitude

$$\mathcal{A}_{\text{eff}} \simeq \left(\frac{4G_F}{\sqrt{2}}\right)^2 \sum_{ij}(\xi_i \xi_j) \int \frac{d^4k}{(2\pi)^4} \left(\bar{d}\gamma_\mu P_L \frac{i}{\slashed{k}-m_i}\gamma_\nu P_L s\right)$$
$$\times \left(\bar{d}\gamma^\nu P_L \frac{i}{\slashed{k}-m_j}\gamma^\mu P_L s\right). \quad (4.34)$$

Now expand in m_i/k under the integral, which is appropriate when examining large k asymptotic of the integral. We see that

$$\mathcal{A}_{\text{eff}} \simeq \left[\sum_{ij}\xi_i\xi_j\right]\left(\frac{4G_F}{\sqrt{2}}\right)^2 \int \frac{d^4k}{(2\pi)^4}\left(\bar{d}\gamma_\mu P_L \frac{i}{\slashed{k}}\gamma_\nu P_L s\right)\left(\bar{d}\gamma^\nu P_L \frac{i}{\slashed{k}}\gamma^\mu P_L s\right)$$

$$+ \text{ mass-dependent terms.} \qquad (4.35)$$

This first integral is quadratically divergent, however the sum over CKM matrix elements vanishes due to Eq. (4.31). This cancellation is called the Glashow–Iliopoulos–Maiani (GIM) mechanism. Notice that if $m_i = m_j$, then the sum always vanishes and the amplitude is exactly zero. The fact that the masses are different will mean that there is a finite part proportional to the mass differences. We will see this explicitly.

Now we would like to go to scales below m_t. Recalling the rule of EFT, we must remove the top quark completely from our theory. This means the GIM mechanism will no longer cancel our divergences, and some work needs to be done. Since we still have $m_c, m_u < \mu$, contributions from these quarks will be explicitly included in the matrix elements; however, to include the top quark, we must perform a matching calculation. Loops that involve a top quark internal line are shrunk to a point, and we are left with a point vertex with coefficient $C^{(2)}$ that must be calculated. By performing the loops involving top quarks we can compute

$$C^{(2)}(m_t) = \frac{G_F^2}{4\pi^2}\left[\xi_t^2 m_t^2 + 2\xi_t\xi_c(m_t^2 + m_c^2) + 2\xi_t\xi_u(m_t^2 + m_u^2)\right]$$

$$= \frac{G_F^2}{4\pi^2}\left[-\xi_t^2 m_t^2 + 2\xi_t\xi_c m_c^2\right] \qquad (4.36)$$

where we set $m_u = 0$ and used Eq. (4.31). Note that these terms come from the *finite part* of the diagram.

Now that we have the value of the coefficient at the scale where we integrate out the top quark, we can proceed to run the theory from m_t to m_c. Once again, without the top quark, there is no GIM mechanism, and the diagrams with virtual u and c quarks will be divergent. If C_1 ($= -4G_F/\sqrt{2}$) is the coefficient of the $\Delta S = 1$ operators (two insertions of which go into the loop-generating $\Delta S = 2$ operator) then we can compute the running of $C^{(2)}$ analogously to what was done before

$$\mu\frac{dC^{(2)}}{d\mu} = \frac{1}{8\pi^2}C_1^2 m_c^2 \xi_c \xi_t \qquad (4.37)$$

Ignoring the scale dependence in C_1 or the masses, we can easily integrate this equation

$$C^{(2)}(m_c) = C^{(2)}(m_t) - \frac{G_F^2}{2\pi^2}\xi_c\xi_t m_c^2 \log\left(\frac{m_t^2}{m_c^2}\right). \qquad (4.38)$$

4.4.2 A more detailed calculation

Let us perform the above calculation more carefully to see what happens. In the Fermi effective theory, $\Delta S = 2$ transitions occur through two insertions of the $\Delta S = 1$ vertex:

$$\mathcal{L}_{\Delta S=1} = C_1 V_{is} V_{jd}^* (\bar{q}_i \gamma^\mu P_L s)(\bar{d} \gamma_\mu P_L q_j) \ . \tag{4.39}$$

This leads to the amplitude

$$i\mathcal{A}_{ij} = C_1^2 \xi_i \xi_j^* \int \frac{d^4k}{(2\pi)^4} \left(\bar{d}\gamma_\mu P_L \frac{i(\slashed{k}+m_i)}{k^2 - m_i^2} \gamma_\nu P_L s \right)$$

$$\times \left(\bar{d}\gamma^\nu P_L \frac{i(\slashed{k}+m_j)}{k^2 - m_j^2} \gamma^\mu P_L s \right) \tag{4.40}$$

$$= C_1^2 \xi_i \xi_j^* \left(\bar{d}\gamma_\mu P_L \gamma_\alpha \gamma_\nu P_L s \right) \left(\bar{d}\gamma^\nu P_L \gamma_\beta \gamma^\mu P_L s \right)$$

$$\times \int \frac{d^4k}{(2\pi)^4} \frac{k^\alpha k^\beta}{(k^2 - m_i^2)(k^2 - m_j^2)} \ ,$$

where we have set all external momenta to zero for the matching calculation, and we have used $P_L \gamma^\mu P_L = 0$ to drop the mass terms in the numerator. The integral is nominally divergent; in the full theory, the GIM mechanism cancels this divergence; in the effective theory, GIM fails and we will need to renormalize. To that end, we employ dimensional regularization. After combining denominators with a Feynman parameter the integral becomes

$$I^{\alpha\beta} = \mu^{4-d} \int_0^1 dx \int \frac{d^dk}{(2\pi)^d} \frac{k^\alpha k^\beta}{(k^2 - \Delta)^2}$$

$$= \mu^{4-d} \int_0^1 dx \frac{-i}{(4\pi)^{d/2}} \frac{g^{\alpha\beta}}{2} \frac{\Gamma(1-d/2)}{\Delta^{1-d/2}}$$

$$= (-i)g^{\alpha\beta} \int_0^1 dx\, \Delta \left[\frac{\mu^{4-d}}{2-d} \frac{\Gamma(2-d/2)}{(4\pi)^{d/2}} \frac{1}{\Delta^{2-d/2}} \right] \ , \tag{4.41}$$

where we used the master identity Equation (C.19) and defined

$$\Delta = x m_i^2 + (1-x) m_j^2 \ . \tag{4.42}$$

Notice the pole at $d = 2$, signifying a quadratic divergence. The quantity in brackets contains the dependence on dimension; letting $d = 4 - 2\epsilon$ gives for the bracketed term

$$-\frac{1}{2(4\pi)^2} \left[\frac{1}{\epsilon} - \gamma_E - \log\left(\frac{\Delta}{4\pi\mu^2}\right) + \mathcal{O}(\epsilon) \right] \ . \tag{4.43}$$

We are now in a position to state a very useful rule: *the coefficient of the $\log(\mu^2)$ is equal to the coefficient of the $1/\epsilon$ pole*. This rule holds generally, and it means that in order to compute anomalous dimensions it is always sufficient to study the residue of the amplitude. In our case, that residue is

$$+\frac{i}{2}\frac{g^{\alpha\beta}}{(4\pi)^2}\int_0^1 dx\,[xm_i^2+(1-x)m_j^2] = +\frac{i}{2}\frac{g^{\alpha\beta}}{(4\pi)^2}\frac{1}{2}(m_i^2+m_j^2)\,. \quad (4.44)$$

Plugging into the total amplitude in Equation (4.41) then gives

$$\mathcal{A}_{ij} = \frac{1}{\epsilon}C_1^2\xi_i\xi_j^*\frac{m_i^2+m_j^2}{4(4\pi)^2}(\bar{d}\gamma_\mu P_L\gamma_\alpha\gamma_\nu P_L s)(\bar{d}\gamma^\nu P_L\gamma^\alpha\gamma^\mu P_L s)+\text{finite}. \quad (4.45)$$

Now we need to reduce the spinor structure. Apply the Fierz identity

$$(\bar{u}_1\gamma_\nu P_L u_2)(\bar{u}_3\gamma^\nu P_L u_4) = (\bar{u}_1\gamma_\nu P_L u_4)(\bar{u}_3\gamma^\nu P_L u_2)\,, \quad (4.46)$$

to rewrite the spinor product as

$$\begin{aligned}(\bar{d}_i\gamma_\mu P_L\gamma_\alpha\gamma_\nu P_L s_i)&(\bar{d}_j\gamma^\nu P_L\gamma^\alpha\gamma^\mu P_L s_j)\\ &= (\bar{d}_i\gamma_\mu P_L\gamma_\alpha\gamma_\nu P_L\gamma^\alpha\gamma^\mu P_L s_j)(\bar{d}_j\gamma^\nu P_L s_i)\\ &= (\bar{d}_i\gamma_\mu P_L(-2\gamma_\nu)\gamma^\mu P_L s_j)(\bar{d}_j\gamma^\nu P_L s_i)\\ &= (\bar{d}_i(+4\gamma_\nu)P_L s_j)(\bar{d}_j\gamma^\nu P_L s_i)\\ &= 4(\bar{d}_i\gamma_\nu P_L s_i)(\bar{d}_j\gamma^\nu P_L s_j)\,.\end{aligned}$$

We can now see the operator of the $\Delta S = 2$ piece of the Lagrangian of Eq. (4.33). The amplitude becomes

$$\mathcal{A}_{ij} = \frac{C_1^2}{\epsilon}\xi_i\xi_j^*\frac{m_i^2+m_j^2}{(4\pi)^2}(\bar{d}_i\gamma_\nu P_L s_i)(\bar{d}_j\gamma^\nu P_L s_j)\,. \quad (4.47)$$

Notice now that the residue of the $C^{(2)}$ coefficient has a very particular form

$$C_1^2\xi_i\xi_j^*f(m_i,m_j), \quad (4.48)$$

where $f(0,0) = 0$ and $f(0,m) = \frac{1}{2}f(m,m)$. Now we can perform the sum over $i,j = u,c$ and use Eq. (4.31) with $m_u = 0$ to get

$$\sum_{i,j=u,c} C_1^2\xi_i\xi_j^*f(m_i,m_j) = C_1^2\left[\xi_u\xi_u^*f(0,0) + 2\xi_u\xi_c f(0,m_c) + \xi_c\xi_c^*f(m_c,m_c)\right]$$

$$= -2C_1^2\xi_c\xi_t f(m_c,0)\,. \quad (4.49)$$

This piece of the coefficient gives us the anomalous dimension for $C^{(2)}$

$$\mu\frac{dC^{(2)}}{d\mu} = +\frac{C_1^2}{8\pi^2}\xi_c\xi_t^*m_c^2\,. \quad (4.50)$$

If we ignore any scale dependence[4] of G_F or m_c, then this equation integrates very easily to

$$C^{(2)}(m_c) = C^{(2)}(m_t) - \frac{G_F^2}{2\pi^2}\xi_c\xi_t^* m_c^2 \log\left(\frac{m_t^2}{m_c^2}\right), \qquad (4.51)$$

precisely what we stated earlier in Eq. (4.38).

Finally, we would like to match at the scale m_c, in which case we must integrate out the charm quark, leaving only the up quark loops in the theory. Our matching condition becomes

$$\begin{aligned}C^{(2)}(m_c - 0) &= C^{(2)}(m_c + 0) + \frac{G_F^2}{4\pi^2}m_c^2(\xi_c^2 + 2\xi_c\xi_u)\\ &= \frac{G_F^2}{4\pi^2}m_c^2(\xi_c^2 + 2\xi_c\xi_u) - \frac{G_F^2}{2\pi^2}\xi_c\xi_t^* m_c^2\log\left(\frac{m_t^2}{m_c^2}\right) \quad (4.52)\\ &\quad + \frac{G_F^2}{4\pi^2}\left[-\xi_t^2 m_t^2 + 2\xi_t\xi_c m_c^2\right].\end{aligned}$$

With this coefficient, the charm and top loop contributions are represented by a point vertex with $C^{(2)}$ as the "coupling constant," plus loop integrals involving up quarks.

4.5 QCD corrections in EFTs

The effective theory of electroweak interactions that were discussed so far in this chapter deal with quark fields. Quarks carry color-charge, which means they are susceptible to QCD interactions. Are those important? If they are, how would one take them into account?

In building effective theories with QCD we need to remember that the basis of operators of a given dimension will be enlarged due to color degrees of freedom for each quark field. For example, operators of dimension 6 built out of four different quark fields q_i

$$\begin{aligned}Q_1 &= (\bar{q}_{1\alpha}\Gamma_\mu q_{2\beta})(\bar{q}_{3\beta}\Gamma_\mu q_{4\alpha}), \text{ and}\\ Q_2 &= (\bar{q}_{1\alpha}\Gamma_\mu q_{2\alpha})(\bar{q}_{3\beta}\Gamma_\mu q_{4\beta}),\end{aligned} \qquad (4.53)$$

with $\alpha, \beta = 1, 2, 3$ being the color indices (summation is assumed), are not the same. Moreover, since gluons carry color quantum numbers, one would expect that perturbative QCD (pQCD) corrections will mix Q_1 into Q_2 and vise versa. Thus, below the scale associated with the W-boson mass we can conjecture that the basis of operators describing electroweak

---
[4]For example, we are ignoring QCD corrections.

transitions contains at least two operators[5]. So, the weak Hamiltonian describing transitions among q_i is

$$\mathcal{H}_{\text{eff}} = \frac{4G_F}{\sqrt{2}} \xi_{CKM} \left(C_1(\mu) Q_1(\mu) + C_2(\mu) Q_2(\mu) \right), \qquad (4.54)$$

where ξ_{CKM} parameterizes relevant CKM parameters. Notice that we introduced an auxiliary scale, μ, that separates (or *factorizes*) contributions from short distance physics (in the Wilson coefficients C_i) from long distance physics (in the operators Q_i). Our terminology suggests that Eq. (4.54) has the form of an operator product expansion.

For example, for a strangeness-changing $\Delta S = 1$ interaction discussed in Eq. (4.24) of section 4.3, the effective Hamiltonian describing decays of strange mesons and baryons is

$$\mathcal{H}_{\text{eff}} = \frac{4G_F}{\sqrt{2}} V_{us} V_{ud}^* \big[\ C_1 \ (\bar{u}_\alpha \gamma^\mu P_L s_\beta)(\bar{d}_\beta \gamma_\mu P_L u_\alpha)$$
$$+ \ C_2 \ (\bar{u}_\alpha \gamma^\mu P_L s_\alpha)(\bar{d}_\beta \gamma_\mu P_L u_\beta) \big]. \qquad (4.55)$$

Notice that QCD corrections to effective electroweak operators are oftentimes more important than the contributions of higher-dimensional operators (as we saw in section 4.3, these are further suppressed by powers of $1/M_W$). This is especially true for the description of low-energy processes, where $\alpha_s(\mu) \gg \mu^2/M_W^2$, and two-loop or even three-loop pQCD corrections give more important contributions than electroweak $1/M_W^4$ effects. Thus, it is important to learn how to calculate pQCD effects that are parameterized by the coefficients C_1 and C_2.

4.5.1 *Matching at one loop in QCD*

Methods for calculating a QCD-corrected effective Lagrangian are not very different from the procedures that we have developed for the leading order operators. That is to say, we would need to calculate transition amplitudes[6] in full and effective theories, $\mathcal{A}_{\text{full}}$ and $\mathcal{A}_{\text{eff}}$, to the desired order in α_s and match them at a chosen energy scale to determine Wilson coefficients

$$\mathcal{A}_{\text{full}}(p_i, m_i, \mu) = \mathcal{A}_{\text{eff}}(p_i, m_i, \mu), \qquad (4.56)$$

where we also need to specify a kinematical point for external momenta p_i and external masses m_i that we choose for matching full and effective theories. Since Nature chose to have several flavors of quarks with different

---
[5] We shall return to this assumption later in this chapter.
[6] To be exact, we need to calculate amputated Green functions.

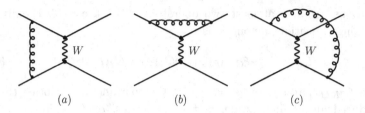

Fig. 4.2 Examples of one loop QCD corrections in "full" EW theory.

masses, the matching procedure might be repeated every time we cross a threshold associated with various quarks' masses. At the very least, this would change the number of active quark flavors that we need to include in the running of α_s.

We shall retrace the steps that gave us Eq. (4.24), but now include perturbative QCD corrections. There are two potential concerns that one can have while implementing this procedure. First of all, as with almost all loop calculations, one can expect the appearance of ultraviolet divergences. These divergences can be roughly divided into two classes, (1) appearing from renormalization of external quark legs, and (2) from renormalization of the composite operators Q_i themselves. We shall deal with these UV divergences by applying standard renormalization theory. Secondly, calculations of transition amplitudes in Eq. (4.56) include computations of matrix elements of four-fermion operators between hadron states. This is a non-trivial and, in general, non-perturbative task. However, since we are only interested in the results for the Wilson coefficients C_i, we can choose any external states for the calculations of matrix elements of those operators. In particular, the simplest choice, perturbative quark states, will do.

In order to perform the matching, we compute one loop QCD corrections in the full electroweak theory. These will involve calculations of six diagrams like those in Fig. 4.2. There are three scales that could affect our results: the mass of the W-boson M_W, renormalization scale scale μ, and the scale associated with external momenta p of the diagrams in Fig. 4.2. In our calculation we shall only keep the results enhanced by the logarithms of the ratios of scales, as those can be large. This corresponds to the so-called *leading logarithmic approximation* (LLA). We also assume that all quarks have the same momentum p and set their masses to zero. The result for

Effective Field Theories of Type I

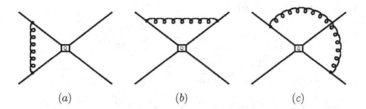

Fig. 4.3 Examples of one loop QCD corrections in effective theory.

$\mathcal{A}_{\text{full}}$ in dimensional regularization is [Buras (1998)]

$$\mathcal{A}_{\text{full}} = \frac{4G_F}{\sqrt{2}} \xi_{CKM} \left[\left(A_1 + \frac{B_1}{\epsilon} \right) \langle Q_2 \rangle^T + A_2 \langle Q_1 \rangle^T \right],$$

$$A_1 = 1 + 2C_F \frac{\alpha_s}{4\pi} \log \frac{\mu^2}{-p^2} + \frac{3}{N_c} \frac{\alpha_s}{4\pi} \log \frac{M_W^2}{-p^2}, \qquad (4.57)$$

$$A_2 = -\frac{3\alpha_s}{4\pi} \log \frac{M_W^2}{-p^2}, \quad B_1 = 2C_F \frac{\alpha_s}{4\pi},$$

where we used a well-known identity for the Gell-Mann matrices

$$2\delta_{\alpha\delta}\delta_{\gamma\beta} = \frac{2}{N_c}\delta_{\alpha\beta}\delta_{\gamma\delta} + \lambda^a_{\alpha\beta}\lambda^a_{\gamma\rho}, \qquad (4.58)$$

where summation over a is implied. $\langle Q_i \rangle^T$ are the tree-level matrix elements of operators Eq. (4.53).

The amplitude Eq. (4.57) is divergent for $\epsilon \to 0$. This divergence is cancelled by QCD loop contributions on the external quark legs, which gives for the fully renormalized amplitude

$$\mathcal{A}^r_{\text{full}} = \frac{4G_F}{\sqrt{2}} \xi_{CKM} \left[A_1 \langle Q_2 \rangle^T + A_2 \langle Q_1 \rangle^T \right]. \qquad (4.59)$$

This is the amplitude that we shall match to the effective amplitude computed with the effective Hamiltonian of Eq. (4.54).

The effective theory contribution $\mathcal{A}_{\text{eff}}$ can be computed from the diagrams of Fig. 4.3. Denoting insertion of the operators Q_1 and Q_2 by squares in Fig. 4.3 and evaluating the diagrams in dimensional regularization yields

$$\mathcal{A}_{\text{eff}} = \frac{4G_F}{\sqrt{2}} \xi_{CKM} \left[C_1 \left(\left(a_1 + \frac{b_1}{\epsilon} \right) \langle Q_1 \rangle^T + \left(a_2 + \frac{b_2}{\epsilon} \right) \langle Q_2 \rangle^T \right) \right.$$
$$\left. + C_2 \left(\left(a_1 + \frac{b_1}{\epsilon} \right) \langle Q_2 \rangle^T + \left(a_2 + \frac{b_2}{\epsilon} \right) \langle Q_1 \rangle^T \right) \right], \qquad (4.60)$$

where the coefficients are given by

$$a_1 = 1 + 2C_F \frac{\alpha_s}{4\pi} \log \frac{\mu^2}{-p^2} + \frac{3}{N_c} \frac{\alpha_s}{4\pi} \log \frac{\mu^2}{-p^2},$$

$$a_2 = -\frac{3\alpha_s}{4\pi} \log \frac{\mu^2}{-p^2}. \qquad (4.61)$$

$$b_1 = \frac{\alpha_s}{4\pi}\left(2C_F + \frac{3}{N_c}\right), \quad b_2 = -\frac{3\alpha_s}{4\pi}.$$

Notice that it is not possible to completely remove the divergences by wave function renormalization factors. This is to be expected, as we are dealing with the renormalization of composite operators. As given in Eq. (B.22), additional renormalization factors are needed,

$$\langle Q_i \rangle^{(0)} = Z_\psi^{-2} Z_{ij} \langle Q_j \rangle. \qquad (4.62)$$

Removing some of the UV divergencies with quark wave function renormalization factors Z_ψ and constructing Z_{ij} to remove the remaining divergences as

$$Z_{ij} = 1 + \frac{\alpha_s}{4\pi}\frac{1}{\epsilon}\begin{pmatrix} 3/N_c & -3 \\ -3 & 3/N_c \end{pmatrix} \qquad (4.63)$$

gives us the renormalized amplitude,

$$\mathcal{A}^r_{\text{eff}} = \frac{4G_F}{\sqrt{2}}\xi_{CKM}\Big[C_1\left(a_1\langle Q_1\rangle^T + a_2\langle Q_2\rangle^T\right)$$
$$+ C_2\left(a_1\langle Q_2\rangle^T + a_2\langle Q_1\rangle^T\right)\Big], \qquad (4.64)$$

This expression does not have any UV divergences left. We shall use Eq. (4.63) in the next section.

In the final step of our drive to determine C_1 and C_2, we shall match Eq. (4.59) and Eq. (4.64). Matching the coefficients of $\langle Q_i \rangle^T$ give the following system of equations

$$\begin{pmatrix} a_1 & a_2 \\ a_2 & a_1 \end{pmatrix}\begin{pmatrix} C_1 \\ C_2 \end{pmatrix} = \begin{pmatrix} A_2 \\ A_1 \end{pmatrix}, \qquad (4.65)$$

which can be solved for C_i, as $\det\hat{a} = 1 + (\alpha_s/2\pi)(2C_F + 3/N_c)\log(\mu^2/(-p^2)) \neq 0$. The solution is

$$\begin{pmatrix} C_1 \\ C_2 \end{pmatrix} = \frac{1}{\det\hat{a}}\begin{pmatrix} a_1 & -a_2 \\ -a_2 & a_1 \end{pmatrix}\begin{pmatrix} A_2 \\ A_1 \end{pmatrix}. \qquad (4.66)$$

Performing the multiplications and keeping only the leading results in α_s we obtain

$$C_1(\mu) = -\frac{3\alpha_s}{4\pi}\log\frac{M_W^2}{\mu^2}, \quad C_2(\mu) = 1 + \frac{3}{N_c}\frac{\alpha_s}{4\pi}\log\frac{M_W^2}{\mu^2}. \qquad (4.67)$$

The derivation of these coefficients was the goal of our exercise in this section. We have shown that perturbative QCD effects indeed generate a new operator, Q_1. Setting α_s to zero returns our original matching result of Eq. (4.24) with $C_1 = 0$. The effect of QCD corrections are significant, as for $\mu \sim 1$ GeV (a typical momentum scale for hadronic processes) the logarithms in Eq. (4.67) are large.

Let us finally note that, in principle, there was no need to employ Z_ψ for both full and effective theory calculations, since they are identical in both theories and therefore cancel in the matching. Yet, their use was convenient in obtaining the matrix Z_{ij} of Eq. (4.63), which we shall employ in the next section.

4.5.2 *Renormalization group improvement and EFTs*

Calculation of QCD corrections to effective operators of Eq. (4.53) produced large logarithms of the form $\alpha_s \log(m_W^2/\mu^2)$ in Eq. (4.67). This makes the convergence of QCD perturbative series questionable, as for low scales the size of the log compensates the smallness of α_s. We might expect that at higher orders in α_s all corrections of the form $\alpha_s^n \log^n(m_W^2/\mu^2)$ would be large, so they must be resummed. This can be achieved with the help of renormalization group equations.

What we saw in the previous section is that the operators Q_1 and Q_2 form a basis that *closes under renormalization*. This basis, however, might not be the most convenient for application of renormalization group (RG) techniques. One problem with this basis is that the matrix Z_{ij} of Eq. (4.63) is not diagonal. If we switch to another basis,

$$Q_\pm = \frac{1}{2}(Q_2 \pm Q_1), \quad C_\pm = C_2 \pm C_1 \tag{4.68}$$

we see that the matrix Z_{ij} in this basis also becomes diagonal,

$$Z_\pm = \text{diag}\left(1 - \frac{2\alpha_s}{4\pi\epsilon}, 1 + \frac{4\alpha_s}{4\pi\epsilon}\right). \tag{4.69}$$

This means that $Q_\pm$ do not mix under renormalization. The large logs appearing in Eq. (4.67) will also affect $C_\pm$, but the resummation of these logs will be easier in the new basis.

Now consider the renormalization group equation for $C_\pm$. Viewing C's as coupling constants in the effective Hamiltonian of Eq. (4.54), it is natural to imagine that they would also get renormalized. Then, in general,

$$C_\pm = Z_\pm^C C_\pm^{(0)}, \tag{4.70}$$

where $C_\pm^{(0)}$ denote "bare" values for $C_\pm$. Let us take a logarithmic derivative of both sides of Eq. (4.70),

$$\mu \frac{dC_\pm}{d\mu} = \left(\mu \frac{dZ_\pm^C}{d\mu}\right) C_\pm^{(0)} + Z_\pm^C \left(\mu \frac{dC_\pm^{(0)}}{d\mu}\right). \tag{4.71}$$

Since "bare" couplings do not depend on renormalization scale μ,

$$\mu \frac{dC_\pm(\mu)}{d\mu} = \gamma_\pm C_\pm(\mu), \tag{4.72}$$

where we defined *anomalous dimensions* for $C_\pm$,

$$\gamma_\pm(g_s) = \frac{\mu}{Z_\pm^C} \frac{dZ_\pm^C}{d\mu} = \frac{d \log Z_\pm^C}{d \log \mu} \tag{4.73}$$

Using the identity derived in Eq. (B.27) of Appendix B and expanding the (diagonal) anomalous dimension matrix $\gamma_\pm$ as

$$\gamma_\pm(g_s) = \frac{d \log Z_\pm}{d \log \mu} = \gamma_\pm^{(0)} \frac{g_s^2}{16\pi^2} + \gamma_\pm^{(1)} \left(\frac{g_s^2}{16\pi^2}\right)^2 + \ldots \tag{4.74}$$

we obtain the needed anomalous dimensions for the Wilson coefficients as

$$\gamma_\pm^{(0)} = \mathrm{diag}\,(4, -8). \tag{4.75}$$

Now, the solution of the RG equation of Eq. (4.72) can be obtained if we recall that $C_\pm(\mu)$ depend on μ both explicitly and via μ-dependence of the running coupling constant g_s,

$$\frac{d}{d\mu} = \frac{\partial}{\partial \mu} + \beta(g_s) \frac{\partial}{\partial g_s}. \tag{4.76}$$

The solution of the RG equation can the be represented as

$$C_\pm(\mu) = U_\pm(\mu, M_W) C_\pm(M_W) = \exp \int_{g(M_W)}^{g(\mu)} dg' \frac{\gamma_\pm(g')}{\beta(g')} C_\pm(M_W). \tag{4.77}$$

Expanding anomalous dimensions in $\alpha_s/4\pi$ we can find a LLA solution,

$$C_\pm(\mu) = \left(\frac{\alpha_s(M_W)}{\alpha_s(\mu)}\right)^{\gamma_\pm/(2\beta_0)} C_\pm(M_W). \tag{4.78}$$

This formula achieves our goal: there are no longer any large logarithms in the expression for $C_\pm$. One can check that, if expanded in α_s, the expression derived in Eq. (4.67) is recovered.

Note that according to Eq. (4.68) $C_+(M_W) = C_-(M_W) = 1$. To complete the derivation of $C_\pm$ let us recall that β_0 depends on the number of flavors via $\beta_0 = 11 - 2n_f/3$, which ties together scale and the number of active flavors. For example, for $\mu = m_b$, where five quark flavors are active we get

$$C_+(m_b) = \left(\frac{\alpha_s(M_W)}{\alpha_s(m_b)}\right)^{6/23}, \quad C_-(m_b) = \left(\frac{\alpha_s(M_W)}{\alpha_s(m_b)}\right)^{-12/23}. \quad (4.79)$$

Numerically, $C_+(5 \text{ GeV}) \simeq 0.85$ is suppressed by QCD running, while $C_-(5 \text{ GeV}) \simeq 1.40$ is enhanced compared to $C_\pm(M_W) = 1$.

As we can see, the number of active flavors changes the values of Wilson coefficients, mainly due to the changing of β_0. Thus, the evolution of $C_\pm$ depends on crossing quark thresholds. For example, evolving down to the scales near the charm quark mass $\mu = m_c$ should be done as

$$C_\pm(\mu) = U_\pm^{n_f=4}(m_c, m_b) U_\pm^{n_f=5}(m_b, M_W) C_\pm(M_W) \quad (4.80)$$

and similarly for the lower scales.

4.5.3 Complete basis. Penguin operators

Our assumption was that there are only two operators that form the basis of dimension six flavor-changing operators. This assumption seemed solid, as we proved that the basis closes under renormalization, i.e. no other operator structures are generated by QCD effects.

This, however, does not preclude the appearance of other operators! If we follow EFT framework, we would have to write all possible dimension six (and even some dimension five) operators in order to describe the non-leptonic low energy transitions of quarks. At very least, we should expect that other Dirac structures will appear in our discussion. These new operators might mix among themselves under QCD renormalization. Also, Q_1 and Q_2 might mix into other operators, but not vise versa. Still, all operator structures must be present, so we have to write them out.

This is exactly what has been done for the description of $\Delta Q = 1$ weak decays of $Q = b, c$, and s quarks. In this case, operator structures other than Q_1 and Q_2 can be generated in the standard model. One example is the so-called penguin operator, which appears upon matching the standard model graph like the one in Fig. 4.4 onto our effective theory. Without dwelling on the origins of such a strange name for an operator, we note that depending on the nature of the propagator connecting the electroweak loop and the rest of the graph, one can classify those operators as *gluonic* or *electroweak* penguins.

Let us now write the full $\Delta Q = 1$ non-leptonic effective Hamiltonian. We shall only concentrate on the operators generated in the standard model.

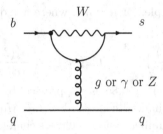

Fig. 4.4 Penguin diagram giving rise to penguin operators.

Any other operator structures that contribute to the amplitude would indicate the presence of physics beyond the standard model, which can be used as a search tool. Concentrating on $\Delta b = 1$ transitions for simplicity

$$\mathcal{H}_{\text{eff}} = \frac{4G_F}{\sqrt{2}} \sum_{q=u,c} \lambda_q \left[C_1 Q_1^q + C_2 Q_2^q + \sum_{i=3,\ldots,10} C_i Q_i + C_{7\gamma} Q_{7\gamma} + C_{8g} Q_{8g} \right]$$
$$+ \text{h.c.}, \qquad (4.81)$$

where $\lambda_q = V_{qb} V_{qs}^*$ denotes the relevant CKM matrix elements. The current-current operators Q_1 and Q_2 for $\Delta b = 1$ transitions take the form

$$Q_1 = (\bar{q}_\alpha b_\beta)_{V-A} (\bar{s}_\beta q_\alpha)_{V-A}, \text{ and}$$
$$Q_2 = (\bar{q}_\alpha b_\alpha)_{V-A} (\bar{s}_\beta q_\beta)_{V-A}. \qquad (4.82)$$

Note that here we explicitly consider $b \to s$ decays and suppressed Dirac structures that are given by $(\bar{q}_1 q_2)_{V\pm A} = (1/2)(\bar{q}_1 \gamma_\mu (1 \pm \gamma_5) q_2)$. Related transitions, such as $b \to d$ can be obtained by direct substitution of the appropriate quark fields and CKM matrix elements. We shall suppress color indices from now on for the structures like $(\bar{q}_\alpha q_\alpha)_{V\pm A} \to (\bar{q}q)_{V\pm A}$.

The gluonic penguin operators Q_{3-6} are given by

$$Q_3 = (\bar{s}b)_{V-A} \sum_q (\bar{q}q)_{V-A}, \quad Q_4 = (\bar{s}_\alpha b_\beta)_{V-A} \sum_q (\bar{q}_\beta q_\alpha)_{V-A},$$
$$Q_5 = (\bar{s}b)_{V-A} \sum_q (\bar{q}q)_{V+A}, \quad Q_6 = (\bar{s}_\alpha b_\beta)_{V-A} \sum_q (\bar{q}_\beta q_\alpha)_{V+A}, \qquad (4.83)$$

while electroweak penguin operators Q_{7-10} are

$$Q_7 = \frac{3e_q}{2}(\bar{s}b)_{V-A} \sum_q (\bar{q}q)_{V+A}, \quad Q_8 = \frac{3e_q}{2}(\bar{s}_\alpha b_\beta)_{V-A} \sum_q (\bar{q}_\beta q_\alpha)_{V+A},$$
$$Q_9 = \frac{3e_q}{2}(\bar{s}b)_{V-A} \sum_q (\bar{q}q)_{V-A}, \quad Q_{10} = \frac{3e_q}{2}(\bar{s}_\alpha b_\beta)_{V-A} \sum_q (\bar{q}_\beta q_\alpha)_{V-A}. \qquad (4.84)$$

Table 4.1 Wilson coefficients for the $\Delta b = 1$ operator basis in Eq. (4.81) for $\mu = m_b(m_b) = 4.40$ GeV and $m_t = 170$ GeV (from [Buchalla et al. (1996)]).

Wilson coefficient	LO	NLO (NDR) $\Lambda^{(5)} = 140$ MeV	LO	NLO (NDR) $\Lambda^{(5)} = 310$ MeV
C_1	−0.273	−0.165	−0.339	−0.203
C_2	1.125	1.072	1.161	1.092
C_3	0.013	0.013	0.016	0.016
C_4	−0.027	−0.031	−0.033	−0.039
C_5	0.008	0.008	0.009	0.009
C_6	−0.033	−0.036	−0.043	−0.046
C_7/α	0.042	−0.003	0.047	−0.001
C_8/α	0.041	0.047	0.054	0.061
C_9/α	−1.264	−1.279	−1.294	−1.303
C_{10}/α	0.291	0.234	0.360	0.288

Finally, there are two dipole operators,

$$Q_{7\gamma} = -\frac{e}{8\pi^2} m_b \, \bar{s}\sigma_{\mu\nu}(1-\gamma_5)F^{\mu\nu}b, \text{ and}$$
$$Q_{8g} = -\frac{e}{8\pi^2} m_b \, \bar{s}\sigma_{\mu\nu}(1-\gamma_5)G^{\mu\nu}b, \quad (4.85)$$

These QED and QCD field strength tensors are defined in Appendix B. In the standard model dipole operators with quarks of opposite helicity are suppressed by factors of the light quark mass, m_s. Therefore, their contribution is usually discarded as being small. There are models of new physics where chirality flip occurs inside the loop diagram that generates the operators $Q_{7\gamma}$ and Q_{8g}. One way to search for those models would be to tag the helicity of the final state photon, which is a challenging measurement.

What about the Wilson coefficients for those operators? Their calculation at leading order (LO) in QCD and renormalization group improvement of the result is done the same way as discussed in sections 4.5.1-4.5.2. The only difference is the size of the basis of the operators, which makes the discussion cumbersome. The details of the calculations can be found in [Buchalla et al. (1996)], we shall not present them here.

Even after resumming the logarithms, the scale dependence of Wilson coefficients remains strong (see Table 4.1). It is then advantageous to go beyond one loop matching. It would then also be consistent to go beyond the leading order in the calculations of anomalous dimensions for the following renormalization group running. The next-to-leading order (NLO) calculation is interesting, as it brings new features in the discussion, such as the dependence of the final result on the renormalization scheme used in the calculation. The discussion of this, however, goes beyond the scope of this book, but can be found in a number of specialized reviews [Buchalla et al. (1996); Buras (1998)]. Here we present numerical results for the Wilson coefficients at LO and NLO (using Naive Dimensional Reduction (NDR) scheme) [Buchalla et al. (1996)], which can be used for practical computations of decay rates.

The discussion of Wilson coefficients for photonics and gluonic dipole operators $Q_{7\gamma}$ and Q_{8g} is similar. Their values are $C_{7\gamma} = 0.299$ and $C_{8g} = 0.143$, which are quite large to affect some decays of B-mesons.

It should be noted that the techniques of calculating QCD corrections for four-fermion operators can be employed for other calculations. For example, analyses of meson mixing in the B-sector ($\Delta b = 2$ operators) can also be done in the same way, as can analyses of kaons and charmed meson decays.

4.6 Chiral perturbation theory

We have been considering effective field theories of weak interactions. The primary goal of building such theories was simplification of calculations of QCD corrections, which is an important development. Yet both the UV completion and the infrared EFT were known and even perturbative. That is to say, matching between those theories did not hide any secrets: both effective operators and Wilson coefficients were calculable.

There are situations when the UV completion is not known. No matching is possible in this case, so Wilson coefficients need to be determined from fits to experimental data. This situation often happens when the standard model is considered as an effective theory, so all new unknown heavy degrees of freedom can be accounted for by expanding the set of operators while maintaining the symmetry structure of the standard model.

We can also encounter yet another situation. It might happen that both UV and IR theories are known, yet direct matching is not possible. This can happen if the matching occurs at a scale where both theories are non-perturbative. Oftentimes, this also means that two theories are written in terms of different degrees of freedom.

One example of such theory is QCD. While perturbative in the UV, quarks and gluons are not the low energy degrees of freedom of strong interactions. The low energy spectrum of QCD should be written in terms of mesons and baryons, as this is what observed in Nature. Can an effective field theory of meson and baryon interactions be built? Would it be useful without a possibility of matching to QCD?

The answers to both of those questions are positive. There is no problem in building a quantum theory of meson interactions. What we have to do is to make sure that the theory has a proper power counting scheme that makes it effective. In order to describe QCD in the low energy regime, it must have the same set of symmetries as QCD. Then, coefficients of the operators could be fixed by experimental measurements. Our low-energy theory would be predictive as long as the chosen power counting scheme assures us that at each order the number of parameters to fix is less than the number of experimental observables. Low energy interactions of those light degrees of freedom is the subject of interest of Chiral Perturbation Theory (χPT).

In the massless limit QCD has a set of new symmetries. The Lagrangian of massless QCD for two lightest quarks u and d,

$$\mathcal{L}_{QCD} = -\frac{1}{4} G^a_{\mu\nu} G^{\mu\nu}_a + \bar{q}(i\slashed{D} - m_q) q, \qquad (4.86)$$

where it is convenient to write q as a doublet

$$q(x) = \begin{pmatrix} u(x) \\ d(x) \end{pmatrix} \qquad (4.87)$$

can be broken into two independent sectors (let us call them *left* and *right*), provided that quark fields are written as

$$q = \frac{1-\gamma_5}{2} q + \frac{1+\gamma_5}{2} q = q_L + q_R \qquad (4.88)$$

and $m_q = 0$. In other words,

$$\mathcal{L}_{QCD} = -\frac{1}{4} G^a_{\mu\nu} G^{\mu\nu}_a + \bar{q}_L (i\slashed{D}) q_L + \bar{q}_R (i\slashed{D}) q_R, \qquad (4.89)$$

The Lagrangian of Eq. (4.89) is invariant under separate $SU(2)$ rotations of q_L and q_R[7], which we shall call $SU(2)_L$ and $SU(2)_R$,

$$q_L \to L q_L, \quad \text{and} \quad q_R \to R q_R. \qquad (4.90)$$

As we know, those symmetries are not exact. At the very least, they are broken by the quark masses, which mix L and R sectors of QCD,

$$\mathcal{L}_m = -m_q (\bar{q}_L q_R + \bar{q}_R q_L). \qquad (4.91)$$

---

[7]Actually, it is invariant under $U(2)_L \times U(2)_R$, but two phases $U(1)_A$ and $U(1)_V$ are irrelevant for our discussion. As it turns out, $U(1)_A$ is broken by chiral anomaly, while unbroken $U(1)_V$ represents baryon number conservation, which is trivial in the meson sector.

They are also broken *spontaneously* by the non-zero expectation values of the scalar operators,

$$\langle 0|\bar{q}_R^\alpha q_L^\beta|0\rangle \equiv \langle \bar{q}_R^\alpha q_L^\beta\rangle = v\delta^{\alpha\beta}, \qquad (4.92)$$

also known as *quark condensates* with $v \sim -(250)$ MeV^3. The quark condensates break the symmetry

$$SU(2)_L \times SU(2)_R \to SU(2)_V, \qquad (4.93)$$

which is realized in Nature. The manifestation of such a symmetry breaking is the triplet of massless particles (goldstone bosons). In reality, there are no massless particles in QCD spectrum, but there are three very light pions. Understanding that chiral symmetry in QCD is also explicitly broken by the quark masses as discussed above, those pions are called *pseudo-Nambu-Goldstone* (PNG) bosons. This symmetry structure must be built into the construction of the low-energy effective theory of QCD.

4.6.1 Goldstone bosons and their properties

A careful examination of experimental data on the spectra of mesons and baryons does not reveal much structure. There is, however, one glaring exception: there are three pseudoscalar meson states that are much lighter than others. Indeed, while the majority of states have masses of approximately around 1 GeV, masses of charged $\pi^\pm$, and neutral π^0 pions are significantly smaller, around 140 MeV.

Before we get into the nitty-gritty of constructing the Lagrangian of χPT, let us first discuss its symmetry structure in more detail, restricting our attention to the the pion sector only. We will do an analysis of the problem along the lines of the CCWZ construction reviewed in Section 2.7. The pions transform according to a representation of $G = SU(2)_L \times SU(2)_R$ group, which means that for each member g of G there exists a mapping $\vec{\pi} \to \vec{\pi}' = f(g, \vec{\pi})$. Chiral transformations in the left and right-handed sectors are denoted L and R, respectively. So, $g = (L, R) \in G$. The fields $\pi^i(x)$ are the coordinates describing the Goldstone fields in the coset space G/H, where $H = SU(2)_V$. Choosing a coset representative $\xi = (\xi_L(\pi), \xi_R(\pi)) \in G$, one can show that their chiral transformations G are non-trivial,

$$\xi_L(\pi) \to L\xi_L(\pi)h^\dagger(\pi, g) \qquad (4.94)$$

and the same for the tight-handed field with $L \to R$. Here we introduced a *compensating* transformation $h \in G$, the same for L and R sectors[8]. It would then be best to choose $\xi_L^\dagger(\pi) = \xi_R(\pi) \equiv \xi(x)$. We can then see that the field

$$U(x) \equiv \xi_L^\dagger(\pi)\xi_R(\pi) \to \xi^2(x) \tag{4.95}$$

has a much simpler transformation properties under G,

$$U(x) \to L U(x) R^\dagger \tag{4.96}$$

i.e. it transforms *linearly* under the chiral transformations. Since for the pions the quotient group G/H is also $SU(2)$, according to the theorem in Chapter 2, the pion field can be represented by a 2×2 matrix field $U(x)$ which belongs to $SU(2)$. Since one needs three parameters to represent elements of $SU(2)$, this matrix $U(x)$ is equivalent to a set of three pion fields, $\pi^i(x)$, where $i = 1,2,3$. Since experimental observables should be independent on the choice of the mathematical formulation of our theory, it is up to us to choose a representation of the pion fields. It is quite convenient to choose the canonical representation,

$$U(x) = \exp\left(\frac{i\sqrt{2}}{f}\pi(x)\right), \quad \text{with} \quad \pi(x) = \sum_{i=1}^{3} \frac{\sigma^i}{\sqrt{2}}\pi^i(x), \tag{4.97}$$

where f is some parameter (of mass dimension 1) that we will determine later. Notice that

$$\pi(x) = \frac{1}{\sqrt{2}}\begin{pmatrix} \pi^3 & \pi^1 - i\pi^2 \\ \pi^1 + i\pi^2 & -\pi^3 \end{pmatrix} \equiv \begin{pmatrix} \frac{\pi^0}{\sqrt{2}} & \pi^+ \\ \pi^- & -\frac{\pi^0}{\sqrt{2}} \end{pmatrix} \tag{4.98}$$

and that $\pi^i(x) = (1/2)\text{Tr } \sigma^i \pi(x)$. The chiral transformation of the pion fields $\pi(x)$, contrary to $U(x)$, is highly nonlinear.

The discussion above can clearly be generalized to three flavors, in which case the symmetry breaking pattern would be

$$SU(3)_L \times SU(3)_R \to SU(3)_V \tag{4.99}$$

and the matrix $U(x)$ takes the form similar to Eq. (4.97), but with $\pi(x) \to \phi(x)$ with

$$\phi(x) = \sum_{i=1}^{3} \frac{\lambda_i}{\sqrt{2}}\phi_i(x) = \begin{pmatrix} \frac{\pi^0}{\sqrt{2}} + \frac{\eta_8}{\sqrt{6}} & \pi^+ & K^+ \\ \pi^- & -\frac{\pi^0}{\sqrt{2}} + \frac{\eta_8}{\sqrt{6}} & K^0 \\ K^- & \overline{K}^0 & -\frac{2\eta_8}{\sqrt{6}} \end{pmatrix}, \tag{4.100}$$

---

[8] This is trivially so, since left and right-handed sectors of QCD can be related to each other by a parity transformation.

where λ_i are the Gell-Mann matrices.

Assigning chiral dimension zero to $U(x)$, and taking into account symmetry properties of $U(x)$, one can build an effective Lagrangian for pion-pion interactions using $U(x)$ and derivatives,

$$\mathcal{L} = \mathcal{L}_0 + \mathcal{L}_2 + \mathcal{L}_4 + \dots. \tag{4.101}$$

Since we are dealing with the matrix-valued fields $U(x)$, and taking into account chiral transformation properties of $U(x)$ given in Eq. (4.96), it is convenient to use traces of the fields and their products, where the cyclic property of the trace makes the field products invariant under chiral transformations. The only possible term of dimension 0 that is consistent with symmetry requirements is

$$\mathcal{L}_0 = a \text{ Tr } UU^\dagger, \tag{4.102}$$

where a is some constant of dimension 4. This term, however, represents a meaningless constant, as $\text{Tr } UU^\dagger = 1$, which does not give us any dynamical information.

Including derivatives acting on the field U one can write the next term in derivative expansion of $\mathcal{L}$. There is again only a single term that can be written,

$$\mathcal{L}_2 = b \text{ Tr } \partial_\mu U \partial^\mu U^\dagger, \tag{4.103}$$

as other possible terms are either total derivatives or can be related to a total derivative and Eq. (4.103). For example, the term $\text{Tr } (\partial^2 U) U^\dagger$ can be rewritten as

$$\text{Tr } [(\partial_\mu \partial^\mu U) U^\dagger] = \partial_\mu \left[\text{Tr } \partial^\mu U U^\dagger \right] - \text{Tr } \partial_\mu U \partial^\mu U^\dagger. \tag{4.104}$$

Since $\partial_\mu \left[\text{Tr } \partial^\mu U U^\dagger \right]$ is a total derivative, it can be dropped from the Lagrangian, so this term is equivalent to Eq. (4.103). Finally, terms containing a product of two traces as also irrelevant, as

$$\text{Tr } \partial_\mu U U^\dagger \text{ Tr } \partial^\mu U U^\dagger = 0, \tag{4.105}$$

because $\text{Tr } \partial_\mu U U^\dagger = 0$. Thus only one term, given by Eq. (4.103), is possible at order p^2, if symmetry breaking terms are excluded. We only need to determine b, which can be done by requiring that the kinetic term for the pion field had the canonical form,

$$\mathcal{L}_2 = \frac{1}{2} \partial_\mu \phi_i \partial^\mu \phi^i + \cdots, \tag{4.106}$$

which implies that $b = f^2/4$. Thus, the leading-order chiral Lagrangian for the pions, kaons and etas can be written as

$$\mathcal{L}_2 = \frac{f^2}{4} \text{ Tr } \partial_\mu U \partial^\mu U^\dagger. \tag{4.107}$$

It is important to note that this Lagrangian is nonlinear in the goldstone fields. That is to say, one and the same constant f defines processes with different number of Goldstone bosons, say, $\pi\pi \to \pi\pi$ and $\pi\pi \to 4\pi$, etc. For example, to calculate $\pi^i \pi^k \to \pi^m \pi^n$ (where i, k, m, n denote isospin indices of the pions), we would need to expand the Lagrangian of Eq. (4.107) in $1/f$ and restrict our attention to its pion sector,

$$\mathcal{L}_2 = \frac{1}{2} \left(\partial_\mu \vec{\pi}\right)^2 + \frac{1}{6f^2} \left[(\partial_\mu \vec{\pi} \cdot \vec{\pi})^2 - \vec{\pi}^2 \left(\partial_\mu \vec{\pi}\right)^2 \right]. \tag{4.108}$$

The scattering amplitude $\mathcal{A} = i(S-1)$ for $\pi^i \pi^k \to \pi^m \pi^n$ would result from the second term in Eq. (4.108),

$$\mathcal{A} = -\frac{1}{f^2} \left[s\, \delta_{ik}\delta_{mn} + t\, \delta_{im}\delta_{kn} + u\, \delta_{in}\delta_{mk} \right], \tag{4.109}$$

where the Mandelstam variables are defined in the usual way, $s = (p_i + p_k)^2$, $t = (p_i - p_m)^2$, and $u = (p_i - p_n)^2$. This is a well-known result. This is great – now we can compare our prediction to experiment and determine f!

So far we only considered a case of *massless* Goldstone bosons. We know that this is not the case – while indeed pions and kaons are much lighter than the other mesons, they are not massless. How should we take into account their masses?

4.6.2 Sources in chiral perturbation theory

While χPT is explicitly constructed to be a representation of QCD at low energies, it is also possible to modify it to consider electroweak decays of kaons and pions. A particularly convenient way to do so is by introducing external source fields, a beautiful technique advocated first by J. Schwinger. To see how this is done, let us first couple the light quark QCD Lagrangian to external source fields,

$$\begin{aligned}\mathcal{L} = &-\frac{1}{4} G^a_{\mu\nu} G^{\mu\nu}_a + \bar{q} i \slashed{D} q \\ &- \bar{q}_L \gamma_\mu l^\mu q_L - \bar{q}_R \gamma_\mu r^\mu q_R \\ &- \bar{q}_L \left(s + ip\right) q_R - \bar{q}_R \left(s - ip\right) q_L,\end{aligned} \tag{4.110}$$

where l^μ, r^μ, s, and p are the left, right, scalar, and pseudoscalar source fields, respectively[9]. Notice that the covariant derivative $\mathcal{D}_\mu$ in the first line of Eq. (4.110) refers to color quark degrees of freedom. We give it this notation to avoid confusion with the chiral covariant derivative defined below.

Let us now require that the Largangian of Eq. (4.110) is invariant under *local* flavor $SU(3)_L \times SU(3)_R$ symmetry, i.e.

$$q_L \to L(x)q_L, \quad q_R \to R(x)q_R, \tag{4.111}$$

where $L(x)$ and $R(x)$ are position-dependent flavor $SU(3)$ rotation matrices. Now, if we require that the source fields transform under flavor $SU(3)_L \times SU(3)_R$ as

$$\begin{aligned}
(s+ip) &\to L(x)\,(s+ip)\,R^\dagger(x), \\
(s-ip) &\to R(x)\,(s-ip)\,L^\dagger(x), \\
l_\mu &\to L(x)l_\mu L^\dagger(x) + i\partial_\mu L(x) L^\dagger(x), \\
r_\mu &\to R(x)r_\mu R^\dagger(x) + i\partial_\mu R(x) R^\dagger(x),
\end{aligned} \tag{4.112}$$

and the derivative in the kinetic term of Eq. (4.110) is promoted to a covariant derivative,

$$\not{D}q_L \to L(x)\not{D}q_L, \quad \not{D}q_R \to R(x)\not{D}q_R \tag{4.113}$$

then the Lagrangian of QCD in Eq. (4.110) will indeed remain invariant. Notice that this covariant derivative has *nothing* to do with $SU(3)_c$.

Employing the same logic as before, let us now build a low energy effective Largangian from hadronic degrees of freedom by requiring that it would retain all symmetries of the QCD Lagrangian with external sources. To do so, let us promote a local derivative of Eq. (4.107) to a *covariant derivative*

$$\partial_\mu U \to D_\mu U = \partial_\mu U + il_\mu U - iUr_\mu. \tag{4.114}$$

It is easy to check that, with this definition, the covariant derivative will transform as

$$D_\mu U \to L(x)\,[D_\mu U]\,R^\dagger(x) \tag{4.115}$$

Now we treat those external sources like external fields. We can introduce "field strength tensors" built out of the source fields

$$\begin{aligned}
L_{\mu\nu} &= \partial_\mu l_\nu - \partial_\nu l_\mu + i\,[l_\mu, l_\nu], \\
R_{\mu\nu} &= \partial_\mu r_\nu - \partial_\nu r_\mu + i\,[r_\mu, r_\nu],
\end{aligned} \tag{4.116}$$

---

[9] Oftentimes, instead of l_μ and r_μ one can equivalently choose vector $v_\mu = l_\mu + r_\mu$ and axial $a_\mu = l_\mu - r_\mu$ sources.

which, as you can check, transform as

$$L_{\mu\nu} \to L(x) L_{\mu\nu} L^\dagger(x),$$
$$R_{\mu\nu} \to R(x) R_{\mu\nu} R^\dagger(x). \tag{4.117}$$

We can also put together a "source" Lagrangian,

$$\mathcal{L}_{ext} = H_1 \text{Tr}\left[L_{\mu\nu} L^{\mu\nu} + R_{\mu\nu} R^{\mu\nu}\right] + H_2 \text{Tr} \chi^\dagger \chi, \tag{4.118}$$

where $H_{1,2}$ are some unknown coefficients and $\chi \propto s + ip$ will be defined more precisely below. The fact that L-terms and R-terms in Eq. (4.118) share a coefficient reflects parity invariance of the theory.

Now we have all the necessary ingredients to build a low energy chiral Lagrangian with sources. Generalization of the kinetic term is straightforward, one simply needs to upgrade derivatives to the covariant ones. In addition to that, we can now build terms that contain only one field U (or $U^\dagger$) and a combination of sources $s \pm ip$. Thus, generalizing the kinetic term and taking into account Eq. (4.112) we can now write a leading-order chiral Lagrangian as

$$\mathcal{L}_2 = \frac{f^2}{4} \text{Tr}\left[D_\mu U D^\mu U^\dagger\right] + \frac{f^2}{4} \text{Tr}\left[\chi U^\dagger + U \chi^\dagger\right], \tag{4.119}$$

where covariant derivatives D_μ were defined in Eq. (4.114), and $\chi = 2B_0(s+ip)$. Thus, $\mathcal{L}_2$ contains all sources introduced above. We can define the action described by the Lagrangian of Eq. (4.119) as

$$S_2 = \int d^4x \, \mathcal{L}_2. \tag{4.120}$$

This action makes it convenient to determine various Noether symmetry currents by taking functional derivatives with respect to the appropriate sources. For example, the left-handed current can be obtained as

$$j_L^\mu(y) = \left.\frac{\delta S_2}{\delta l_\mu(y)}\right|_{l_\mu = r_\mu = 0} = \frac{i}{2} f^2 \left(D_\mu U^\dagger\right) U$$
$$= \frac{1}{\sqrt{2}} f D_\mu \phi - \frac{i}{2} [\phi(D^\mu \phi) - (D^\mu \phi)\phi] + ..., \tag{4.121}$$

where we used a familiar identity $\delta j(x)/\delta j(y) = \delta^4(x-y)$ for both l_μ and r_μ and then expanded the results in terms of the meson fields (or $1/f$). Similar result for the right handed current is

$$j_R^\mu(y) = -\frac{1}{\sqrt{2}} f D_\mu \phi - \frac{i}{2} [\phi(D^\mu \phi) - (D^\mu \phi)\phi] + ... \tag{4.122}$$

The introduction of sources allows us not only to describe electroweak transitions with PNG bosons, but also lets us determine unknown coefficients in Eq. (4.119). Low energy QCD without weak interactions has $l_\mu = r_\mu$ and $p = 0$ with

$$s \equiv M = \begin{pmatrix} m_u & 0 & 0 \\ 0 & m_d & 0 \\ 0 & 0 & m_s \end{pmatrix}, \qquad (4.123)$$

as can be seen from Eq. (4.110). We note, in passing, that, in principle, we could have obtained the second term in Eq. (4.119) by using spurion methods described in Chapter 2 assuming that M transforms as

$$M \to RML^\dagger. \qquad (4.124)$$

Let us now determine all of the constants in the leading order chiral Lagrangian. First, in the same approximation $l_\mu = r_\mu$, $p = 0$ let us take a variational derivative of the action S_2 with respect to the source matrix s^{ij},

$$\frac{\delta S_2}{\delta s^{ij}} = 2B_0 \frac{f^2}{4} \text{Tr}\left[\left(U^\dagger\right)^{ij} + U^{ij}\right]. \qquad (4.125)$$

We note that s^{ij} is nothing but the mass matrix M^{ij}. In Eq. (4.125) we computed the scalar current in χPT.

The same quantity can be also computed in QCD. Conveniently rewriting the mass term of the QCD Lagrangian

$$\mathcal{L}_{QCD} = -\bar{q}_{iR} M^{ij} q_{jL} - \bar{q}_{iL} M^{\dagger ij} q_{jR}, \qquad (4.126)$$

the variational derivatives of the QCD action $S_{QCD} = \int d^4x\, \mathcal{L}_{QCD}$ with respect to M^{ij} and $M^{\dagger ij}$ would take the form

$$\frac{\delta S_{QCD}}{\delta M^{ij}} = \bar{q}_R^i q_L^j \quad \text{and} \quad \frac{\delta S_{QCD}}{\delta M^{\dagger ij}} = \bar{q}_L^i q_R^j. \qquad (4.127)$$

For the mass matrix given in Eq. (4.123) and in the lowest energy (vacuum) state, where $U = U^\dagger = 1$ (that is, Kronecker delta δ^{ij}), matching of the vacuum matrix elements of the scalar currents of Eqs. (4.125) and (4.127) yields

$$\langle \bar{q}^i q^j \rangle = -f^2 B_0 \delta^{ij}, \text{ or } B_0 = -\frac{1}{3f^2} \langle \bar{q}q \rangle, \qquad (4.128)$$

where we used the fact that $\langle 0|(\bar{q}_L q_R + \bar{q}_L q_R)|0\rangle = \langle \bar{q}q \rangle$, the usual QCD quark vacuum condensate. Thus, B_0 is directly related to the order parameter of spontaneous breaking of chiral symmetry in QCD.

Similarly we can determine the parameter f. Let us calculate the amplitude of probability to create a (charged) pion state acting on the vacuum with an external axial current. This amplitude is usually parameterized in QCD as

$$\langle 0|j_{A\pi}^{\mu}|\pi\rangle = i\sqrt{2}f_{\pi}p^{\mu}. \tag{4.129}$$

Here $f_{\pi} = 92.4$ MeV is the pion decay constant that can be measured experimentally in the leptonic decay of the pion, $\pi^+ \to \ell^+ \bar{\nu}_\ell$.

The same amplitude can be computed in χPT. Since $j_A^\mu = j_L^\mu - j_R^\mu$ we can get from Eq. (4.119) after expanding in $\phi(x)$,

$$j_A^\mu(x) = \sqrt{2}f\partial_\mu \phi(x). \tag{4.130}$$

Calculating the matrix element of the above matrix current for π^+ in momentum spaces immediately identifies, at the leading order in chiral expansion, $f = f_\pi$.

What have we achieved so far? Just like in section 4.3 we built an infrared effective field theory for which the UV completion is known. There is, however a twist: while the UV completion is perturbative, perturbative matching between QCD and χPT is not possible, as there is no region of the parameter space where both theories are perturbative. Moreover, the EFT in this case is even written in terms of different degrees of freedom than its UV completion.

This, however, has not deterred us from deriving χPT. Since we are dealing with data, we don't really need to match to another theory to work with Eq. (4.119), we might as well use data. Which is exactly what we have done by fixing f. As long as there are enough experimental observables to fix all the parameters at given order in chiral expansion, we are in good shape. So, what predictions can be made with χPT effective Lagrangian?

4.6.3 Applications: Gell-Mann-Okubo relation

The Chiral Lagrangian of Eq. (4.119) has much to say about meson masses. To study those, let us recall that the mass terms in scalar theories come from the coefficients of the relevant operators ϕ^2. Once again setting $l_\mu = r_\mu = p = 0$, and $s = M$, and expanding the second term of Eq. (4.119) (let us call it $\mathcal{L}_m$), and retaining the terms quadratic in $\phi(x)$ yields

$$\mathcal{L}_m = -B_0 \text{Tr}(\phi^2(x)M) + ..., \tag{4.131}$$

where we dropped the meaningless constants. We can now compute the trace to get

$$-\frac{1}{B_0}\mathcal{L}_m = 2\frac{m_u + m_d}{2}\pi^+(x)\pi^-(x) + (m_u + m_s)K^+(x)K^-(x)$$
$$+ (m_d + m_s)\overline{K}^0(x)K^0(x) + \frac{m_u + m_d}{2}(\pi^0(x))^2 \qquad (4.132)$$
$$+ \frac{1}{\sqrt{3}}(m_u - m_d)\pi^0(x)\eta_8(x) + \frac{1}{6}(m_u + m_d + 4m_s)\eta_8^2(x)$$

For simplicity we shall work in the isospin limit, where $m_u = m_d$, so the term that mixes π^0 and η_8 vanishes. Also, it is common to write $m_u + m_d$ as $2\hat{m}$, so the PNG masses can be identified as

$$m_{\pi^\pm}^2 = m_{\pi^0}^2 = 2\hat{m}B_0,$$
$$m_{K^\pm}^2 = m_{K^0}^2 = (\hat{m} + m_s)B_0, \qquad (4.133)$$
$$m_{\eta_8}^2 = \frac{2}{3}(\hat{m} + 2m_s)B_0,$$

where we have to remember that π^0 and η_8 are self-conjugated fields, so their mass terms have an extra factor of $1/2$. These relations usually go under the name of Gell-Mann-Oaks-Renner (GMOR) relations; notice that they are relations between the quark masses and the meson mass-squares. It is interesting to note that while Eq. (4.133) relates PNG and quark masses, extraction of the quark masses is not possible without independent information about B_0 because one can always perform a simultaneous rescaling $m_q \to m_q/\lambda$ and $B_0 \to \lambda B_0$. Nevertheless, the relations in Eq. (4.133) are quite interesting. First, one can note that

$$4m_K^2 - m_\pi^2 = 4(\hat{m} + m_s)B_0 - 2\hat{m}B_0 = 2(\hat{m} + 2m_s)B_0 = 3m_{\eta_8}^2 \qquad (4.134)$$

Thus, PNG masses are related to each other, irrespectively of the values of the quark masses and B_0. Eq. (4.134) is called the *Gell-Mann-Okubo* (GMO) relation; it has been derived long before χPT using methods of chiral algebra.

Secondly, as mentioned above, we cannot obtain absolute quark masses from Eq. (4.133). However, since all PNG masses are proportional to B_0 we can obtain the next best thing: their ratios. In particular,

$$\frac{m_s}{\hat{m}} = \frac{2m_K^2}{m_\pi^2} - 1 \approx 25.6 \qquad (4.135)$$

This relation can be further improved if we figure out how to deal with quantum loops and higher-order terms in the chiral expansion.

4.6.4 Power counting. Chiral loops and higher orders in χPT

The development of chiral perturbation theory would not be complete without a correct treatment of quantum loops. Let us introduce p as some representative momentum that describes a physical process under consideration. For a part of the effective chiral Lagrangian, counting the powers of this momentum scale is equivalent to counting dimension of the operators (see Table 4.2). However, non-linear representations of PNG fields in the chiral Lagrangian complicates the power counting. It is also not clear how to power count quantum loops and terms containing quark mass insertions if only the operator dimension is taken into account.

In order to properly power-count loops, let us introduce a notion of the *chiral dimension* D for each graph, defined as a power related to the response of a given matrix element $\mathcal{M}(p_i, m_q)$ under linear rescaling of its external momenta $p_i \to \lambda p_i$ and *quadratic* rescaling of all masses $m_q \to \lambda^2 m_q$,

$$\mathcal{M}(\lambda p_i, \lambda^2 m_q) = \lambda^D \mathcal{M}(p_i, m_q), \qquad (4.136)$$

Note that the need for quadratic rescaling of the quark masses follows from the GMOR relations in Eq. (4.133), where PNG masses are rescaled linearly.

For any graph with V_i vertices of different types with d_i derivatives each, L loops, and I internal PNG lines, the overall degree of divergence D would be

$$D = \sum_i V_i d_i - 2I + 4L. \qquad (4.137)$$

Using the standard topological relation connecting the number of loops, internal lines and vertices,

$$L = I - \sum_i V_i + 1, \qquad (4.138)$$

to remove the number of internal lines from the expression for D, we can find the overall degree of divergence as

$$D = \sum_i V_i(d_i - 2) + 2L + 2. \qquad (4.139)$$

As we can see from Eq. (4.119), one loop contributions with vertices described by the leading order χPT Lagrangian $\mathcal{L}_2$ scale as $\mathcal{O}(p^4)$. This can be seen examining all possible one loop graphs. For example, for the simplest one-loop tadpole correction with $V_2 = 1$, $d_2 = 2$, and $L = 1$ we

Table 4.2 Power counting and symmetry properties of building blocks of $\mathcal{L}$ of Eq. (4.101).

Operator	Chiral power	Chiral properties	C	P
U	$\mathcal{O}(p^0)$	$RUL^\dagger$	U^T	$U^\dagger$
$D_\mu U$	$\mathcal{O}(p^1)$	$RD_\mu U L^\dagger$	$(D_\mu U)^T$	$(D_\mu U)^\dagger$
χ	$\mathcal{O}(p^2)$	$R\chi L^\dagger$	χ^T	$\chi^\dagger$

explicitly get $D = 0 + 2 + 2 = 4$. Thus, any UV divergences that appear at one loop level would have to be cancelled by counter terms that scale as $\mathcal{O}(p^4)$, for which $L = 0$, $V_4 = 1$, and $d_4 = 4$. This is precisely what makes chiral perturbation theory a consistent EFT. This also means that we need to write out the next term in Eq. (4.101), $\mathcal{L}_4$ that scales as $\mathcal{O}(p^4)$.

In order to write $\mathcal{L}_4$ we need to do the same procedure as in writing $\mathcal{L}_2$: write the most general set of operators that scale as $\mathcal{O}(p^4)$ consistent with the set of chiral and internal symmetries and use equations of motion and operator identities to arrive at the minimal set of operators. For $\mathcal{L}_4$ this procedure is well described in [Donoghue et al. (1992); Scherer and Schindler (2005)], so we shall not repeat it here. The commonly accepted Lagrangian (including sources) is given by

$$\begin{aligned}
\mathcal{L}_4 = &\, L_1 \mathrm{Tr}(D_\mu U^\dagger D^\mu U)^2 + L_2 \mathrm{Tr}(D_\mu U^\dagger D_\nu U)\mathrm{Tr}(D^\mu U^\dagger D^\nu U) \\
&+ L_3 \mathrm{Tr}(D_\mu U^\dagger D^\mu U D_\nu U^\dagger D^\nu U) \\
&+ L_4 \mathrm{Tr}(D_\mu U^\dagger D^\mu U)\mathrm{Tr}(U^\dagger \chi + \chi^\dagger U) \\
&+ L_5 \mathrm{Tr}(D_\mu U^\dagger D^\mu U (U^\dagger \chi + \chi^\dagger U)) \\
&+ L_6 \mathrm{Tr}(U^\dagger \chi + \chi^\dagger U)^2 + L_7 \mathrm{Tr}(U^\dagger \chi - \chi^\dagger U)^2 \\
&+ L_8 \mathrm{Tr}(\chi^\dagger U \chi^\dagger U + U^\dagger \chi U^\dagger \chi) \\
&- i L_9 \mathrm{Tr}(R^{\mu\nu} D_\mu U D_\nu U^\dagger + L^{\mu\nu} D_\mu U^\dagger D_\nu U) \\
&+ L_{10} \mathrm{Tr}(U^\dagger R^{\mu\nu} U L_{\mu\nu}).
\end{aligned} \quad (4.140)$$

In addition to the operators on Eq. (4.140) one can add two more terms,

$$\mathcal{L}_{4S} = H_1 \mathrm{Tr}(R_{\mu\nu} R^{\mu\nu} + L_{\mu\nu} L^{\mu\nu}) + H_2 \mathrm{Tr}(\chi^\dagger \chi). \quad (4.141)$$

These source terms, however, do not contain PNG terms and therefore $H_{1,2}$ are not directly measurable in low energy reactions with pseudoscalar mesons.

4.6.5 Naive dimensional analysis

The chiral Lagrangian is a derivative (energy) expansion. The theory does a good job predicting meson behavior at relatively low energies, but we know that sooner or later, it is going to break down. The question is: when will this happen? As a rule, the energy expansion will continue to work until the sub-leading terms become as important as the leading terms. That is, contributions to matrix elements from $\mathcal{L}_2$ should dominate the contributions from $\mathcal{L}_4$.

Type-I effective theories have a scale at which this no longer happens. For example, we know that Fermi's theory of weak interactions that we presented earlier will work fine as long as we are at energies below the W-boson mass, and the smaller the energy, the more accurate the theory's results. Unfortunately, for theories such as the chiral Lagrangian, there is no obvious scale like there was in Fermi's theory. The problem is that the chiral Lagrangian does not come from integrating out something as clear-cut as a particle from the full QCD world of quarks and gluons, but rather we are matching to a different "phase" of QCD, where the chiral symmetry is broken. Without something as straightforward as a mass, we do not seem to have any idea how to decide the place where the chiral Lagrangian fails.

However, if we think of the problem as looking for the energy scale where $\mathcal{L}_4$ contributions start to matter as much as $\mathcal{L}_2$ contributions, that can let us know where to stop trusting our theory. This method goes under the general name of *naive dimensional analysis*, or NDA. It works for many effective theories of strongly-interacting degrees of freedom, such as technicolor and superconductivity, so it is worth seeing how the method works.

We need to quantify how much the terms in $\mathcal{L}_4$ are suppressed relative to $\mathcal{L}_2$. The coefficients in $\mathcal{L}_4$ are all dimensionless (denoted by L_i in Eq. (4.140)), while the coefficient of $\mathcal{L}_2$ was (ignoring factors of 2) f^2. Therefore, the ratio of these coefficients is given by

$$\frac{L_i}{f^2} \simeq \frac{1}{\Lambda_\chi^2}. \qquad (4.142)$$

where Λ_χ is called the *chiral symmetry breaking scale*; the energy expansion is then an expansion in the ratio E/Λ_χ, so when the energy of the process approaches this scale, we know that our chiral EFT will no longer give us a good answer.

So what is that scale? To answer that question, let us consider $\pi\pi$ scattering. The tree-level, lowest order amplitude for this process was computed in Eq. (4.109) without the contributions from $\mathcal{L}_4$. If we add these higher dimensional contributions, as well as compute the one-loop corrections and gloss over the details, we would find, after renormalization,[10]

$$\mathcal{A} \sim \frac{p^2}{f^2} + \frac{p^4}{f^4}\left(\frac{1}{16\pi^2}\log(p/\mu) + L(\mu)\right), \qquad (4.143)$$

where any divergences from the loops can be cancelled by counterterms, at the expense of introducing RG scale dependence into the renormalized $\mathcal{L}_4$ coupling. Now our final answer cannot depend on μ, which means that any changes in this scale must be compensated for by a change in $L(\mu)$. This means that if we change μ by an $\mathcal{O}(1)$ amount (say $\mu \to e\mu$), we must also let

$$L(\mu) \longrightarrow \frac{f^2}{\Lambda_\chi^2}\left(1 + \frac{\Lambda_\chi^2}{16\pi^2 f^2}\right), \qquad (4.144)$$

where the f^2/Λ_χ^2 factor came from enforcing the relation in Eq (4.142).

Now changing the RG scale can never have any effect on the answer, but changing it by an $\mathcal{O}(1)$ factor should have a minimal direct effect on the couplings as well. At worst, we would expect that the coupling L will also shift by an $\mathcal{O}(1)$ factor. For that to be the case, we must have

$$\Lambda_\chi \sim 4\pi f \qquad (4.145)$$

In the case of the chiral Lagrangian, this implies that $\Lambda_\chi \sim 1$ GeV, which is near the mass of other particles that are not included in the chiral Lagrangian, such as the ρ (770 MeV) and the proton (938 MeV). It also happens to be the scale at which we expect asymptotic freedom shifts the degrees of freedom back to the quarks and gluons, and we can use ordinary QCD again. Once again, our effective theory gives us the tools to tell us where to expect new physics to crop up, even though we know nothing about QCD or how to match to it.

Naive dimensional analysis, as we mentioned above, is another very powerful tool in the Effective Field theorist's "Bag of Tricks." The logic that led to Eq. (4.145) applies to any theory that becomes strongly coupled in the UV, where you don't have any other way to decide where the theory should be cut off.

---

[10] In this expression, $L(\mu)$ is some (renormalized) combination of the coupling constants in Eq. (4.140). We are only interested in the rough details, so we do not include any finite parts or precise coefficients.

4.6.6 Baryons and chiral perturbation theory

In the previous sections of this chapter we developed chiral perturbation theory – a theory of interacting Goldstone bosons. It is quite successful in describing very low energy interactions of pions, kaons and etas. This is, of course, not the end of the story. Pions and kaons interact with nucleons and nuclei, so, in principle, we should be able to understand those interactions at low energies with help of chiral perturbation theory.

Let us see how this works. For simplicity we will consider interactions of pions and nucleons. Just like before, the pions will be described by the field $U(x)$ given by Eq. (4.97) that transforms under $SU(2)_V$ as described in Eq. (4.96). The nucleons are best described by a doublet,

$$N = \begin{pmatrix} n \\ p \end{pmatrix}, \qquad (4.146)$$

The lowest-order Lagrangian describing interactions of pions and nucleons that is invariant under chiral symmetry is [Gasser et al. (1988)]

$$\mathcal{L}_{\pi N} = \overline{N}\left(i\slashed{D}' - m_N + ig_A \slashed{\Delta}\gamma_5\right)N, \qquad (4.147)$$

where we introduced g_A, a nucleon axial coupling. We also introduced a covariant derivative,

$$D'_\mu = \partial_\mu + \Gamma_\mu, \qquad (4.148)$$

where the connection Γ_μ is given by [Gasser et al. (1988)]

$$\Gamma_\mu = \frac{1}{2}\left[u^\dagger, \partial_\mu u\right] - iu^\dagger\left(v_\mu + a_\mu\right)u - iu\left(v_\mu - a_\mu\right)u^\dagger. \qquad (4.149)$$

Here, v^μ and a^μ were defined in the footnote below Eq. (4.110). Also, we defined

$$\Delta_\mu = \frac{1}{2}u^\dagger D_\mu U u^\dagger \qquad (4.150)$$

Here D_μ (no prime!) is the covariant derivative acting on the pion fields that is given by Eq. (4.114). Let us expand the pion matrices

$$\mathcal{L}_{\pi N} = \overline{N}\left(i\slashed{\partial} - m_N\right)N + g_A \overline{N}\gamma^\mu \gamma_5 \frac{\sigma^i}{2}N\left[\frac{i}{f_\pi}\partial_\mu \pi^i + 2a^i_\mu\right]$$

$$- \frac{1}{4f_\pi^2}\epsilon_{ijk}\overline{N}\gamma^\mu \sigma^i N\, \pi^j \partial_\mu \pi^k + \ldots \qquad (4.151)$$

This Lagrangian yields the Goldberger-Treiman relation,

$$f_\pi g_{\pi NN} = m_N g_A. \qquad (4.152)$$

The lowest-order Lagrangian of Eq. (4.147) can lead to a number of successful predictions [Gasser et al. (1988)]. There are, however, several issues with this description. In particular, exchanges of low-energy pions leave nucleons pretty much static, so a derivative acting on a nucleon field would bring down a factor of the nucleon mass, m_N. This is not a small number, in fact, $m_N > \Lambda_\chi$. It is therefore not immediately clear how to go beyond the static case. In fact, adding the nucleon field brings another dimensionful parameter, m_N, into play, so it is also not immediately clear if the calculation of loop diagrams with Eq. (4.147) will bring terms that are suppressed compared to the tree level case, and results proportional to $m_N/(16\pi f)$ are not obviously ruled out.

While it is possible to deal with nucleons in the formalism described above [Gasser et al. (1988)], other methods will prove to be superior in dealing with light quark baryons. We shall learn these methods in the next chapter.

4.7 Notes for further reading

Calculations of QCD corrections to effective operators in low-energy electroweak interactions are important. The state of the art involves next-to-leading order calculations done by many groups, mostly with help of computer programs that help to evaluate multi loop Feynman integrals (see appendix C for some details). The techniques for those calculations beyond the leading order described in this book are described in the review of A. Buras [Buras (1998)] (see also [Buchalla et al. (1996)]).

Chiral perturbation theory is now a mature subject, heavily used in predictions of many low-energy processes. Some early insights can be found in H. Leutwyler's review [Leutwyler (1991)], while reviews of modern status of this field can be found in A. Pich's review [Pich (1998)]. A very nice description can also be found in [Georgi (1984)]. More information about chiral Lagrangians and their applications to QCD with exercises can be found in [Scherer and Schindler (2005)].

Problems for Chapter 4

(1) **Matching in a model effective field theory**
    Determine the coefficient C_1 of Eq. (4.10) by matching Green functions in full and effective theories.

(a) Start with the Lagrangian of Eq. (4.1). Derive Feynman rules for the propagators and vertices. Calculate the amplitude of fermion-antifermion ($\psi\bar{\psi}$) production close to kinematic threshold and expand it in the fermion pair's energy E, which sets up the energy scale of the process $\mu \sim E$. Pay particular attention to the signs of the amplitude. If in trouble, consult [Peskin and Schroeder (1995)].

(b) Derive Feynman rules for the effective theory given by Eq. (4.10) and calculate the same amplitude as in (a).

(c) Match the amplitudes to determine the Wilson coefficient C_1 and show that $C_1 = g^2$, as obtained in the path integral method.

(2) **Fermi model at next-to-leading order**

Determine the Wilson coefficients C_i, $i = 1, 2, 3$ of next-to-leading order expansion of the Fermi model given in Eq. (4.27).

(3) **Chiral Lagrangian and source fields**

Show that all terms of the Lagrangian of Eq. (4.110) are invariant under local flavor $SU(3)$ transformations given by Eq. (4.112), provided that the derivative is substituted by the covariant derivative defined in Eq. (4.113).

Chapter 5

Effective Field Theories of Type II. Bound States with One Heavy Quark

5.1 Introduction

Bound states play an important role in our life. From mesons and baryons of QCD to atoms and molecules of QED, we are surrounded by them and made of them. The problems of particles bound in some potential are among the first ones studied by students in the beginning quantum mechanics courses. In field theories, it is very often the case that theory parameters can be tuned in such a way that a bound state become possible. Why would one then want to use EFT techniques to describe bound states? In light of the situations described in the previous chapter, this is a perfectly valid question, especially since in many cases a full theory describing the components of a bound states is well known. It is sometimes even perturbative! What is going on here?

A description of non-relativistic bound states constitutes a considerably difficult enterprise even in perturbative field theories. The root of the problem is that bound states represent intrinsically non-perturbative objects, so any attempt to describe them order-by-order in perturbation theory is doomed to failure: the wave function of a bound state would automatically only differ by a little from the free state. This is clearly something that does not happen in Nature! Any consistent calculation would therefore have to attempt to resum the whole perturbation series. This can be done using, for example, the Bethe-Salpeter equation. A solution of this equation is, by no means, simple. Thus, any possibilities for simplifications of calculations are welcomed.

A modern way to approach description of bound state physics relies on a firm understanding of what energy scales are involved in the description of a bound state. Let us first look at a heavy-light bound state, such as a B-meson in QCD. This state is characterized by two essential energy scales: the heavy quark mass, m_Q, and the scale of non-perturbative physics, $\Lambda_{QCD} \sim 400$ MeV. All light degrees of freedom are relativistic and have energy and momentum of order Λ_{QCD}, which also characterizes the bound state energy. Thus, the scale m_Q is irrelevant for a consistent description of a bound state. If degrees of freedom that are only relevant at this scale can be integrated out at the Lagrangian level, the resulting *effective Lagrangian* would consistently describe all of the important physics. This is exactly the same framework that we employed in previous chapters. Thus, an application of EFT is possible to heavy-light mesons, as was championed in [Eichten and Hill (1990); Georgi (1990)]. This effective theory, which goes under the name *Heavy Quark Effective Theory* or HQET, is described in Sec. 5.2.

With this observation at hand, we can apply EFT techniques to a more complicated situation, a bound state of two nearly-equal mass objects, such as quarkonium states, such as Υ or χ_b, in QCD. These states actually possess a multitude of scales: m_Q (mass of the constituents), $p \sim m_Q v$ (average momentum of a heavy quark), $E \sim m_Q v^2$ (energy of the bound state), and Λ_{QCD}. Moreover, depending on the mass of the heavy constituent (c or b-quark), the hierarchy of scales ($p > E > \Lambda_{QCD}$) might not exist! Nevertheless, a consistent application of EFT techniques, separating various scales at the Lagrangian level, leads to a much simplified description of these bound states. The resulting effective theories, non-relativistic QCD (NRQCD) or non-relativistic QED (NRQED), will be described in the next chapter.

As we shall see later in this chapter, analytical calculations of Feynman diagrams and other relevant quantities are simpler with the introduction of effective field theories, both because the diagrams are simpler and because EFT could be invariant under a much larger symmetry group than the original theory, at least in some limit. Thus, application of EFT methods allows us to better understand the problem of bound states.

5.2 Heavy quark effective theory

It is always important to "keep your eyes on the ball," i.e. to remember the physics of the system we are trying to describe when deriving effective field theory. The EFT derived in this section will be dealing with bound states of heavy and light particles. There are plenty of examples of such systems in

Nature, from the hydrogen atom and heavy-light-mesons to more exotic or even outright fantastic systems as heavy-light molecular states or so-called R-hadrons. However, the most popular application of this EFT was the description of bound states of heavy b (or even c)-quarks and light antiquarks u, d, and s, where it goes under the name of heavy-quark effective theory (HQET). The physics of such system can be easily understood in the rest frame of the heavy quark, where the meson closely resembles the hydrogen atom.

5.2.1 Quantum mechanics of heavy particles

Before we delve deep into discussion of field-theoretic implications of the heavy-mass limit, let us consider a familiar case of relativistic quantum mechanics. We shall discuss a spin-1/2 particle Q, whose relativistic dynamics is described by the Dirac equation,

$$\left(i\slashed{D} - m_Q\right)\Psi_Q(x) = 0, \tag{5.1}$$

where $\Psi_Q(x)$ is a coordinate space wave function of a Dirac particle Q. Let us move to the frame where the particle Q is at rest, which we can always do for a massive particle. Also, to simplify the following discussion, let us for a second forget about the gauge part of the covariant derivative. In this case it is not hard to guess the form of the particle's wave function, which would simply be proportional to the exponential $\exp(-im_Q t)$. Thus, similarly to the case of nonrelativistic quantum mechanics where we often separate out a known asymptotic part of the wave function to find a perturbative solution, we make the following redefinition of the wave function,

$$\Psi_Q(x) = e^{-im_Q t}\psi_Q(x), \tag{5.2}$$

where we separated out the large mechanical part. This wave function can be inserted back into Eq. (5.1),

$$\begin{aligned}(i\gamma^\mu\partial_\mu - m_Q)\Psi_Q(x) &\approx \left(i\gamma^0\partial_0 - m_Q\right)e^{-im_Q t}\psi_Q(x)\\ &= \gamma^0 m_Q e^{-im_Q t}\psi_Q(x) - m_Q e^{-im_Q t}\psi_Q(x) \quad (5.3)\\ &= m_Q e^{-im_Q t}\left(\gamma^0 - 1\right)\psi_Q(x) = 0,\end{aligned}$$

where we only retained the leading terms in the $m_Q \to \infty$ limit, thus ignoring derivatives acting on $\psi_Q(x)$. The equation above is equivalent to the condition on the spinor ψ_Q,

$$\mathcal{P}^0_-\psi_Q = 0. \tag{5.4}$$

The reader noticed that we introduced a *projection operator*,

$$\mathcal{P}_\pm^0 = \frac{1 \pm \gamma^0}{2}, \qquad (5.5)$$

where 1 is a 4×4 unit matrix. It is easy to check that all properties of a projection operators, such as $\left(\mathcal{P}_\pm^0\right)^2 = \mathcal{P}_\pm^0$, and $\mathcal{P}_+^0 \mathcal{P}_-^0 = \mathcal{P}_-^0 \mathcal{P}_+^0 = 0$ are satisfied. To clarify the role of that projection operator, let's look at its explicit form

$$\mathcal{P}_+^0 = \begin{pmatrix} \hat{1} & \hat{0} \\ \hat{0} & \hat{0} \end{pmatrix}, \qquad \mathcal{P}_-^0 = \begin{pmatrix} \hat{0} & \hat{0} \\ \hat{0} & \hat{1} \end{pmatrix}, \qquad (5.6)$$

where $\hat{1}$ is a 2×2 unit matrix and $\hat{0}$ is a matrix of zeros of the same dimension. Acting with these operators on Dirac bi-spinor ψ_Q

$$\mathcal{P}_+^0 \psi_Q = \begin{pmatrix} \hat{1} & \hat{0} \\ \hat{0} & \hat{0} \end{pmatrix} \begin{pmatrix} \psi \\ \chi \end{pmatrix} = \begin{pmatrix} \psi \\ 0 \end{pmatrix} \qquad (5.7)$$

And similarly for the $\mathcal{P}_-^0$. Here we used a short-hand notation for the 2-component spinors

$$\psi = \begin{pmatrix} \psi_1 \\ \psi_2 \end{pmatrix}, \qquad \chi = \begin{pmatrix} \chi_1 \\ \chi_2 \end{pmatrix}. \qquad (5.8)$$

Since ψ describe positive energy (particle) degrees of freedom, the role of the projection operator $\mathcal{P}_+^0$ is to project out negative energy (antiparticle) solutions from the theory.

Before we discuss implications of this observation for non relativistic effective field theory, let us make one further generalization: we introduce a 4-velocity vector, v^μ. In the frame of reference where the particle is at rest this vector takes the form $v = (1, \vec{0})$, so $v \cdot x = t$. It is now easy to generalize the formulas of Eqs. (5.2-5.5) to the frames of reference where the particle is moving nonrelativistically, i.e. where the momentum of the particle can be written as

$$p_Q = m_Q v + k, \qquad (5.9)$$

where k is the so-called residual momentum, with $|k| \ll m_Q$, as the particle is presumed to be nearly at rest, and v is the *velocity* of the particle. Note that it is convenient to impose the constraint $v^2 = 1$, which follows from the on-shell condition on Q. Then the residual momentum k parameterizes any slight off-shellness of the fermion.

The formalism described in this chapter is most useful when the mass scale of the considered particle is high compared to other scales present in the problem. In this physical situation it is often convenient to take a limit $m_Q \to \infty$, in which case it is the particle's *velocity* that corresponds to a useful conserved quantity that can be used to label the particle.

We can also generalize Eq. (5.5) into a velocity-dependent projection operator,

$$\mathcal{P}^v_\pm = \frac{1 \pm \slashed{v}}{2}. \qquad (5.10)$$

We leave it to the reader to check that all properties of projection operators are, once again, satisfied, provided that $v^2 = 1$, and a useful trick,

$$\slashed{v}\slashed{v} = v_\mu v_\nu \gamma^\mu \gamma^\nu = \frac{1}{2} v_\mu v_\nu \left(\gamma^\mu \gamma^\nu + \gamma^\nu \gamma^\mu \right) = \frac{1}{2} v_\mu v_\nu 2 g^{\mu\nu} = 1 \qquad (5.11)$$

is employed. Similarly, we can prove that

$$\slashed{v} \mathcal{P}^v_\pm = \pm \mathcal{P}^v_\pm, \qquad (5.12)$$

which we shall use in the next section.

5.2.2 From quantum mechanics to field theory: HQET. Field redefinitions

We now shift to the discussion fields describing heavy particles. Consider a field $\psi_Q(x)$ that describes a heavy fermion[1]. For QCD bound states, the heavy quark's mass scale is much higher than any other scale present in the bound state, so even though the problem is still highly nonperturbative in the QCD coupling, it is possible to expand in the ratio of Λ/m_Q, where Λ represents all other scales present in the problem.

The physics of the system dictates the power counting. For QCD bound states, in the limit of an infinitely heavy quark, the quark simply serves as a source of gluonic fields, while itself remaining (nearly) static. For instance, the fermion's kinetic energy, $K = p_Q^2/(2m_Q)$, is represented by a power-suppressed operator. The power counting is then extremely easy: we simply count inverse powers of inverse heavy quark mass, which grows with each operator's dimension. As in previous chapters, we shall augment it by a mass-independent subtraction scheme.

The equations of motion for the heavy fermion field can be derived from the usual Dirac Lagrangian,

$$\mathcal{L} = \overline{\psi}_Q(x) \left(i\slashed{D} - m_Q \right) \psi_Q(x). \qquad (5.13)$$

---

[1] In this section we will mainly be talking about heavy quarks in a meson or a baryon states, where all other quarks are light, $m_q \sim \Lambda_{QCD}$. Such systems are sometimes called "hydrogen atoms of QCD" due to their relative simplicity.

In order to take a nonrelativistic limit in complete analogy with the example in relativistic quantum mechanics considered in the previous section, let us separate a large mechanical part of the field

$$\psi_Q(x) = e^{-im_Q v \cdot x}\tilde{\psi}_Q(x) = e^{-im_Q v \cdot x}\left[\mathcal{P}_+^v \tilde{\psi}_Q(x) + \mathcal{P}_-^v \tilde{\psi}_Q(x)\right]$$
$$\equiv e^{-im_Q v \cdot x}\left(h_v(x) + H_v(x)\right), \qquad (5.14)$$

where we labelled the fields h_v and H_v by their velocity, and used the fact that $1 = \mathcal{P}_+^v + \mathcal{P}_-^v$. Note that because of Eq. (5.12), the fields h_v and H_v satisfy the relations

$$\slashed{v} h_v = h_v, \quad \text{and} \quad \slashed{v} H_v = -H_v. \qquad (5.15)$$

Now, what would happen if we insert Eq. (5.14) into Eq. (5.9)? The derivative acting on the exponential takes down factors of $-im_Q v_\mu$, so, opening the brackets,

$$\mathcal{L} = \overline{h}_v i \slashed{D} h_v + \overline{H}_v i \slashed{D} H_v + 2m_Q \overline{H}_v H_v + \overline{h}_v i \slashed{D} H_v + \overline{H}_v i \slashed{D} h_v \qquad (5.16)$$

So far we have not done much besides factoring out a part of the fermion's field for which we supposedly know the solution of the equations of motion. In fact, both fields h_v and H_v are still present in the theory. Moreover, they are mixed up by the last two terms in the Lagrangian above. Before we diagonalize the Lagrangian above, it might be useful to use Eq. (5.15) to get rid of the gamma matrices in the first two terms. As a result, the Lagrangian simplifies a bit

$$\mathcal{L} = \overline{h}_v iv \cdot D h_v - \overline{H}_v \left[iv \cdot D - 2m_Q\right] H_v$$
$$+ \overline{h}_v i \slashed{D}_\perp H_v + \overline{H}_v i \slashed{D}_\perp h_v, \qquad (5.17)$$

where we introduced the "perpendicular" component of the derivative $D_\perp$. This notation is quite useful. For any 4-vector a^μ we can define a "perp component" based on the condition $a_\perp \cdot v = 0$, therefore

$$a_\perp^\mu = a^\mu - (a \cdot v)v^\mu. \qquad (5.18)$$

The reader can easily check that in the rest frame $v^\mu = (1, \vec{0})$ the perp component corresponds to the 3-vector $\vec{a}$.

We can already see emerging features of the theory that we are trying to derive. We see that the Lagrangian contains two fields, a massless one h_v and the "heavy" one H_v with the mass $2m_Q$. In the limit $m_Q \to \infty$ the field H_v describes infinitely heavy particles. Thus, just like in the previous chapter, we can integrate that degree of freedom out of our theory. Now, the proper way of doing so would be to write an effective action in terms of the functional integral and properly integrate out this field. However, this might be a bit of overkill for a problem at hand: everything can be done using equations of motion derived from Eq. (5.17) using, once again, Eq. (5.14).

$$(i\slashed{D} - m_Q) \psi_Q(x) \Rightarrow i\slashed{D} h_v + (i\slashed{D} - 2m_Q) H_v = 0. \qquad (5.19)$$

Now, applying projectors $\mathcal{P}^v_\pm$ from the left of Eq. (5.10) and commuting them with derivatives, we obtain the sought-for equation,

$$(iv \cdot D + 2m_Q) H_v = i\slashed{D}_\perp h_v \quad \text{or} \quad H_v = (2m_Q + iv \cdot D)^{-1} i\slashed{D}_\perp h_v, \qquad (5.20)$$

where $(2m_Q + iv \cdot D)^{-1}$ is the operator inverse to $2m_Q + iv \cdot D$, which we assume to exist. The question of existence of this operators could be bothersome for some readers, so let us define it in terms of Taylor series. Since in deriving Eq. (5.17) we separated out parts of the fermion's field that correspond to large, $\mathcal{O}(m_Q)$ momenta, any time a derivative is acting on h_v, it brings down a momentum that is of order $\mathcal{O}(\Lambda_{QCD})$, since that is the only other scale left. Thus, the expansion

$$(2m_Q + iv \cdot D)^{-1} = \frac{1}{2m_Q} \sum_{n=0}^{\infty} (-1)^n \left(\frac{iv \cdot D}{2m_Q}\right)^n \qquad (5.21)$$

is convergent and we can use it to derive effective Lagrangians of increasing field dimension! In fact, we are almost done deriving the HQET Lagrangian to order $1/M_Q^2$. Inserting Eq. (5.20) into Eq. (5.16), and using the fact that $h_v = \mathcal{P}^v_+ h_v$ and

$$\mathcal{P}^v_+ i\slashed{D}_\perp i\slashed{D}_\perp \mathcal{P}^v_+ = \mathcal{P}^v_+ \left[(i\slashed{D}_\perp)^2 + \frac{g_s}{2}\sigma_{\mu\nu}G^{\mu\nu}\right]\mathcal{P}^v_+ \qquad (5.22)$$

we finally obtain the sought result,

$$\mathcal{L}_{\text{eff}} = \overline{h}_v iv \cdot D h_v + \frac{1}{2m_Q}\overline{h}_v (i\slashed{D}_\perp)^2 h_v$$
$$+ C_g \frac{g_s}{4m_Q}\overline{h}_v \sigma_{\mu\nu} G^{\mu\nu} h_v + \mathcal{O}(1/m_Q^2). \qquad (5.23)$$

This is the HQET Lagrangian to order $\mathcal{O}(1/m_Q^2)$ [Eichten and Hill (1990); Georgi (1990)]. As we discussed in the beginning of the section, we simply count powers of $1/m_Q$ to assign the importance of the operators in the above Lagrangian. Thus, it is obvious that only the first term, $\overline{h}_v iv \cdot D h_v$, survives in the limit $m_Q \to \infty$. Note that in writing the above expression, we introduced a (Wilson) coefficient $C_g = 1$ for the operator $\overline{h}_v \sigma_{\mu\nu} G^{\mu\nu} h_v$. We shall discuss this coefficient in Sec. 5.2.5.

A careful reader might have noticed that we essentially introduced, in a quantum-mechanical system, a field with definite position and velocity. Given a well known Heisenberg uncertainty relation connecting momentum and coordinate, and the fact that momentum is directly proportional to velocity, one might wonder if this construction is actually allowed quantum-mechanically. Excellent point! However, a simple argument shows that the construction of Eq. (5.23) does not overturn basic notions of quantum mechanics,

$$[v^\mu, x^\nu] = i\hbar \frac{g^{\mu\nu}}{m_Q} \to 0 \qquad (5.24)$$

in the limit heavy quark limit $m_Q \to \infty$, which does indeed allows to measure both position and velocity of the field introduced above. Another thing that immediately captures our attention is the fact that the leading-order Lagrangian

$$\mathcal{L}_{\text{eff}}^\infty = \bar{h}_v iv \cdot D h_v \qquad (5.25)$$

does not have any Dirac matrices, even though it is used to describe fermions! This simplifies the Feynman rules that follow from it (see Fig. (5.1)). More importantly, it enlarges the spin symmetry group of the resulting effective theory. We shall talk about that in the next section.

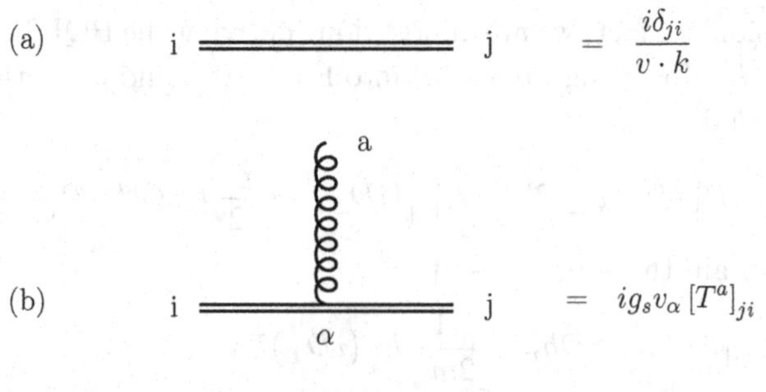

(a) $\quad i \;=\!=\!=\!=\!=\; j \quad = \quad \dfrac{i\delta_{ji}}{v \cdot k}$

(b) $\quad i \;=\!=\!=\!=\!=\; j \quad = \quad ig_s v_\alpha [T^a]_{ji}$
  with gluon a, α

Fig. 5.1 Feynman rules for the heavy fields (a) propagator and (b) vertex.

The final observation that we make in this section concerns *redefinitions of fields* in HQET. We can derive the equations of motion for the fields h_v and $\bar{h}_v$ using Eq. (5.23),

$$iv \cdot D h_v = -\frac{1}{2m_Q} \left[(i\slashed{D}_\perp)^2 + C_g \frac{g_s}{2} \sigma_{\mu\nu} G^{\mu\nu} \right] h_v. \qquad (5.26)$$

While the left-hand side of this equation is $\mathcal{O}(1)$, the right-hand one is of the higher order in $1/m_Q$. This means that one can always redefine the field

h_v such that some of the terms in the Lagrangian are absorbed into the definition of h_v – the equations of motions will only change by an operator that contributes at higher orders in $1/m_Q$.

A good example of this field redefinition technique can be seen in the construction of the Lagrangian of Eq. (5.23) itself: one can notice that terms of the type

$$\mathcal{L}' = \frac{1}{2m_Q} \bar{h}_v \left(iv \cdot D\right)^2 h_v \qquad (5.27)$$

are absent. The reason this kind of operator is not included, even though it should be allowed, is that a field redefinition

$$h_v \to \left[1 - \frac{(iv \cdot D)^2}{4m_Q}\right] h_v \qquad (5.28)$$

will remove the operator $\mathcal{L}'$ from the picture completely! Similar field redefinitions can also be done at higher orders in $1/m_Q$. In practice, this means that any operator that is proportional to the lowest order equation of motion will not contribute. This technique will be quite helpful for us in Sec. 6.1.

5.2.3 Spin symmetry and its consequences

One might already suspect that the absence of Dirac matrices in $\mathcal{L}_{\text{eff}}^\infty$ noted above simplifies manipulation with spin degrees of freedom. The situation is even better! Let us see what happens if we perform an infinitesimal spin rotation of the h_v field, i.e. consider

$$\delta \mathcal{L}_{\text{eff}}^\infty = \bar{h}_v' iv \cdot Dh_v' - \bar{h}_v iv \cdot Dh_v, \qquad (5.29)$$

with $h_v' = \left(1 + i\vec{\alpha} \cdot \vec{S}\right) h_v$ and $\vec{S}$ being the usual fermion spin operators,

$$S_i = \frac{1}{2} \begin{pmatrix} \sigma_i & \hat{0} \\ \hat{0} & \sigma_i \end{pmatrix}, \qquad [S_i, S_k] = i\epsilon_{ijk} S_k. \qquad (5.30)$$

The parameter α_i above parameterizes the angle of infinitesimal spin rotation. Since $\vec{S}$ commutes with γ^0 and $v \cdot D$ contains no Dirac matrices, it follows

$$\delta \mathcal{L}_{\text{eff}}^\infty = 0. \qquad (5.31)$$

Since the spin transformation belongs to $SU(2)$, we immediately conclude that the Lagrangian $\mathcal{L}_{\text{eff}}^\infty$ possesses an additional $SU(2)$ spin symmetry not present in the original Lagrangian of Eq. (5.13). But we are not done yet!

Notice that we implicitly talked about QCD with quark of a single type (or flavor). It is totally appropriate since, flavor-changing interactions are absent in full QCD. This is the property that any effective field theory of QCD must respect as well. However, there are several flavors of heavy quark in Nature (top, beauty and, to some degree, charm), so we can sum $\mathcal{L}_{\text{eff}}^\infty$ over flavors of all heavy quarks. If the number of heavy quarks is N_f, then the total symmetry group of heavy quark effective theory would be $SU(2N_f)$! This makes HQET a powerful effective theory for calculations of physical properties of heavy quark transitions.

Another interesting observation about the Lagrangian of Eq. (5.23) can be made regarding the symmetry-breaking (or $1/m_Q$-suppressed) terms. It is best to switch to the particle's rest frame to see what physics they represent. The first operator,

$$\mathcal{O}_{kin} = \frac{1}{2m_Q}\overline{h}_v \left(i\slashed{D}_\perp\right)^2 h_v \;\;\Rightarrow\;\; \frac{1}{2m_Q}\overline{h}_v \left(\vec{D}\right)^2 h_v, \tag{5.32}$$

which clearly represents kinetic energy of heavy quark motion inside the hadron. The second operator,

$$\mathcal{O}_{mag} = \frac{g_s}{4m_Q}\overline{h}_v \sigma_{\mu\nu} G^{\mu\nu} h_v \;\;\Rightarrow\;\; -\frac{1}{m_Q}\overline{h}_v \vec{S}\cdot\vec{B} h_v \tag{5.33}$$

represents its interaction with chromomagnetic field B^i present inside of the heavy hadron. As usual, this field can be defined in terms of the stress tensor,

$$B^i = -\frac{1}{2}\epsilon^{ijk} G^{jk}. \tag{5.34}$$

The physics of higher order terms is less transparent, yet they are easy to derive using the expansion method presented here. We shall return to the question of higher-order terms in our discussion of other non-relativistic effective theories in Sec. 6.1.

Spin symmetry of the heavy quark Lagrangian has immediate implications for spectroscopy! If we consider B_q and B_q^* meson states, their quark composition is very similar; both of them are made of a heavy $\bar{b}$ antiquark and a light u (or d or s) quark. The only difference is the spin state of those quarks – they are in relative singlet and a triplet states, respectively. A crucial observation is that since the spin interaction in HQET is suppressed by a power of a heavy quark mass, those two states will be degenerate in the heavy quark limit! This is indeed what is seen experimentally: $m_{B^*} - m_B = 45.0 \pm 0.4$ MeV, while mass differences between B and higher states are several hundred MeV [Olive et al. (2014)].

Recall that a similar situation occurs in the physics of nucleons: the mass difference between a proton and a neutron is very small. Then, as far as strong interactions are concerned, it makes sense to talk about a *nucleon* state, which is a doublet of isospin. By analogy, it also makes sense to talk about a *heavy superfield* that includes both B_q and B_q^* states. We shall construct such states in Section 5.3.

5.2.4 *More symmetry: reparameterization invariance*

A careful reader might have noticed that while the momentum definition of Eq. (5.9) is very natural, it is, nevertheless, not unique. In particular, if we modify *both* velocity and residual momentum as

$$v' = v + \frac{q}{m_Q} \quad \text{and} \quad k' = k - q \qquad (5.35)$$

everything will still be unchanged provided that q satisfies the condition $(v + q/m_Q)^2 = 1$ as

$$v^2 = v'^2 = \left(v + \frac{q}{m_Q}\right)^2 = 1. \qquad (5.36)$$

This more or less trivial observation, called reparameterization invariance, has quite powerful consequences for the operators in this and other non-relativistic effective theories. In the case of heavy quark effective theory, it encodes the Lorentz invariance of the full theory, QCD, so that the effective and full theories share the same set of symmetries, as it is often required for construction of EFT. In particular, it allows us to fix the form of subleading operators [Luke and Manohar (1992); Heinonen *et al.* (2012)], since velocity labels effective fields.

Let us see how this works for the heavy fermion fields h_v. Since this field satisfies velocity-dependent conditions Eq. (5.15) and Eq. (5.14), we need to make sure that these conditions are also satisfied by the fields labelled by shifted velocity v'. This can be achieved by employing Lorentz boost operators,

$$\Lambda(v,w) = \exp\left(i\theta J_{\alpha\beta} w^\alpha v^\beta\right), \quad \text{with} \quad J_{\alpha\beta} = -\frac{1}{2}\sigma_{\alpha\beta}, \qquad (5.37)$$

which, in general, boosts spinor fields in the $v - w$-velocity plane. Here θ is the boost parameter. Then, a reparameterization transformation for the fermion fields would be

$$h_{v'}(x) = e^{iq \cdot x}\Lambda(v', p/m_Q)\Lambda(v, p/m_Q)^{-1}h_v(x), \qquad (5.38)$$

where we suppressed the anti-fermion contribution and defined $v' = v + q/m_Q$. Here p is the total momentum that is invariant under reparameterization. It follows from Eq. (5.38) that the combination

$$\widetilde{h}_v(x) = \Lambda(v, p/m_Q) h_v(x) = \left(1 + \frac{\slashed{D}}{2m_Q}\right) h_v(x) + \mathcal{O}(1/m_Q^2) \qquad (5.39)$$

transforms covariantly under reparameterization,

$$\widetilde{h}_{v'}(x) = e^{iq \cdot x} \widetilde{h}_v(x). \qquad (5.40)$$

The Eq. (5.39) implies particular form of reparameterization-invariant bilinears, e.g.

$$\overline{\widetilde{h}_v}(x) \widetilde{h}_v(x) = \overline{h}_v(x) h_v(x),$$
$$\overline{\widetilde{h}_v}(x) v^\mu \widetilde{h}_v(x) = \overline{h}_v(x) \left[v^\mu + \frac{1}{m_Q} i D^\mu\right] h_v(x), \qquad (5.41)$$

and similarly for other Dirac structures. Note that we only expanded the fields to order $\mathcal{O}(1/m_Q)$.

There is a much deeper reason for this symmetry. As we noted in the previous chapters, one of the requirements for constructing an effective field theory is that it shares the same set of symmetries as the original theory. One of the symmetries of the full QCD is invariance under Lorentz transformations, yet in HQET we have picked out a special 4-vector, namely the velocity v^μ. Reparameterization invariance restores the Lorentz invariance of QCD in HQET.

5.2.5 HQET Green's functions. Radiative corrections

The idea that EFT techniques are beneficial for the calculations of processes with heavy particles was in part based on the premise that those calculations would be simpler to perform than in the corresponding full theories. Let us illustrate how that happens in HQET. In particular, we shall see that use of EFT techniques is advantageous for the calculations of radiative corrections. This has an immediate application to decays of heavy beauty-flavored mesons and baryons, where QCD radiative corrections are needed for interpretation of precision measurements done at the heavy flavor factories.

To set up a stage, let us first consider Green's functions in HQET at tree level and then prove a theorem relating full and effective Green's functions including radiative corrections.

- **Two-point function.** Let us derive HQET two-point functions from full QCD

$$G^{(2)}(p_Q) = \frac{i(m_Q \slashed{v} + \slashed{k} + m_Q)}{(m_Q v + k)^2 - m_Q^2} = \frac{1+\slashed{v}}{2} \frac{i}{v \cdot k} + \mathcal{O}(1/m_Q), \quad (5.42)$$

where $p_Q = m_Q v + k$ and we expanded to the leading order in the off-shell momentum k. Since the two-point function is nothing but a propagator, we simply checked that our derivation of the Feynman rule for the propagator in Fig. (5.1) is correct up to the projection operator.

- **Three-point function.** The three-point functions can also be obtained from the full-QCD result

$$G^{(3)a}(p_Q, q)_\mu = \frac{i}{\slashed{p}_Q - m_Q} (-i g_s T^b \gamma_\nu) \frac{i}{\slashed{p}_Q + \slashed{q} - m_Q} \Delta^{ba}_{\nu\mu}(q), \quad (5.43)$$

where $\Delta^{ba}_{\nu\mu}(q)$ is the standard gluon propagator with momentum q. Expanding the propagators as in Eq. (5.42) and commuting the projection operator to the left using the identity

$$\mathcal{P}^v_+ \gamma_\nu \mathcal{P}^v_+ = \mathcal{P}^v_+ v_\nu, \quad (5.44)$$

one obtains the desired result,

$$G^{(3)a}_\mu(p_Q, q) = \frac{1+\slashed{v}}{2} \frac{i}{v \cdot k} (-i g_s T^b v_\nu) \frac{i}{v \cdot (k+q)} \Delta^{ba}_{\nu\mu}(q) + \mathcal{O}(1/m_Q), \quad (5.45)$$

from which one can read off the HQET vertex function derived earlier and presented in Fig. (5.1).

As can be easily seen, the use of the projection operator is inconsequential, as we are dealing with the fields h_v satisfying the relation $h_v = \mathcal{P}^v_+ h_v$.

The same exercise can be performed for higher Green's functions, but the tendency is already clear: at tree level all Green's functions in HQET ($\widetilde{G}^{(n)}(k; q_1, \ldots)$) and in full QCD ($G^{(n)}(p_Q; q_1, \ldots)$) satisfy the following matching condition

$$G^{(n)}(p_Q; q_1, \ldots) = \widetilde{G}^{(n)}(k; q_1, \ldots) + \mathcal{O}(1/m_Q). \quad (5.46)$$

In other words, HQET is equivalent to QCD in the limit of large m_Q – it is simply constructed so. Once can, however, worry that this relation does not hold when the radiative corrections are included. This worry is justified, as some Green's functions would have different renormalization properties. Thus, it is not surprising that in general the relation of Eq. (5.46) should be modified. The essence of this modification is expressed in the following (factorization) theorem [Grinstein (1995)].

Theorem 5.1. *A matching condition for any one-particle irreducible amputated Green's function in full QCD and HQET can be represented as*

$$G^{(n)}(p_Q; q_1, \ldots) = C_n\left(m_Q/\mu, g_s\right) \widetilde{G}^{(n)}(k; q_1, \ldots) + \mathcal{O}(1/m_Q), \quad (5.47)$$

where the dimensionless coefficient function C_n can only depend on (the logarithm of) the ratio of the heavy quark mass and momentum scale at which the HQET Green's function is computed.

Proof. We shall provide a proof of theorem 5.1 up to one loop in QCD. At tree level, when gluons are directly connected to the heavy quark line, this relation holds with $C_n = 1$, as we can directly expand the QCD Green's functions in powers of $1/m_Q$ and employ the relation of Eq. (5.44) similarly to the examples discussed above.

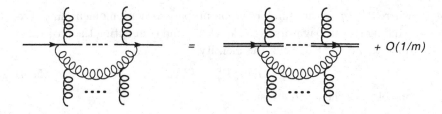

Fig. 5.2 Feynman diagrams for Green's function matching in QCD and HQET.

Now let us move to the one-loop case. Here the situation is not as obvious as in the tree-level case considered above because the integrals over loop momentum could be divergent, so we have to be careful when expanding in $1/m_Q$. To set up the stage, let us first consider $n = m + 2$, where m is the number of external gluons. In the case $m \geq 2$, represented pictorially in Fig. 5.2, the Green's function in QCD would be proportional to

$$G^{(n)}(p_Q; q_1, \ldots) \sim \int \frac{d^4\ell}{(2\pi)^4} \left(\frac{1}{\ell^2}\right)^{m_g+1} \left(\frac{i}{\not{p}_Q + \not{\ell} - m_Q}\right)^{m_q+1}, \quad (5.48)$$

where for the purpose of power counting we set $q_i = 0$ on the right hand of the equation above. Here, m_q counts the number of gluons on the heavy quark line, while m_g counts the number of gluons on the gluon line such that $m_q + m_g = m$.

Let us calculate the superficial degree of divergence $d_G^{(n)}$ for the diagrams in Fig. 5.2. From Eq. (5.48),

$$d_G^{(n)} = 1 - 2m_g - m_q. \quad (5.49)$$

Clearly, for any $m_q \geq 1$ and $m_g \geq 1$ such that $m \geq 2$ the superficial degree of divergence is $d_G^{(n)} < 0$ for any one-particle irreducible amputated Green's function. Thus, the integral in Eq. (5.48) is convergent, as the overlapping divergences are absent for the one-loop diagrams. Also, similar integrals on the HQET side would also be convergent. Thus, we can freely interchange taking the integral and taking limit $m_Q \to \infty$, expanding the quark propagators *before* calculation of the integral. Thus, as in the tree-level case, the theorem is proven.

The tricky part is the case of $m < 2$. But here we only have two possibilities, a three-point and a two-point functions. Let us consider the three-point function; the discussion of the two-point function is completely analogous. The calculations of the three-point function in both QCD and HQET involve gluon emission from quark and gluon lines. While the diagram with gluon emission from the gluon line is finite, it is clear from Eq. (5.48) that the superficial degree of divergence for the diagram with gluon emission from the quark line is $d_G = 0$, so the diagram is logarithmically divergent. Thus, we must regularize the integrals before proceeding any further. Indeed, the argument that we used before fails, if we try to remove the regulator. Note, however, that it still works if we take a derivative of the three-point function with respect to external momenta k_μ or q_μ, i.e.

$$\frac{\partial}{\partial k_\mu} G^{(3)}(p_Q; q) = C_3\left(m_Q/\mu, g_s\right) \frac{\partial}{\partial k_\mu} \widetilde{G}^{(3)}(k; q) + \mathcal{O}(1/m_Q),$$

$$\frac{\partial}{\partial q_\mu} G^{(3)}(p_Q; q) = C_3\left(m_Q/\mu, g_s\right) \frac{\partial}{\partial q_\mu} \widetilde{G}^{(3)}(k; q) + \mathcal{O}(1/m_Q). \quad (5.50)$$

The counter-terms that need to be introduced to remove divergencies are such that their difference is q- and k-independent (in the sense that the difference is proportional to $a G_{tree}^{(3)}(p_Q; q) - b C_3 \widetilde{G}_{tree}^{(3)}(k; q)$ with a and b independent of k and q), so subtracting the divergence does not affect the relation Eq. (5.47), which proves the theorem. As we pointed out before, the proof holds for the two-point function as well – except for the fact that two differentiations are needed to render the function finite. □

We would like to finish this section by making some comments about the coefficient C_g introduced earlier in Eq. (5.23). One can show, using reparameterization invariance, that to all orders in perturbation theory, the coefficient of the kinetic energy operator is exactly one – i.e. it is not renormalized. The magnetic operator is an entirely different story – and should be treated as any other operator in EFT that can receive perturbative corrections upon matching to full QCD.

5.2.6 External currents and external states

As an example of the techniques described above, let us consider calculation of renormalization of external heavy-to-light flavor-changing currents [Ji and Musolf (1991); Neubert (1994b); Falk and Grinstein (1990)]. These currents are indeed external to QCD (which has no flavor-changing operators), generated by electroweak interactions at leading order. They have huge practical significance, as they drive leptonic, semileptonic and radiative decays of heavy mesons and baryons, which in turn are used for experimental extraction of the fundamental parameters of the standard model Lagrangian.

For simplicity, let us only discuss the construction of the short-distance expansion for the heavy-light vector current $V^\mu = \bar{q}\gamma^\mu Q$. Recall that as part of the definition of HQET, we chose a mass-independent renormalization prescription (dimensional regularization with modified minimal subtraction, or $\overline{MS}$). It is known that this prescription is ambiguous about how to treat the Dirac matrix γ_5 beyond four dimensions. If we stick to using a scheme with anti commuting γ_5, then our evaluation would also be applicable to the operator expansion on the axial current, $A^\mu = \bar{q}\gamma^\mu\gamma_5 Q$ upon the replacement of $\bar{q} \to -\bar{q}\gamma_5$ in all operators on the HQET side. This is because we can always anticommute γ_5 such that it appears next to the light quark field.

Suppressing the exponential that will disappear once the matrix elements of the operators on both sides are taken, we can write that up to corrections of order $\mathcal{O}(1/m_Q^2)$,

$$V^\nu = \sum_i C_i(\mu) J_i^\nu(\mu) + \frac{1}{2m_Q} \sum_k B_k(\mu) O_k^\nu(\mu)$$
$$+ \frac{1}{2m_Q} \sum_l A_l(\mu) T_l^\nu(\mu) + \mathcal{O}(1/m_Q^2). \quad (5.51)$$

We introduced several operators on the HQET side. At the leading order the situation is very simple: there are only two operators that contribute,

$$J_1^\nu = \bar{q}\gamma^\nu h_v, \qquad J_2^\nu = \bar{q}v^\nu h_v \quad (5.52)$$

Note that the only requirement on the operators on the right hand side is that they have the same quantum numbers as the current that we are trying to match. Both J_1^μ and J_2^μ transform as vectors in HQET. No operators with more powers of v are required due to Eqs. (5.11) and (5.15). At tree-level, as one can see by direct matching at the momentum scale $\mu = m_Q$,

$C_1^{(0)}(m_Q) = 1$ and $C_2^{(0)}(m_Q) = 0$. Both of those Wilson coefficients will become non-zero after taking into account perturbative QCD corrections, which we shall discuss later in this section.

The subleading, $1/m_Q$-suppressed corrections are represented by local operators O_i^ν and non-local time-ordered products T_i^ν. A basis of operators O_i^ν can be obtained by writing out all possible operators of dimension four and using leading-order equations of motion (or field redefinitions) to reduce the set to minimal. The conventionally-selected operators are

$$O_1^\nu = \bar{q}\gamma^\nu i\slashed{D}h_v, \qquad O_4^\nu = \bar{q}\left(-iv\cdot \overleftarrow{D}\right)\gamma^\nu h_v,$$
$$O_2^\nu = \bar{q}v^\nu i\slashed{D}h_v, \qquad O_5^\nu = \bar{q}\left(-iv\cdot \overleftarrow{D}\right)v^\nu h_v, \qquad (5.53)$$
$$O_3^\nu = \bar{q}iD^\nu h_v, \qquad O_6^\nu = \bar{q}\left(-i\overleftarrow{D}^\nu\right)h_v.$$

Note that we consistently omitted operators that vanish by the leading-order equation of motion: this is an application of field redefinition discussed earlier.

Since our goal in this chapter is to talk about matching the QCD and HQET results including perturbative QCD corrections, one needs to include all operators that can in general mix into each other under renormalization. The set of operators of Eq. (5.53) closes under renormalization. However, it is convenient to introduce additional non-local operators T_i that result from forming time-ordered products of the leading order operators J_i^ν and kinetic and chromomagnetic energy operators of the effective Lagrangian defined in Eqs. (5.32) and (5.33)

$$T_1^\nu = i\int dx\, \text{T}\left[J_1^\nu(0), \mathcal{O}_{kin}(x)\right], \quad T_2^\nu = i\int dx\, \text{T}\left[J_2^\nu(0), \mathcal{O}_{kin}(x)\right],$$
$$T_3^\nu = i\int dx\, \text{T}\left[J_1^\nu(0), \mathcal{O}_{mag}(x)\right], \quad T_4^\nu = i\int dx\, \text{T}\left[J_2^\nu(0), \mathcal{O}_{mag}(x)\right].$$
$$(5.54)$$

As we shall see later, these operators would mix into the operators of Eq. (5.53) under renormalization, but not vice versa.

As we know, reparameterization invariance defines a unique combination of operators whose Wilson coefficients renormalize the same way to all orders in QCD perturbation theory. Thus, before we look into the calculation of the Wilson coefficients C_i, B_i, and A_i, let us make a couple of comments.

First, as was noted in Eq. (5.39), the combination $(1 + i\slashed{D}/(2m_Q))h_v$ transforms covariantly under a reparameterization transformation. Thus, inserting it into J_1 we find that the combination

$$\bar{q}\gamma^\nu h_v \to \bar{q}\gamma^\nu \left(1 + \frac{i\slashed{D}}{2m_Q}\right) h_v = J_1^\nu + \frac{1}{2m_Q} O_1^\nu \qquad (5.55)$$

must have the same Wilson coefficient to all orders in QCD, i.e. $B_1(\mu) = C_1(\mu)$. Similarly, from Eq. (5.41) one can read that it is the combination

$$\bar{q}v^\nu h_v \to J_2^\nu + \frac{1}{2m_Q}(O_2^\nu + 2O_3^\nu) \qquad (5.56)$$

that must have the same Wilson coefficient. Thus, $B_2(\mu) = B_3(\mu)/2 = C_2(\mu)$.

Second, since operators T_i^ν are simply time-ordered products, their Wilson coefficients are products of Wilson coefficients of the operators that form them,

$$A_1(\mu) = C_1(\mu), \qquad A_2(\mu) = C_2(\mu),$$
$$A_3(\mu) = C_1(\mu)C_g(\mu), \qquad A_4(\mu) = C_2(\mu)C_g(\mu). \qquad (5.57)$$

The coefficient C_g was defined in Eq. (5.23). Armed with those observations, we can perform matching of HQET and QCD currents to any order in perturbative QCD. Only $C_1 = A_1 = A_3 = B_1 = 1$ are non-zero at the leading order in pQCD.

We now look at the matching at higher orders in pQCD and see how this is done step-by-step. First, we shall look at how to calculate perturbative QCD corrections to the leading HQET operators, $J_{1,2}$. Then, using the results of the discussion above, we show how to extend our result to include $1/m_Q$ corrections discussed previously. We shall see that the results that we obtain will contain large logarithms of ratios of scales, which need to be resumed, if we are to get a stable result. We shall discuss how this can be done first for $J_{1,2}$ and then for the whole set of operators with $1/m_Q$ corrections with leaving logarithmic (LL) and then next-to-leading logarithmic (NLL) precision.

Let us start by looking at the leading (in $1/m_Q$) order operators. As we should expect, one-loop perturbative QCD corrections to heavy-to-light current operators would renormalize HQET currents,

$$J_i \simeq Z_{ij}^{-1} Z_q Z_h \, \bar{q}\Gamma_j h_v. \qquad (5.58)$$

Here Z_q and Z_h are wave function renormalization factors for q and h_v, respectively, Z_{ij} is the matrix of renormalization factors for the current operator, and $\bar{q}\Gamma_j h_v$ are the leading-order current operators given in Eq. (5.52). We should expect that at the end all Wilson coefficients would look like

$$C_i(m_Q) = C_i^{(0)} + c_i \frac{\alpha_s}{2\pi} + \cdots. \qquad (5.59)$$

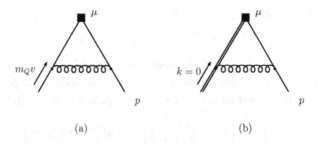

Fig. 5.3 One-loop vertex corrections for the matching of heavy-light currents.

In reality, the currents would be seen experimentally in terms of their matrix elements between some hadronic states, so one would need to match the entire matrix elements. However, since pQCD corrections only involve large, $\mu \sim m_Q$ scales, matching can be performed for any states, including perturbative quark ones! In other words [Ji and Musolf (1991); Neubert (1994b); Falk and Grinstein (1990)], $\langle V_\mu \rangle = \bar{u}_q \Gamma_\mu u_Q$, with $\slashed{v} u_Q = u_Q$. Now, while at the tree level on the QCD side of the matching $\Gamma_\mu = \gamma_\mu$, one-loop pQCD corrections would induce the non-zero coefficients for both γ_μ and v_μ. Including the factors $Z_q^{1/2}$ and $Z_Q^{1/2}$ for the light and heavy quarks that can be found in [Peskin and Schroeder (1995)] and calculating the one-loop graphs depicted in Fig. 5.3a, we get

$$\Gamma^\mu_{\text{full}} = \left[1 + \frac{\alpha_s}{2\pi}\left(\log\frac{m_Q^2}{\lambda_g^2} - \frac{11}{6}\right)\right]\gamma^\mu + \frac{2}{3\pi}\alpha_s v^\mu. \quad (5.60)$$

Here we introduced a finite mass for the gluon λ_g, which is convenient way to regulate IR divergencies. Note that the result is UV finite, as it should be, as it is a conserved current in QCD. Yet, it does have an IR divergence associated with the use of perturbative external states [Neubert (1994b)].

Calculating the graph depicted in Fig. 5.3b and, just like above, multiplying it by $Z_q^{1/2}$ and $Z_h^{1/2}$, we get

$$\Gamma^\mu_{\text{HQET}} = \left[1 + \frac{\alpha_s}{2\pi}\left(\frac{1}{\bar{\epsilon}} + \log\frac{m_Q^2}{\lambda_g^2} - \frac{5}{6}\right)\right]\Gamma^\mu. \quad (5.61)$$

for either $\Gamma^\mu = (\gamma^\mu, v^\mu)$. This current is UV divergent! This might be surprising, but it is not unexpected. In EFT we removed high frequency modes of the heavy fields, so even though in full theory the current operator corresponds to a conserved quantity (and, thus, must be UV finite), we are no longer working in the full theory. Thus, we need to renormalize currents in Eq. (5.52) by choosing Z-factors of Eq. (5.58) to be

$$Z_{ii} = 1 + \frac{\alpha_s}{2\pi}\frac{1}{\bar\epsilon}, \quad Z_{ik\ (i\neq k)} = 0, \qquad (5.62)$$

which follows from Eq. (5.61) and corresponds to $\overline{MS}$ subtraction.

Once the UV infinities are subtracted, we can write a HQET current which can then be used to match to the full QCD result in Eq. (5.60),

$$\Gamma^\mu_{\text{HQET}} = \left[1 + \frac{\alpha_s}{2\pi}\left(\log\frac{m_Q^2}{\lambda_g^2} - \frac{5}{6}\right)\right][C_1(\mu)\gamma^\mu + C_2(\mu)v^\mu]. \qquad (5.63)$$

When compared to Eq. (5.60), this gives us the Wilson coefficients

$$C_1(\mu) = 1 + \frac{\alpha_s}{\pi}\left(\log\frac{m_Q}{\mu} - \frac{4}{3}\right), \quad C_2(\mu) = \frac{2\alpha_s}{3\pi}. \qquad (5.64)$$

Notice that IR-divergent log has disappeared. This is to be expected, since IR physics is the same in QCD and HQET. Looking at the equation above, we can immediately notice two things. First, the coefficient C_2 is not zero any more, in line with our expectations expressed in Eq. (5.59). Second, we notice that $C_1(\mu)$ acquired a log term. For $\mu \sim 1$ GeV one can see that $\alpha_s \log(m_Q/\mu)/\pi \sim 0.5$, which makes the perturbation series converging quite slowly, as we expect to get powers of logs at higher orders in α_s. It would therefore be nice to resum all the logs![2] We shall see how this can be done after we discuss inclusion of $1/m_Q$ corrections.

A calculation of α_s corrections for the whole basis of Eq. (5.51) is very similar to the calculation that we just described. The only difference is that in the evaluation of Feynman diagrams of Fig. 5.3 one needs to set $p_Q = m_Q v + k$ and $p_q = p$ and keep terms linear in k and p. The result, as well as a description of the calculation, can be found in [Neubert (1994b)]. The results are

$$C_1(\mu) = 1 + \frac{\alpha_s}{\pi}\left(\log\frac{m_Q}{\mu} - \frac{4}{3}\right),$$

$$C_2(\mu) = B_2(\mu) = \frac{1}{2}B_3(\mu) = \frac{2\alpha_s}{3\pi},$$

$$B_4(\mu) = \frac{4\alpha_s}{3\pi}\left(3\log\frac{m_Q}{\mu} - 1\right), \qquad (5.65)$$

$$B_5(\mu) = -\frac{4\alpha_s}{3\pi}\left(2\log\frac{m_Q}{\mu} - 3\right),$$

$$B_6(\mu) = -\frac{4\alpha_s}{3\pi}\left(\log\frac{m_Q}{\mu} - 1\right).$$

---

[2] Even though $\log(m_Q/\mu)$ is largely cancelled by 4/3 in Eq. (5.64), in general we do not expect this to hold at higher orders in α_s.

While this gives the matching coefficients for the HQET expansion of vector and axial currents, as we already pointed out above, the coefficients of effective operators would differ substantially at different scales $\mu < m_Q$ due to the large logarithms present in Eq. (5.65). Those logarithms will need to be resummed. This can be done using renormalization group methods developed in Chapter 3.

To illustrate how to perform renormalization group improvement of this matching, let us concentrate for now, again, on the leading order operators J_1^ν and J_2^ν [Ji and Musolf (1991)]. We shall show how to generalize the results to include $1/m_Q$ corrections later on. Let us take a logarithmic derivative with respect to scale μ of both sides of Eq. (5.51). The quantity on the left-hand side is scale independent, so its derivative should be zero. However, the quantity on the right-hand side is a product of Wilson coefficients $C_i(\mu)$, which are scale-dependent, and the operators $J_i^\nu(\mu)$, which are also scale-dependent[3]

$$\mu \frac{d}{d\mu} V^\mu = \left[\mu \frac{d}{d\mu} C_i(\mu) \right] J_i^\nu(\mu) + C_i(\mu) \left[\mu \frac{d}{d\mu} J_i^\nu(\mu) \right]. \quad (5.66)$$

Now, if we multiply both sides by $J_i^{-1}(\mu)$ from the right, we get the familiar renormalization group equation for Wilson coefficients,

$$\left[\mu \frac{d}{d\mu} - \hat{\gamma}_i \right] C_i(m_Q/\mu, \alpha(\mu)) = 0. \quad (5.67)$$

Notice that we explicitly wrote out where the μ-dependence of the Wilson coefficients comes from. Since we are dealing with two Wilson coefficients of the operators $J_{1,2}^\mu$, it is convenient to form a column vector composed of C_1 and C_2. In that case Eq. (5.67) can be viewed as a 2×2 matrix equation. That makes $\hat{\gamma}_i$ a matrix of anomalous dimensions of the HQET operators J_i^ν. Note that this matrix is diagonal, i.e. $\hat{\gamma} = \text{diag}(\gamma_1, \gamma_2)$.

In order to solve this equation we recall that

$$\mu \frac{d}{d\mu} = \mu \frac{\partial}{\partial \mu} + \beta(g_s) \frac{\partial}{\partial g_s}. \quad (5.68)$$

Recall that the QCD beta function, which describes scale dependence of the renormalized coupling constant, can be expanded in powers of g_s with

$$\beta(g_s) = -g_s \left[\beta_0 \frac{g_s^2}{16\pi^2} + \beta_1 \left(\frac{g_s^2}{16\pi^2} \right)^2 + \ldots \right], \quad (5.69)$$

---

[3] In fact, the scale dependence of the operators must compensate the scale dependence of the Wilson coefficients! The vector current on the LHS is not renormalized in full theory because it corresponds to a conserved quantity.

where β_i are given by

$$\beta_0 = 11 - \frac{2}{3}n_f, \quad \beta_1 = 102 - \frac{38}{3}n_f. \quad (5.70)$$

Here n_f is the number of light flavors, i.e., the number of flavors with masses below m_Q. Similarly, anomalous dimension matrix $\hat{\gamma}$ can also be expanded in powers of g_s,

$$\hat{\gamma}(g_s) = \hat{\gamma}_0 \frac{g_s^2}{16\pi^2} + \hat{\gamma}_1 \left(\frac{g_s^2}{16\pi^2}\right)^2 + \ldots \quad (5.71)$$

To calculate $\hat{\gamma}_i$ we need to know Z_{ij}. As we are working with $\overline{MS}$ subtraction scheme, the coefficients of anomalous dimension matrix can be obtained from the coefficients of $1/\bar{\epsilon}$ terms in expansion

$$Z_{ij} = 1 + \sum_k \frac{1}{\bar{\epsilon}^k} Z_{ij}^{(k)}(g_s) \quad \rightarrow \quad \hat{\gamma} = -g_s \frac{\partial Z_{ij}^{(1)}}{\partial g_s}. \quad (5.72)$$

At the leading order in $1/m_Q$, the 2×2 anomalous dimension matrix can be obtained from Eq. (5.62),

$$\hat{\gamma}_0 = \begin{pmatrix} -4 & 0 \\ 0 & -4 \end{pmatrix}. \quad (5.73)$$

Thus, at the leading order in $1/m_Q$, the only non-zero entries in the anomalous dimensions matrix correspond to $\gamma_{11} = \gamma_{22} = -4$.

With these expansions the formal solution of Eq. (5.67) can be obtained at the leading or next-to-leading order depending on how many terms in the expansions of Eqs. (5.69,5.71) one is willing to consider. The first non-trivial result ("leading-log approximation") can be found truncating those series just after the leading terms in g_s,

$$C_i(m_Q/\mu, \alpha_s(\mu)) = \exp\left(-\int_{g_s(\mu)}^{g_s m_Q} \frac{\gamma(g_s)}{\beta(g_s)} dg_s\right) \times C_i(1, \alpha(m_Q))$$

$$= \left(\frac{\alpha(m_Q)}{\alpha(\mu)}\right)^{\gamma_0/(2\beta_0)} C_i(1, \alpha(m_Q)) \quad (5.74)$$

It is convenient to introduce $x = \alpha(\mu)/\alpha(m_Q)$. As we can see from Eq. (5.74), we need both β_0 and γ_0 obtained above. The LL results for C_1 and C_2 are

$$C_1(\mu) = x^{2/\beta_0}, \quad C_2(\mu) = 0. \quad (5.75)$$

We can also include $1/m_Q$ corrections to the current. To do so we need to employ very similar techniques, but work with higher-dimensional matrices to accommodate the whole basis of Eq. (5.51). Putting everything together, we can obtain the leading-log results [Neubert (1994b); Falk and Grinstein (1990)],

$$C_1(\mu) = B_1(\mu) = x^{2/\beta_0},$$
$$C_2(\mu) = B_2(\mu) = B_3(\mu) = 0,$$
$$B_4(\mu) = \frac{34}{27}x^{2/\beta_0} - \frac{4}{27}x^{-1/\beta_0} - \frac{10}{9} + \frac{16}{3\beta_0}x^{2/\beta_0}\log x, \quad (5.76)$$
$$B_5(\mu) = -\frac{28}{27}x^{2/\beta_0} + \frac{88}{27}x^{-1/\beta_0} - \frac{20}{9},$$
$$B_6(\mu) = -2x^{2/\beta_0} - \frac{4}{3}x^{-1/\beta_0} + \frac{10}{3}.$$

To check our result we can expand the formulas in Eq. (5.76) for $\mu \sim m_Q$ to recover all coefficients in front of the logs in Eq. (5.65). Since the goal of the leading logarithmic approximation is to resume all large logs (as they are assumed to dominate the result), we will not recover the constant terms in Eq. (5.65). To get those we need to go to next-to-leading order results.

The next-to-leading solution of Eq. (5.67) is

$$C_i(m_Q/\mu, \alpha_s(\mu)) = \exp\left(-\int_{g_s(\mu)}^{g_s m_Q} \frac{\gamma(g_s)}{\beta(g_s)}dg_s\right) \times C_i(1, \alpha(m_Q))$$
$$= C_i(1, \alpha(m_Q))\left(\frac{\alpha(m_Q)}{\alpha(\mu)}\right)^{\gamma_0/(2\beta_0)} \quad (5.77)$$
$$\times \left[1 + \frac{\alpha_s(m_Q)}{4\pi}\left(c_i + \frac{\gamma_0}{2\beta_0}\left(\frac{\gamma_1}{\gamma_0} - \frac{\beta_1}{\beta_0}\right)\right)\right].$$

Thus, as we can see, in order to have C_i to first order in α_s, one needs the *two-loop* anomalous dimension γ_1, the two-loop beta-function β_1 and the first-order matching coefficient c_i. The calculations of anomalous dimensions present the hard part of the problem, as the operators at subleading orders in $1/m_Q$ can now mix. Yet, even though the calculation is tedious, it is tractable and can be achieved by the methods discussed previously. In fact,

$$\gamma_1 = -\frac{254}{9} - \frac{56}{27}\pi^2 + \frac{20}{9}n_f. \quad (5.78)$$

The expressions for subleading coefficients are even more complicated, so we will not display them here. But they can be found in original papers [Neubert (1994b); Amoros et al. (1997)]. A full next-to-leading order matching of the heavy-to-light currents can be found in [Becher et al. (2001)].

5.3 Different degrees of freedom: heavy mesons

Can one build an effective Lagrangian describing heavy *meson* states? This question is not unreasonable, especially if we are concerned with the situation when heavy mesons (or baryons) interact with light mesons that are members of the pseudo-goldstone multiplet of pions, kaons and etas. If those pseudo-goldstones remain soft in the interaction, it is possible to apply the language of chiral perturbation theory that was set up in previous chapters to describe physical processes involving those particles.

The techniques developed in previous sections of this chapter deal with quark degrees of freedom. Yet, several observations that were made could be useful even if we deal with heavy flavored mesons and baryons! For example, the fact that the masses of pseudoscalar B and vector B^* mesons are degenerate in the heavy quark limit allows us to think about them as members of the same multiplet, similarly to $SU(2)$ doublet of protons and neutrons.

5.3.1 *Heavy meson states. Tensor formalism*

Let us concentrate on heavy mesons in this section. Before we build an effective chiral Lagrangian describing heavy mesons interacting with pseudo-goldstones, let us build the heavy meson states themselves. To build a state, recall that the heavy meson state contains one heavy quark and one light antiquark[4], so it must transform as a triplet under $SU(3)_V$ chiral symmetry transformation. In order to build the heavy meson states, it will be convenient to first do so in the meson rest frame where $v = (1, \vec{0})$.

The eigenstates of the spin operator of Eq. (5.30) can be chosen as follows. For the heavy quark the spin part of the quantum field would be

$$\text{Spin up: } b_\alpha^{(\Uparrow)} = \begin{pmatrix} 1 \\ 0 \\ 0 \\ 0 \end{pmatrix} \equiv \delta_{1\alpha},$$

$$\text{Spin down: } b_\alpha^{(\Downarrow)} = \begin{pmatrix} 0 \\ 1 \\ 0 \\ 0 \end{pmatrix} \equiv \delta_{2\alpha} \quad (5.79)$$

---

[4] Due to historical reasons, this means that we are describing D (or $c\bar{q}$) or $\bar{B}$ (or $b\bar{q}$) states.

and, similarly, for the light antiquarks as $\bar{q}_\alpha^{(\Downarrow)} = -\delta_{3\alpha}$ and $\bar{q}_\alpha^{(\Uparrow)} = -\delta_{4\alpha}$. It is then relatively straightforward to construct the meson spin states. For the pseudoscalar state

$$B_{\alpha\beta}^q \propto b_\alpha^{(\Uparrow)}\bar{q}_\beta^{(\Downarrow)} + b_\alpha^{(\Downarrow)}\bar{q}_\beta^{(\Uparrow)} = -\begin{pmatrix}\hat{0} & \hat{1}\\ \hat{0} & \hat{0}\end{pmatrix} = -\frac{1+\gamma^0}{2}\gamma_5. \qquad (5.80)$$

Here the index q goes over from one to three for $q = u, d$ and s. It is now clear how this should be generalized to any frame that is moving with velocity v – in fact, we have already done that in the beginning of this chapter in Eq. (5.10)! Doing so, we can write for the pseudoscalar state in coordinate representation,

$$H_q^{(ps)} = -\frac{1+\slashed{v}}{2}\bar{B}_q\gamma_5. \qquad (5.81)$$

Similarly, we can construct spin wave function of the vector B^* state. This time, it should contain three polarization states, as we are dealing with a massive vector state. Let us define polarization vectors as

$$\epsilon_B^{(\pm)} = (0, 1, \pm i, 0),$$
$$\epsilon_B^0 = (0, 0, 0, 1), \qquad (5.82)$$

and consider each polarization (helicity) state, once again, in the rest frame of the heavy field. Just like in non relativistic quantum mechanics we construct

$$B_{\alpha\beta}^{q,(+)} \propto b_\alpha^{(\Uparrow)}\bar{q}_\beta^{(\Uparrow)} = \frac{1}{\sqrt{2}}\begin{pmatrix}\hat{0} & \sigma_1+i\sigma_2\\ \hat{0} & \hat{0}\end{pmatrix} = -\frac{1+\gamma^0}{2}\slashed{\epsilon}_B^{(+)},$$

$$B_{\alpha\beta}^{q,(0)} \propto b_\alpha^{(\Uparrow)}\bar{q}_\beta^{(\Downarrow)} - b_\alpha^{(\Downarrow)}\bar{q}_\beta^{(\Uparrow)} = \begin{pmatrix}\hat{0} & \sigma_3\\ \hat{0} & \hat{0}\end{pmatrix} = -\frac{1+\gamma^0}{2}\slashed{\epsilon}_B^0, \qquad (5.83)$$

$$B_{\alpha\beta}^{q,(-)} \propto b_\alpha^{(\Downarrow)}\bar{q}_\beta^{(\Downarrow)} = \frac{1}{\sqrt{2}}\begin{pmatrix}\hat{0} & \sigma_1-i\sigma_2\\ \hat{0} & \hat{0}\end{pmatrix} = -\frac{1+\gamma^0}{2}\slashed{\epsilon}_B^{(-)}.$$

Similarly to the case of pseudoscalar mesons, this result can be generalized to an arbitrary frame of reference,

$$H_q^{(vect)} = \frac{1+\slashed{v}}{2}\slashed{\epsilon}_B \qquad (5.84)$$

for all three helicity states in the momentum representation. In the coordinate representation we would have to replace $\epsilon_B^\mu \to \bar{B}_q^{*\mu}$. Since we are dealing with the fields in the heavy quark limit, the fields $\bar{B}_q$ and $\bar{B}_q^{*\mu}$ do not create antiparticles. Also notice that the heavy vector states satisfy the condition $v_\mu \bar{B}_q^{*\mu} = 0$.

We have argued before that in the heavy quark limit the pseudoscalar and vector states are degenerate. Thus, it is convenient to introduce a supermultiplet state that includes both of them, which in coordinate representation would read

$$H_q(v) = \frac{1+\slashed{v}}{2} \left(\bar{B}_q^{*\mu} \gamma_\mu - \bar{B}_q \gamma_5 \right). \tag{5.85}$$

This tensor representation of the states in the heavy quark limit will be used to build a chiral Lagrangian for interactions of heavy and light fields. Just like any effective heavy field, this field has the following properties,

$$\slashed{v} H_q(v) = H_q(v), \qquad H_q(v)\slashed{v} = -H_q(v), \tag{5.86}$$

which could be easily derived using Eq. (5.12). It also transforms under heavy quark spin symmetry the same way as the heavy quark field it contains, i.e.

$$H_q(v) \rightarrow S H_q(v) = e^{i\vec{\alpha}\cdot\vec{S}} H_q(v), \tag{5.87}$$

where $\vec{S}$ is defined in Eq. (5.30). We should also define a conjugated field $\overline{H}$,

$$\overline{H}_i(v) = \gamma^0 H_i^\dagger(v) \gamma^0 = \left(\bar{B}_q^{*\dagger\mu} \gamma_\mu + \bar{B}_q^\dagger \gamma_5 \right) \frac{1+\slashed{v}}{2}, \tag{5.88}$$

which transforms under the spin symmetry transformation as $\overline{H} \rightarrow \overline{H} S^{-1}$.

An introduction of this *tensor representation* for the heavy meson field significantly simplifies the calculation of matrix elements of various operators between heavy meson states and allows for a compact way of writing of effective Lagrangians for heavy meson states.

For example, the calculation recipe for the matrix elements is as follows. First, substitute the bra and ket-vectors with either Eq. (5.80) or (5.83). Second, substitute the operator with all possible tensors transforming appropriately under the spin symmetry. Put those two ingredients together and multiply by an unknown coefficient that would have to be determined from experimental data. The object that we are building must be invariant under heavy quark spin symmetry, which means that we must take a trace over spin indices. That is it! We now show how it works for the simplest case of normalization of the pseudoscalar and vector heavy meson states,

$$\langle \bar{B}_q(v) | \bar{B}_q(v) \rangle = a \, \text{Tr} \left[B^{q\dagger} B^q \right] = a \, \text{Tr} \left[\gamma_5 \frac{1+\slashed{v}}{2} \frac{1+\slashed{v}}{2} \gamma_5 \right] = 2a,$$

$$\langle \bar{B}_q^*(v) | \bar{B}_q^*(v) \rangle = a \, \text{Tr} \left[\slashed{\epsilon}_B^* \frac{1+\slashed{v}}{2} \frac{1+\slashed{v}}{2} \slashed{\epsilon}_B \right] = 2a. \tag{5.89}$$

Note that both of these states should have the same normalization. Also, we implicitly eliminated all other possible tensors built out of $\not{v}$ due to Eqs. (5.12). If we require to normalize the external states to one, $a = 1/2$.

The Lagrangian we are about to derive can be used to describe properties of both B and D mesons, so we shall call them collectively H in the following discussion.

5.3.2 Leading-order Lagrangian

To build a Lagrangian that involves interactions of heavy and light PNG mesons, we need to require its ingredients to have certain transformation properties under chiral $SU(3)_L \times SU(3)_R$. In particular, recall that U transforms as

$$U \to LUR^\dagger. \tag{5.90}$$

Because in our definition $\bar{B}$ and D-mesons contain light antiquarks, we require the heavy field $H \equiv H_q(v)$ defined above and its covariant derivative transform under $SU(3)_L \times SU(3)_R$ chiral symmetry as

$$H \to HV^\dagger, \qquad (D^\mu H) \to (D^\mu H)V^\dagger, \tag{5.91}$$

where the matrix V is in vectorial $SU(2)$ or $SU(3)$. The conjugated field transforms as

$$\overline{H} \to V\overline{H}. \tag{5.92}$$

The matrix V is introduced to describe transformational properties of ξ (recall that $U = \xi^2$),

$$\xi \to L\xi V^\dagger = V\xi R^\dagger. \tag{5.93}$$

We defined the chiral covariant derivatives in Chapter 4. The corresponding vector and axial currents are

$$\begin{aligned}
D^\mu_{ab} &= \delta_{ab}\partial^\mu + V^\mu_{ab}, \\
V^\mu_{ab} &= \frac{1}{2}\left(\xi^\dagger \partial^\mu \xi + \xi \partial^\mu \xi^\dagger\right)_{ab}, \\
A^\mu_{ab} &= \frac{i}{2}\left(\xi^\dagger \partial^\mu \xi - \xi \partial^\mu \xi^\dagger\right)_{ab}.
\end{aligned} \tag{5.94}$$

The axial current is defined to transform as $A \to VAV^\dagger$. With these ingredients it is easy to build the most general effective Lagrangian satisfying heavy quark spin and light quark chiral symmetries,

$$\begin{aligned}
\mathcal{L} = &- \operatorname{Tr}\left[\overline{H}_a(v)\, iv \cdot D_{ba} H_b(v)\right] \\
&+ g \operatorname{Tr}\left[\overline{H}_a(v) H_b(v) \slashed{A}_{ba} \gamma_5\right],
\end{aligned} \tag{5.95}$$

where the trace is taken over spin indices, and flavor indices a, b are summed over. Eq. (5.95) gives the leading-order interaction Lagrangian of heavy and light mesons. Note that while the coefficient of the first term is fixed by the equations of motion, the coefficient g of the second term is a parameter that needs to be fixed by experimental data. Both chiral-symmetry and heavy-quark symmetry-breaking corrections can be added to Eq. (5.95). We shall discuss those in the next section.

One important thing to remember about the chiral Lagrangian for heavy and light mesons is that the heavy fields have unusual canonical dimension, 3/2, which follows from examination of the kinetic term in Eq. (5.95). This is because we absorbed factors of $\sqrt{m_H}$ into the definitions of those fields. In practical calculations one must always remember to multiply scattering amplitudes by factors $\sqrt{m_H}$ for each external heavy meson (both pseudoscalar and vector) to restore the correct relativistic normalization. We shall determine numerical value of g using decay $D^* \to D\pi$ in sect. 5.3.4.

5.3.3 Subleading Lagrangians

The leading order Lagrangian of HHχPT of Eq. (5.95) describes interactions of heavy and light mesons in the limit of combined heavy quark and chiral symmetries. Indeed, those symmetries are only approximate: corrections of order $\sim m_s/\Lambda$ or Λ/m_c for some hadronic scale $\Lambda \sim 1$ GeV could be quite significant! So, it is important to address those corrections.

Let us first discuss non-leading terms that violate heavy quark spin symmetry, but preserve chiral $SU(3)_V$. Those terms enter at order $1/M_H$ and can be parameterized by

$$\mathcal{L}_{1/m_Q} = \frac{C_1}{m_Q} \text{Tr}\left[\overline{H}_a(v)\sigma^{\mu\nu} H_a(v)\sigma_{\mu\nu}\right] + \frac{C_2}{m_Q} \text{Tr}\left[\overline{H}_a(v)\slashed{A}_{ab}\gamma_5 H_b(v)\right]. \tag{5.96}$$

Note that it does not matter at this order if we write $1/m_Q$ or $1/M_H$, as the difference is between heavy quark and heavy mesons masses will only enter at higher orders in $1/m_Q$. The corrections described by Eq. (5.96) are especially important when we consider observables in the charm sector.

Since the first term in Eq. (5.96) only includes heavy meson fields, its effect is to introduce splitting between the members of the heavy meson supermultiplet,

$$m_{H^*} - m_H = -\frac{8C_1}{m_Q}. \tag{5.97}$$

While the heavy quark limit might be a good approximation when we are talking about B mesons, chiral symmetry breaking could be important when comparing decays of B_s and B_d mesons. This flavor $SU(3)_V$ breaking can be conveniently parameterized by introducing light quark mass matrix $M_q = \text{diag}(m_u, m_d, m_s)$. There is a slick way to build a flavor-symmetry-violating Lagrangian that is related to the "spurion" methods defined in Chapter 2. We assume that M_q transforms as

$$M_q \to L M_q R^\dagger, \tag{5.98}$$

and use it as one of the building blocks of our Lagrangian. The lowest-order correction would enter at the first order in m_q, so we would insert M_q only once,

$$\begin{aligned}\mathcal{L}_{m_q} &= \lambda_0 \text{Tr}\left[M_q U + M_q U^\dagger\right] \\ &+ \lambda_1 \text{Tr}\left[\overline{H}_a(v) H_b(v) \left(\xi M_q \xi + \xi^\dagger M_q \xi^\dagger\right)_{ba}\right] \\ &+ \lambda_1' \text{Tr}\left[\overline{H}_a(v) H_a(v) \left(M_q U + M_q U^\dagger\right)_{bb}\right]\end{aligned} \tag{5.99}$$

Calculating perturbative corrections to these terms can generate contributions that break both heavy quark spin and flavor symmetries.

5.3.4 Calculations with HHχPT

In order to do calculations with HHχPT we need two things. First of all, we should derive Feynman rules. This would greatly simplify our calculations. Next, we need to determine all unknown constants defined in the previous section in order to get numerical predictions to be compared to experiment. Just like in other effective field theories one needs to consistently calculate those constants to required order. Some of the constants can be easily computed even without the need to derive Feynman rules. For example, the constant g of Eq. (5.95) can be determined from the experimentally-observed strong $D^* \to D\pi$ decay[5]. To evaluate g, let us expand A_μ in terms of component fields

$$A_\mu = -\frac{1}{f}\partial_\mu \mathcal{M} + \mathcal{O}(\mathcal{M}^3), \tag{5.100}$$

and and compute the relevant traces to obtain

$$\mathcal{L}_{D^*D\pi} = -2\frac{g}{f} D_q^{*\mu} \partial_\mu D_q^\dagger + \text{h.c.} \tag{5.101}$$

This results in the decay width

$$\Gamma(D^{*+} \to D^0 \pi^+) = \frac{1}{6\pi}\frac{g^2}{f^2}|\vec{p}_\pi|^3, \tag{5.102}$$

---

[5]Note that the corresponding B-decay does not exists due to phase-space suppression.

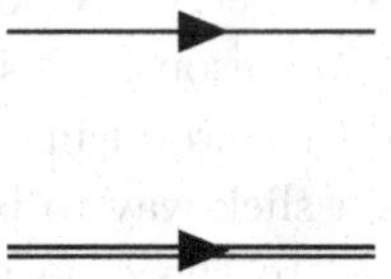

Fig. 5.4 Propagators of the heavy meson fields (a) spin-0, and (b) spin-1.

$$\text{(a)} \quad \longrightarrow \quad = \frac{i}{2v\cdot k + i\epsilon}$$

$$\text{(b)} \quad =\!=\!=\!\Rightarrow\!=\!=\!= \quad = \frac{-i\,(g_{\mu\nu} - v_\mu v_\nu)}{2v\cdot k + i\epsilon}$$

with $\Gamma(D^{*+} \to D^0\pi^+) = 2\Gamma(D^{*+} \to D^+\pi^0)$. Measurements done by experimental collaborations CLEO at Cornell and BaBar at SLAC determined that $g \approx 0.6$ [Anastassov et al. (2002); Lees et al. (2013)].

It is now easy to list the relevant Feynman rules. The propagators can be determined by writing the kinetic part of the HHχPT Lagrangian separately in terms of its pseudoscalar and vector components and inverting them in the usual way. We present those in Fig. 5.4. Notice that we used double line notation to denote heavy vector mesons, while for heavy pseudo scalars we reserved a single line notation. Since we never use heavy quarks and mesons in the same diagram, this should not lead to any confusion. Light meson propagators are the same as in Chapter 4.

The vertices can be obtained by expanding the Lagrangian Eq. (5.95) in the number of PNG fields. For example, the one-PNG vertex can be obtained from expansion of Eq. (5.100) and taking the traces. It is easy to see that there are no vertices connecting heavy particles of the same spin and one PNG, which follows from its derivative coupling and also explains its momentum dependence. The values of the coefficients $C^M_{qq'}$ are in Table 5.1 and the vertices are in Fig. (5.5). Notice that reversing the flow of momentum in the light PNG line and switching vector and scalar lines results in the extra overall minus sign for the vertex. The same is true if one considers antiparticles instead of particles in the vertices.

Expanding the axial current further we can get the expression for the terms in the Lagrangian that describe interactions of two heavy mesons and two PNG fields. These are also presented in Fig. (5.5).

To give a slightly less trivial example than the one in Eq. (5.102), let us consider diagrams contributing to wave function renormalization of the B-meson. These graphs are important in the calculations of chiral corrections to decay constants, $B^0\overline{B}^0$ mixing and other important phenomenological quantities. Note that the contribution of these graphs cancel in the ratio of decay constants (see Problem 3 at the end of this chapter).

In order to define a wave function renormalization constant $\sqrt{Z_H}$, let us

Table 5.1 Coefficients of the vertices in HHχPT Feynman rules

q	q'	M	$C^M_{qq'}$
u	u	π^0	1
u	u	η	$1/\sqrt{3}$
u	d	π^-	$\sqrt{2}$
u	s	K^-	$\sqrt{2}$
d	d	π^0	-1
d	d	η	$1/\sqrt{3}$
d	u	π^+	$\sqrt{2}$
d	s	$\overline{K}^0$	$\sqrt{2}$
s	s	η	$-2/\sqrt{3}$
s	u	K^+	$\sqrt{2}$
s	s	K^0	$\sqrt{2}$

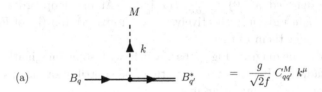

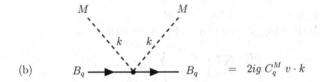

Fig. 5.5 Vertices with one (a), and two (b) pseudogoldstone bosons.

recall that the full propagator $i\Delta(\omega)$ of a heavy meson state can be written as

$$i\Delta(\omega) = i\Delta_0(\omega) + i\Delta_0(\omega)\left[-iM(\omega)\right]i\Delta_0(\omega) + ..., \qquad (5.103)$$

where $M(\omega)$ is the one-particle irreducible graph contributing to wave function renormalization and mass renormalization. We introduced $\omega = v \cdot p$ for the propagator with momentum p.

At one loop order in the chiral and $1/m_B$ expansions the wave function renormalization is given by two diagrams depicted in Fig. (5.6). Expanding

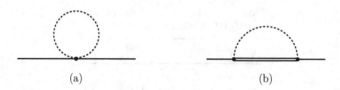

Fig. 5.6 Feynman diagrams contributing to wave function renormalization of heavy mesons.

$M(\omega)$ in a Taylor series around $\omega = 0$

$$M(\omega) = M(0) + \omega \left.\frac{\partial}{\partial \omega} M(\omega)\right|_{\omega=0} + ... \qquad (5.104)$$

we find that the part that contributes to the wave function renormalization is

$$Z_H = 1 + M'(0), \qquad (5.105)$$

where we denoted $M'(0) = \left.\frac{\partial}{\partial \omega} M(\omega)\right|_{\omega=0}$. At one loop order $M(\omega) = M_a(\omega) + M_b(\omega)$ given, respectively, by diagrams (a) and (b) of Fig. (5.6). Let us consider them in turn.

Consider the diagram Fig. (5.6a), which, with some imagination, can be called a tadpole. It is easy to show that the contribution of this diagram vanishes due to Lorentz invariance,

$$M_a(\omega) = 2im_B \sum_{M=\pi^0,\eta} C_u^M \int \frac{d^4k}{(2\pi)^4} \frac{v \cdot k}{k^2 - M^2 + i\epsilon}, \qquad (5.106)$$

as the integrand is an odd function of k.

The contribution of diagram (b) can be written as

$$M_b(\omega) = \frac{im_B}{4} \sum_M \left(\frac{g C_{qq'}^M}{\sqrt{2} f}\right)^2 (g_{\mu\nu} - v_\mu v_\nu)$$

$$\times \int \frac{d^4k}{(2\pi)^4} \frac{k^\mu k^\nu}{(-v \cdot k + \omega + i\epsilon)(k^2 - M^2 + i\epsilon)}. \qquad (5.107)$$

While we can calculate this integral directly using formulas from Appendix C, let is do it explicitly. Calculating $M_b'(0)$ needed for Eq. (5.105),

$$M_b'(0) = -\frac{im_B}{4} \sum_M \left(\frac{g C_{qq'}^M}{\sqrt{2} f}\right)^2 (g_{\mu\nu} - v_\mu v_\nu) I^{\mu\nu}, \qquad (5.108)$$

where the integral is

$$I^{\mu\nu} = \int \frac{d^4k}{(2\pi)^4} \frac{k^\mu k^\nu}{(-v \cdot k + i\epsilon)^2 (k^2 - M^2 + i\epsilon)}$$

$$= -\frac{i}{16\pi^2 \bar{\epsilon}} g^{\mu\nu} (M^2)^{1-\epsilon}, \qquad (5.109)$$

where we only retained the part that is non-zero upon contraction with $(g_{\mu\nu} - v_\mu v_\nu)$. Putting everything together we obtain, for a pseudoscalar B-meson (setting $H = B$),

$$Z_B = 1 - \frac{3}{4} \frac{m_B}{16\pi^2 \bar{\epsilon}} \sum_M \left(\frac{g C_{qq'}^M}{\sqrt{2} f}\right)^2 (M^2)^{1-\epsilon}. \tag{5.110}$$

This result can be used in many calculations, including B^* width at next-to-leading order.

5.4 Light baryons in heavy particle formalism

We already discussed the effective theory describing transitions between different baryon states in chapter 4. As it turns out, the formalism that we developed here can help us solve several problems that we encountered there.

Recall that when we developed the EFT for the light quark baryons we made a note that baryons are *not* massless – or even light – in the chiral symmetry limit. What this means is that every time we have a derivative acting on a baryon field, we bring down a power of momentum $p_B \sim 1$ GeV, which is by no means small compared to Λ_χ. Therefore, the power counting in a theory with baryons is non-trivial; a simple derivative expansion does not work.

A solution comes from the realization that, assuming restricted kinematics, the momentum transfer between baryons via pion or kaon exchange is small, $|q| \ll m_B$, where m_B, like before, denotes the mass of a baryon. This makes the situation very similar to the one described earlier in this chapter. If we consider baryon as a static source of light mesons, we can use the techniques that we developed here to describe low-energy interactions of baryons.

5.4.1 *Leading-order Lagrangian*

As we know from chapter 4, there is a plethora of light quark baryons. There, we classified them according to their spin and the way they transform under flavor $SU(3)$. While all baryon multiplets have interesting phenomenology, for the purpose of simplicity, let us only consider an octet of spin-1/2 baryons in this chapter. In order to build an effective Lagrangian we can parameterize baryon momentum p_B as

$$p_B^\mu = m_B v^\mu + k^\mu, \qquad (5.111)$$

with v^μ being the baryon velocity, and k being a "residual" momentum. Just like before, $v \cdot k$ would parameterize the "off-shellness" of the baryon field. Using the familiar procedure of scaling out the "mechanical" static solution, we can now define a velocity-dependent baryon field,

$$B_v = e^{im_B v \cdot x} P_+^v B(x), \qquad (5.112)$$

where P_+^v is a projector introduced in Eq. (5.10).

Fantastic! What did it buy us? Let us consider equations of motion of the new field B_v. The Dirac equation for the baryon field

$$(i\slashed{\partial} - m_B) B(x) = 0 \qquad (5.113)$$

will now be used to derive an equation of motion for B_v,

$$i(-i)m_B e^{-im_B v \cdot x} \slashed{v} P_-^v B_v + i e^{-im_B v \cdot x} P_-^v \slashed{\partial} B_v \\ - m_B e^{-im_B v \cdot x} P_-^v B_v = 0, \qquad (5.114)$$

or, after cancellations and using projector properties of the operator P_-^v,

$$i\slashed{\partial} B_v = 0 \qquad (5.115)$$

That's it! Now we can expect that any derivative acting on B_v will bring down a momentum that is of order k, not p_B, so each new term with a derivative acting on a baryon field would be suppressed by k/Λ_χ, not p_B/Λ_χ. This implies that derivative terms would get progressively smaller and we would be able to develop simple and consistent power counting scheme. Combined with the usual symmetry requirements for the construction of the effective theory for baryons, we can now build a convenient effective Lagrangian.

To do so, let us recall that B_v fields, just like their "full" counterparts, can be arranged in a matrix form,

$$B_v = \sum_{a=1}^8 \frac{\lambda^a}{\sqrt{2}} B^a = \begin{pmatrix} \Sigma_v^0/\sqrt{2} + \Lambda_v/\sqrt{6} & \Sigma_v^+ & p_v \\ \Sigma_v^- & -\Sigma_v^0/\sqrt{2} + \Lambda_v/\sqrt{6} & n_v \\ \Xi_v^- & \Xi_v^0 & -2\Lambda_v/\sqrt{6} \end{pmatrix}, \qquad (5.116)$$

and still have the same transformation properties under chiral $SU(3)_L \times SU(3)_R$,

$$B_v \to V B_v V^\dagger. \qquad (5.117)$$

Great simplification occurs because of the spin operators that satisfy

$$v \cdot S_v = 0, \quad S_v^2 B_v = -\frac{3}{4} B_v \qquad (5.118)$$

with the following commutation relations

$$\{S_v^\mu, S_v^\nu\} = -\frac{1}{2}\left(g^{\mu\nu} - v^\mu v^\nu\right), \quad [S_v^\mu, S_v^\nu] = i\epsilon^{\mu\nu\alpha\beta} v_\alpha S_{v\beta}. \qquad (5.119)$$

If we want to apply the techniques of building of an effective Lagrangian developed in the previous sections, we first need to construct the "building blocks" that have certain transformational properties in EFT. We can conveniently build the baryon effective Lagrangian, if we first see what happens to all fermion bilinears when we switch to the "heavy baryon" formulation. First of all,

$$\overline{B}_v \equiv \gamma^0 B_v^\dagger \gamma^0 = \gamma^0 \left[\frac{1+\slashed{v}}{2} B_v\right]^\dagger \gamma^0$$

$$= [\gamma^0 B_v^\dagger \gamma^0]\gamma^0 \left[\frac{1+\slashed{v}}{2}\right]^\dagger \gamma^0 = \overline{B}_v \frac{1+\slashed{v}}{2}, \qquad (5.120)$$

where we appropriately inserted $1 = \gamma^0 \gamma^0$. Thus, we can always substitute $\overline{B}_v \to \overline{B}_v P_+^v$ and $B_v \to P_+^v B_v$. Using Eq. (5.120) and the fact that B_v is an eigenstate of the projection operator, $B_v = P_+^v B_v$, we can see what happens to all baryon bilinears,

$$\overline{B}_v \gamma_5 B_v = 0, \quad \overline{B}_v \gamma_\mu B_v = v^\mu \overline{B}_v B_v,$$
$$\overline{B}_v \gamma^\mu \gamma_5 B_v = 2\overline{B}_v S_v^\mu B_v, \quad \overline{B}_v \sigma^{\mu\nu} B_v = 2\epsilon^{\mu\nu\alpha\beta} v_\alpha \overline{B}_v S_{v\beta} B_v, \qquad (5.121)$$
$$\overline{B}_v \sigma^{\mu\nu} \gamma_5 B_v = 2i \left[v^\mu \overline{B}_v S_v^\nu B_v - v^\nu \overline{B}_v S_v^\mu B_v\right].$$

We can now see a number of simplifications: besides the vanishing pseudoscalar combination, all other combinations lost their γ-matrices!

As part of the proof of those relations, let us derive the first identity on Eq. (5.121),

$$\overline{B}_v \gamma_5 B_v = \overline{B}_v P_+^v \gamma_5 P_+^v B_v = \overline{B}_v P_+^v P_-^v \gamma_5 B_v = 0 \qquad (5.122)$$

with the rest following analogously. Armed with Eqs. (5.117), (5.121), the effective chiral Lagrangian now looks like [Jenkins and Manohar (1991); Bernard et al. (1992)]

$$\mathcal{L}_B = \text{Tr } \overline{B}_v iv \cdot DB_v + 2D \text{ Tr } \overline{B}_v S_v^\mu \{A_\mu, B_v\}$$
$$+ 2F \text{ Tr } \overline{B}_v S_v^\mu [A_\mu, B_v] + \text{meson part}, \qquad (5.123)$$

where we also used the covariant derivative

$$D^\mu B_v = \partial^\mu B_v + [V^\mu, B_v], \qquad (5.124)$$

and vector and axial currents introduced in Eq. (5.94). One thing to note in Eq. (5.123) is the absence of the mass term for the baryon fields, just like in HQET.

It is also possible to show that, for the processes with one incoming and one outgoing baryon, a well-defined power counting can be developed based on the formula

$$D = 2L + 1 + \sum_n (n-2) V_{MM}^{(n)} + \sum_m (m-1) V_{MB}^{(m)}. \qquad (5.125)$$

Here we defined the chiral dimension D. As usual, L denotes the number of loops, $V_{MM}^{(n)}$ denotes the number of vertices from the $\mathcal{O}(k^n)$ chiral Lagrangian for the light PNG bosons from chapter 4, while $V_{MB}^{(m)}$ denote the number of vertices of order $\mathcal{O}(k^m)$ from the meson-baryon Lagrangian developed here. This effective Lagrangian can now be used to compute various experimentally-measured quantities using formalism of this effective theory.

5.5 Notes for further reading

Given that the first papers with analyses of heavy hadrons appeared in 1980's, heavy quark theory, as a field, is quite mature. There is plenty of material that should be assimilated for further study of effective theories of particles with one heavy quark. We purposely omitted extensive studies of heavy-to-heavy transitions started with pioneering papers by Voloshin and Shifman [Shifman and Voloshin (1988)] and Isgur and Wise [Isgur and Wise (1989)]. These serve as prime phenomenological applications of effective theory techniques to experimental problems and are extensively covered in a number of excellent reviews [Neubert (1994a); Shifman (1995); Bigi et al. (1997)] and books [Manohar and Wise (2000); Donoghue et al. (1992); Shifman (1999)].

A better understanding of techniques would include studies of higher-order perturbative QCD corrections in effective Lagrangians for heavy quarks [Buchalla et al. (1996)]. A derivation of the HQET Lagrangian from QCD using path integral formalism can be found in [Mannel et al. (1992)].

Important phenomenological applications include studies of inclusive B-decays and their lifetimes [Neubert and Sachrajda (1997); Beneke et al. (1999); Ciuchini et al. (2002); Gabbiani et al. (2004)] studies of $B^0 \overline{B}^0$ mixing [Lenz and Nierste (2007)], semileptonic [Bigi et al. (1993); Chay et al. (1990); Benson et al. (2003)] and rare heavy meson and baryon decays

[Buchalla et al. (1996)]. More on baryons within HQET can be found in [Mannel et al. (1991)].

Important applications of chiral techniques in HQET, starting with original papers [Burdman and Donoghue (1992); Wise (1992)] are extensively reviewed as well. Finally, lattice QCD methods enjoy applications of effective field theories [Kronfeld (2002)]. We invite interested reader to read them to gain deeper understanding of effective field theories with one heavy quark.

Problems for Chapter 5

(1) **Reparametrization**
In Eq. (5.41) we introduced two most-useful reparameterization-invariant bilinears. Find all reparameterization-invariant bilinears. Check your result with [Luke and Manohar (1992)].

(2) **Isgur-Wise function**
Consider flavor-changing transitions between heavy quark states that are driven by a heavy-to-heavy current $Q_1 \Gamma Q_2$, where Γ could be any Dirac matrix, $\Gamma = 1, \gamma_5, \gamma^\mu, \ldots$. An example of such transition can be exclusive weak decay $B \to D \ell \bar{\nu}$ or $B \to D^* \ell \bar{\nu}$, in which case $\Gamma = \gamma^\mu (1 + \gamma_5)$. Depending on the spin state of initial and final states, proper description of this transition would involve a number of form factors (six to describe both B-decays). Show that, in the heavy quark limit, all those form factors are related to only *one* form factor whose normalization can be determined model-independently. In order to accomplish that, consider

$$\langle H(v')|\overline{h}_{v'}^{Q_2} \Gamma h_v^{Q_1}|H(v)\rangle, \quad \langle H^*(v')|\overline{h}_{v'}^{Q_2} \Gamma h_v^{Q_1}|H(v)\rangle \quad (5.126)$$

for all possible Γ. Here $h_v^{(Q_i)}$ represent effective heavy fields in HQET for Q_i quarks. The actions of the heavy fields on the states can be represented as a Dirac matrix

$$\sqrt{m_{H_{Q_1}} m_{H_{Q_2}}} \, \overline{H}_{Q_2}(v') \Gamma H_{Q_1}^{ps}(v), \quad (5.127)$$

where index ps means that we are only interested in a pseudoscalar component of the meson superfield. This matrix must be coupled to the 4×4 matrix representing light degrees of freedom $\mathcal{M}$.

(a) Show that this matrix can be represented by

$$\mathcal{M} = A + B\slashed{v} + C\slashed{v}' + D\slashed{v}\slashed{v}'. \quad (5.128)$$

(b) Show that this means that all matrix elements can be written as
$$\frac{\langle H^{(*)}(v')|\overline{h}_{v'}^{Q_2}\Gamma h_v^{Q_1}|H(v)\rangle}{\sqrt{m_{H_{Q_1}} m_{H_{Q_2}}}} = \text{Tr}\left[\overline{H}_{Q_2}(v')\Gamma H_{Q_1}^{ps}(v)\mathcal{M}\right] \quad (5.129)$$
(c) Show that the properties of $H_{Q_i}(v)$ imply that the result of Eq. (5.129) would be the same if
$$\mathcal{M} = A + B + C + D = \xi(v \cdot v') \quad (5.130)$$
Here $\xi(v \cdot v')$ is called Isgur-Wise function and it is implied that all constants are multiplied by a 4×4 unit matrix in Dirac space.
(d) Show that $\xi(v \cdot v) = 1$.

(3) **Chiral corrections to decay constants**
Consider leptonic decays of heavy-flavored mesons, $D \to \ell\bar{\nu}$ and $D_s \to \ell\bar{\nu}$. The experimental studies of the ratio of these decay rates yields the ratio of D-decay constants, f_{D_s}/f_D. Show that this ratio receives contributions from "chiral logarithms,"
$$\frac{f_{D_s}}{f_D} = 1 + \kappa \frac{m_K^2}{16\pi^2 F^2} \log \frac{m_K^2}{\mu^2} + ... \quad (5.131)$$
In order to accomplish that, compute four diagrams in Fig. (5.7) with π, K, and η mesons in the loop (denoted by dashed lines). Show that

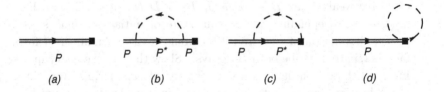

Fig. 5.7 Feynman diagrams for f_{D_s}/f_D calculations.

the contribution from the diagram (c) vanishes. Neglecting the mass of the pion and using the GMO relation $m_\eta^2 = 4m_K^2/3$ show that
$$\kappa = -\frac{5}{6}(1 + 3g^2), \quad (5.132)$$
where g is a pseudogoldstone-heavy meson coupling from the Lagrangian
$$\mathcal{L} = \frac{ig}{2}\text{Tr}\,\overline{H}^{(Q)a} H_b^{(Q)} \gamma_\nu \gamma_5 \left[\xi^\dagger \partial^\nu \xi - \xi \partial^\nu \xi^\dagger\right]_a^b. \quad (5.133)$$
If in trouble, consult [Grinstein et al. (1992)] for help.

(4) **Baryon bilinears**
Prove the relations displayed in Eq. (5.121).

Chapter 6

Effective Field Theories of Type-II. Bound States with Two Heavy Quarks

6.1 Introduction. Non-relativistic QED and QCD

Effective theories with one heavy particle discussed in the previous chapter remind us that physical scales present in the problem at hand are of the utmost importance to the building of a proper effective theory. In a way, Heavy Quark Effective Theory discussed in Chapter 5 is a relatively simple theory: there are only two scales present, the large scale, m_Q, and the small scale, Λ_{QCD}. Expanding in the ratio of these two scales simplifies QCD in the heavy quark limit. Let us now consider a situation when we need to deal with several physical scales.

We do not need to think hard to find an example: a bound state of two heavy quarks provides an excellent representation of a multi scale system. Provided the Coulomb interaction governs the physics of this bound state, the virial theorem implies that the heavy quark velocity in such bound state scales as

$$v^2 \sim \frac{\alpha_s}{m_Q r}. \tag{6.1}$$

The position-momentum uncertainty relation implies that the radius of the bound state scales inversely to the three-momentum of its components, $r \sim 1/m_Q v$, which together with the above equation implies that $v \sim \alpha_s$. The smallness of the velocity v guarantees the presence of several scales. Indeed, formation of the two-quark bound state takes place at a scale $m_Q v$, which is much smaller than m_Q, but might still be larger than the hadronic scale Λ_{QCD}[1]. Further, heavy quark energy introduces yet another scale, $m_Q v^2$, which might or might not be larger than Λ_{QCD}. We are clearly

---
[1]This statement, of course, depends on the mass of the heavy quark. The average relative velocity of the quark and antiquark can be estimated to be $v^2 \sim 0.3$ for the charmonium and $v^2 \sim 0.1$ for the bottomonium states.

seeing a multitude of scales here, which tells us that dealing with the two-quark system would not be easy. To proceed, let us denote the scale m_Q as *hard*, the scale $m_Q v$ as *soft*, and the scale $m_Q v^2$ as *ultrasoft*.

How would the presence of several scales affect our construction of an effective field theory? To answer this question let us look more closely at the diagram in Fig. 6.1. This diagram will certainly come up in the calculations

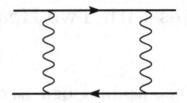

Fig. 6.1 Box diagram with two heavy quarks

of a spectrum of bound states of two heavy particles. The gauge particles in Fig. 6.1 could be photons or gluons: it does not matter for us at the moment. The integral I that we would need to compute when dealing with this diagram in HQET would look something like this,

$$I \sim \int \frac{d^d q}{(2\pi)^d} \frac{1}{q^0 + i\epsilon} \frac{1}{-q^0 + i\epsilon} \frac{1}{(q+k)^2 + i\epsilon} \frac{1}{(q-k)^2 + i\epsilon}, \quad (6.2)$$

where we suppressed everything appearing in the numerator and set $v = (1, \vec{0})$. In calculating this integral we usually analytically continue from Minkowski to Euclidian space to properly identify IR and UV behavior of the integral. This is most easily achieved by a Wick rotation in the q^0 space. However, when we look at the analytic structure of this integral, we notice that it has two poles at $q^0 = \pm i\epsilon$ coming from the heavy particle propagators with the integration contour passing right in between those poles. This is the case of a famous pinch singularity: we cannot deform the integration contour without crossing one of the poles, which is an illegal procedure. In order to remove the pinch one needs to "separate" the poles by introducing terms other than q^0 in the heavy particle propagator. This, however, is also illegal, as it is forbidden by the power counting rules of HQET. We have no other choice but to admit that we need to modify our power counting scheme in the case where two heavy particles are present. In other words, we need to change the way we build our effective field theory. The resulting construction goes under the name "Non-Relativistic QCD" or NRQCD, if we are dealing with strong interactions, or "Non-relativistic

QED" or NRQED if we work with electromagnetism. In this book we shall extensively discuss NRQCD.

6.2 NRQCD Lagrangian at the scale m_Q

In order to build an effective Lagrangian for NRQCD we would employ variations of the same techniques as the ones that we developed in previous chapters. First, we need to identify the degrees of freedom which would be used to construct the EFT. Since we are building a non-relativistic effective theory it would be very convenient to deal directly with two-component Pauli spinors for quark and anti-quark fields. This is so because heavy quark-antiquark pair creation is not possible for energies below heavy quark mass, just like in the case of HQET discussed in the previous chapter. Therefore, it is convenient to work with a Pauli spinor $\psi(x)$ that annihilates heavy quarks and a spinor $\chi(x)$ that creates heavy antiquarks, and build effective operators composed of ψ, χ, and their (covariant) derivatives. The matching coefficients for the NRQCD Lagrangian can then be obtained by matching it to the "full theory," which is ordinary QCD.

It would seem, however, that the issues related to the pinch singularities discussed in the previous section might make the matching procedure a very nontrivial affair. In HQET, since the heavy quark propagator is independent of the scale m_Q, power counting of relevant Green's functions came simply from counting factors of $1/m_Q$ in the vertex factors. Matching involved expanding Green's functions to the needed orders in $1/m_Q$ and matching the results.

We argued in the previous section, however, that the HQET propagator is not suitable for NRQCD, as we must keep extra terms to avoid pinch singularities in calculating the matching integrals. But a propagator with power-suppressed terms is bad for matching calculations: each propagator has explicit factors of $1/m_Q$, wrecking simple power counting of Green's functions! What should we do?

The solution was found in [Manohar (1997)], where it was proposed to expand the propagator, treating the extra terms as a perturbation, just like in HQET discussed in the previous chapter. It was argued that while positive powers of masses can appear when using a full NRQCD propagator, they do not appear if a propagator is treated as an infinite series, where the relevant integrals are performed first and the then the series is resummed. The disappearance of the dangerous terms with positive powers of m_Q can be traced to a non-analyticity of those terms in dimensional regularization. Since all hadronization effects happen at scales lower than m_Q, matching can be done for any states, including the perturbative state of two heavy quarks.

The NRQCD Lagrangian is traditionally written in the usual NRQCD frame $v = (1, \vec{0})$. Derivatives of the QCD gluon potentials can be expressed in terms of (chromo)electric and (chromo)magnetic fields. At the end, simply tracking the factors of $1/m_Q$ an effective Lagrangian of NRQCD (NRQED) can be written as follows [Manohar (1997)],

$$\mathcal{L}_\psi = \psi^\dagger \Bigg[iD^0 - c_2 \frac{\vec{D}^2}{2m_Q} + c_4 \frac{\vec{D}^4}{8m_Q^3} + c_F g \frac{\sigma \cdot B}{2m_Q} + c_D g \frac{\vec{D} \cdot \vec{E}}{8m_Q^2}$$

$$+ i c_s g \frac{\vec{\sigma} \cdot (\vec{D} \times \vec{E} - \vec{E} \times \vec{D})}{8m_Q^2} + c_{W1} g \frac{\{\vec{D}^2, \vec{\sigma} \cdot \vec{B}\}}{8m_Q^3}$$

$$- c_{W2} g \frac{D^i \vec{\sigma} \cdot \vec{B} D_i}{4m_Q^3} + c_{p'p} g \frac{\vec{\sigma} \cdot \vec{D} \vec{B} \cdot \vec{D} + \vec{D} \cdot \vec{B} \vec{\sigma} \cdot \vec{D}}{8m_Q^3} \quad (6.3)$$

$$+ i c_M g \frac{\vec{D} \cdot (\vec{D} \times \vec{B}) + (\vec{D} \times \vec{B}) \cdot \vec{D}}{8m_Q^3} + c_{A1} g^2 \frac{\vec{B}^2 - \vec{E}^2}{8m_Q^3}$$

$$- c_{A2} g^2 \frac{\vec{E}^2}{16m_Q^3} + c_{A3} g^2 \mathrm{Tr}\left(\frac{\vec{B}^2 - \vec{E}^2}{8m_Q^3}\right) - c_{A4} g^2 \mathrm{Tr}\left(\frac{\vec{E}^2}{16m_Q^3}\right)$$

$$+ i c_{B1} g^2 \frac{\vec{\sigma} \cdot (\vec{B} \times \vec{B} + \vec{E} \times \vec{E})}{8m_Q^3} - i c_{B2} g^2 \frac{\vec{\sigma} \cdot (\vec{E} \times \vec{E})}{8m_Q^3} \Bigg] \psi,$$

This is the usual NRQCD/NRQED Lagrangian defined at the hard scale m_Q. Here indices F, S, and D denote Fermi, spin-orbit, and Darwin terms in the Lagrangian. The fields ψ are the two-component Pauli spinor fields for heavy quarks. .The covariant derivative for NRQCD is defined as $D^\mu = \partial^\mu + i g t^a A^{a\mu} = (D^0, -\vec{D})$. A much simpler NRQED Lagrangian can be obtained from Eq. (6.3)[2]. Similar term can be written down for the heavy antiquark fields that we denote by χ. The coefficients c_i can be determined by matching to full QCD Lagrangian which, just like in previous chapters, can be done by comparing amputated Green's functions in full and effective theories.

Let us sketch how the matching of QCD and NRQCD takes place. First of all, reparameterization invariance implies that there are six linear rela-

---
[2]For the electron set $g \to -e$. Also, A^μ is no longer a matrix in NRQED

tions among the Wilson coefficients [Manohar (1997)],

$$c_2 = c_4 = 1, \quad c_S = 2c_F - 1$$
$$c_{W2} = 2c_{W1} - 1, \quad c_{p'p} = c_F - 1, \quad 2c_M = c_F - c_D. \tag{6.4}$$

At tree level, matching coefficients can be obtained by simply performing a non-relativistic expansion of the QCD Lagrangian,

$$c_F = c_D = c_S = c_{W1} = 1,$$
$$c_{W2} = c_{p'p} = c_M = 0. \tag{6.5}$$

One loop corrections in the two-quark sector of the NRQCD Lagrangian can be done by computing one-loop corrections to various Green's functions.

Let us sketch how this can be done. As in previous chapters, matching of QCD and NRQCD is done by matching the amputated Green's functions. The matching scheme can be approximately expressed as

Diagrams in QCD = Diagrams in NRQCD
$$+ c_i \times \text{(matrix elements of NRQCD operators)}$$

What should we expect from the NRQCD side of this equation? As can be seen from Appendix C, scaleless integrals vanish in dimensional regularization. Thus, all diagrams on the NRQCD side must have the form

$$A_{\text{eff}} \times \left(\frac{1}{\epsilon_{UV}} - \frac{1}{\epsilon_{IR}}\right), \tag{6.6}$$

where A is some combination of parameters, and we introduced separate regulators for IR ϵ_{IR} and UV ϵ_{UV} in dimensional regularization, just like in Appendix C. Recall that it is appropriate to set $\epsilon_{UV} = \epsilon_{IR}$ in dimensional regularization. Thus, the NRQCD side of the matching would look like

$$A_{\text{eff}} \times \left(\frac{1}{\epsilon_{UV}} - \frac{1}{\epsilon_{IR}}\right) + c_i, \tag{6.7}$$

On the QCD side the integrals have an internal mass scale m_Q. Thus, in general, the QCD side of the matching condition would have a form,

$$\frac{A}{\epsilon_{UV}} + \frac{B}{\epsilon_{IR}} + (A+B)\log\frac{\mu}{m_Q} + D. \tag{6.8}$$

We remind the reader that the form of Eq. (6.8) comes from our use of dimensional regularization.

Since full and effective theories are, in general, written in terms of different degrees of freedom, their UV properties are expected to be different. Thus, $A \neq A_{\text{eff}}$. This, however, has no effect on our results whatsoever

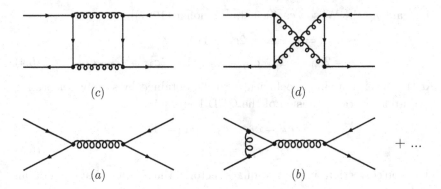

Fig. 6.2 A subset of diagrams that represents tree-level matching for annihilation operators $\mathcal{L}_{4F}$.

because, just like in Chapter 4, we elect to remove all UV divergencies by renormalization of the appropriate parameters of both full and effective theories. However, QCD and NRQCD must have the same infrared properties (as one is the low-energy approximation of the other), so $B = -A_{\text{eff}}$. Taking these two facts into account we obtain the matching condition by comparing Eqs. (6.7) and (6.8) [Manohar (1997)],

$$c_i = (A+B)\log\frac{\mu}{m_Q} + D. \tag{6.9}$$

All one-loop corrections to the matching coefficients can be obtained this way. Their exact form is a bit cumbersome, but can be readily found in [Manohar (1997)].

While high-frequency modes have been removed from the theory, we must still account for the annihilation of quark-antiquark pairs in heavy quarkonium states[3]. In order to make sure that our NRQCD Lagrangian correctly reproduces quark-antiquark annihilation that is present in QCD, we need to introduce local operators that describe that physics. In particular, quark-antiquark annihilation will be accomplished by local four fermion operators

$$\mathcal{L}_{4F} = \frac{f_1(^1S_0)}{m_Q^2}O_1(^1S_0) + \frac{f_1(^3S_1)}{m_Q^2}O_1(^3S_1)$$
$$+ \frac{f_8(^1S_0)}{m_Q^2}O_8(^1S_0) + \frac{f_8(^3S_1)}{m_Q^2}O_8(^3S_1), \tag{6.10}$$

---

[3] We remind the reader that we did not "remove antiparticle degrees of freedom" when we switched to non-relativistic effective theory. Both particles and antiparticles exist in the theory, but the pair creation is suppressed by inverse powers of m_Q.

where the operators themselves are

$$O_1(^1S_0) = \psi^\dagger \chi \chi^\dagger \psi,$$
$$O_1(^3S_1) = \psi^\dagger \vec{\sigma} \chi \chi^\dagger \vec{\sigma} \psi, \quad (6.11)$$
$$O_8(^1S_0) = \psi^\dagger t^a \chi \chi^\dagger t^a \psi,$$
$$O_8(^3S_1) = \psi^\dagger \vec{\sigma} t^a \chi \vec{\sigma} t^a \chi^\dagger \psi.$$

The Wilson coefficients can be obtained by matching diagrams of the type displayed in Fig. 6.2. Notice that only $f_8(^3S_1)$ is non-zero at tree level (Fig. 6.2a). Other coefficients will receive non-zero corrections at one- or two-loop levels

$$f_1(^1S_0) = \alpha_s^2 C_F \left(\frac{C_A}{2} - C_F\right)(2 - 2\log 2 + i\pi)$$

$$f_1(^3S_1) = 0,$$

$$f_8(^1S_0) = \frac{\alpha_s^2}{2}\left(-\frac{3C_F}{2} + 4C_F\right)(2 - 2\log 2 + i\pi), \quad (6.12)$$

$$f_8(^3S_1) = -\pi\alpha_s \left[1 + \frac{\alpha_s}{\pi}\left(\frac{n_f T_F}{3}\left(2\log 2 - \frac{5}{3} - i\pi\right) - \frac{8T_F}{9}\right) + \frac{109 C_A}{36} - 4C_F\right],$$

where the group theory factors C_F, C_A and T_F are defined in Appendix A. Notice that these coefficients have imaginary parts: this will be important later. With these Wilson coefficients determined we can now perform several practical calculations, if we know how to properly calculate matrix elements of four-fermion operators.

6.3 Going down: non-perturbative scales $m_Q v$ and $m_Q v^2$

Let us now consider the case when all NRQCD scales, except for the scale m_Q, are non-perturbative. That is to say, while the hierarchy of scales $m_Q \ll m_Q v < m_Q v^2 < \Lambda_{QCD}$ still exists, the scales $m_Q v$ and $m_Q v^2$ are too low to perform perturbative matching. This happens often when we deal with bound states of charm quarks. How should one deal with this situation?

To resolve this, let us recall what we did when we performed matching in HQET. To start off with a concrete example, let us concentrate on an example of inclusive annihilation of a charmonium state H_c. The total width can be written in the form of a factorization theorem,

$$\Gamma(H_c \to X) = \sum_n \frac{C_n}{m_Q^{d_n-4}} \langle H_c | \mathcal{O}_n^{H_c}(^{2S+1}L_J) | H_c \rangle. \qquad (6.13)$$

Since the scale m_Q is always perturbative for heavy quarks, Eq. (6.13) simply states that all short distance effects ($\mu > m_Q$) are encoded in the short-distance coefficients C_n, while all scales below m_Q are contained in the matrix elements $\langle H_c | \mathcal{O}_n | H_c \rangle$. This is similar to the situation discussed in Sect. 6.2.

We can now establish a power counting scheme for the *matrix elements* in Eq. (6.13). This can be done most simply from the following arguments [Lepage *et al.* (1992)]. Consider the number operator

$$\int d^3x \ \psi^\dagger(x)\psi(x). \qquad (6.14)$$

For the leading, "baseline," Fock state this operator counts the number of heavy quarks inside the bound state. Since we removed heavy quark pair creation at leading order this number should be very close to one. Now, since the heavy quarkonium size is determined by the heavy quarks' momenta (uncertainty principle), $\int d^3x \sim 1/(m_Q v)^3$. Thus,

$$\psi^\dagger(x)\psi(x) \sim (m_Q v)^3, \quad \text{or} \quad \psi \sim (m_Q v)^{3/2}. \qquad (6.15)$$

Similar arguments allow us to determine power counting of derivatives and other elements. For example, we should expect that kinetic energy operator of a heavy quark should scale like kinetic energy, i.e.

$$\int d^3x \ \psi^\dagger(x) \frac{\vec{D}^2}{2m_Q} \psi(x) \sim m_Q v^2, \qquad (6.16)$$

which immediately provides power counting for a derivative, $\vec{D} \sim m_Q v$, as expected. Note that time derivative power counts as $D_t \sim m_Q v^2$ because of the leading order equation of motion, $(iD^0 + \vec{D}/(2m_Q))\psi = 0$. Opening up the definitions of covariant derivative we can power count QCD/QED potentials and fields. The results can be seen in Table 6.1.

A convenient way to power-count other matrix elements is to set up their v scaling relative to the "baseline value" described above. This can be done in two steps. First, one needs to assign v counting of the operators. To do so, let us first define *baseline* operators as the operators that contain quark bilinears that project out the *predominant* $Q\bar{Q}$ component of the quarkonium state[4]. For example, the baseline bilinears for the η_c

---
[4] We define $Q\bar{Q}$ state as predominant component when its spin and angular momentum state mimics that of the quarkonium state.

Table 6.1 NRQCD/NRQED power counting of fields, derivatives and potentials (in Coulomb gauge).

Operator	Description	Scaling
ψ	Quark field	$(m_Q v)^{3/2}$
χ	Antiquark field	$(m_Q v)^{3/2}$
D^0	Gauge-covariant time derivative	$m_Q v^2$
$\vec{D}$	Gauge-covariant space derivative	$m_Q v$
gA_0	Scalar potential[a]	$m_Q v^2$
$g\vec{A}$	Vector potential[a]	$m_Q v^3$
$g\vec{E}$	(Chromo)electric field	$m_Q^2 v^3$
$g\vec{B}$	(Chromo)magnetic field	$m_Q^2 v^4$

[a]In Coulomb gauge.

state ($^1S_0$) would be $\psi^\dagger \chi$, the ones for the J/ψ state ($^3S_1$) would be $\psi^\dagger \vec{\sigma}\chi$, etc. This allows us to define scaling rules for the four-fermion operators. Factorizing matrix elements of four-fermion operators into the products of quark bilinears we can immediately classify operators according to their v-scaling. For example,

$$\langle \eta_c | \mathcal{O}_1^{\eta_c}(^1S_0)|\eta_c\rangle = \langle \eta_c|\psi^\dagger\chi|0\rangle\langle 0|\chi^\dagger\psi|\eta_c\rangle + \ldots$$
$$\langle J/\psi|\mathcal{O}_1^{\eta_c}(^3S_1)|J/\psi\rangle = \langle J/\psi|\psi^\dagger\vec{\sigma}\chi|0\rangle\langle 0|\chi^\dagger\vec{\sigma}\psi|J/\psi\rangle + \ldots \quad (6.17)$$

define the baseline operators for the s-wave states J/ψ and η_c. Insertion of the scalar operator $\vec{D}^2/m_c^2$ on either side of the bilinear operator would result in additional v^2 suppression of the resulting matrix elements. Similarly,

$$\langle h_{m_J}|\mathcal{O}_1^h(^1P_1)|h_{m_J}\rangle = \langle h_{m_J}|\psi^\dagger\chi|0\rangle\langle 0|\chi^\dagger\psi|h_{m_J}\rangle + \ldots$$
$$\langle \chi_0|\mathcal{O}_1^{\chi_0}(^3P_0)|\chi_0\rangle = \frac{1}{3}\langle \chi_0|\psi^\dagger\left(-i\vec{\sigma}\cdot\overleftrightarrow{\vec{D}}\right)\chi|0\rangle\langle 0|\chi^\dagger\left(-i\vec{\sigma}\cdot\overleftrightarrow{\vec{D}}\right)\psi|\chi_0\rangle + \ldots$$
(6.18)

define baseline operators for the p-wave states. In Eq. (6.18) we defined $\overleftrightarrow{\vec{D}} = (1/2)\left(\vec{\vec{D}} - \overleftarrow{\vec{D}}\right)$. Just like above, insertion of of the scalar operator $\vec{D}^2/m_c^2$ on either side of the bilinear operator would result in additional v^2 suppression of the resulting matrix elements. Additionally, matrix elements of Eq. (6.18) are roughly suppressed by v^2 compared to the matrix elements of Eq. (6.17). Notice that the ellipses in Eq. (6.17) and Eq. (6.18) denote terms suppressed by v^2.

One important point about heavy quarkonium is that the *states* can also be expanded in velocity. Since the heavy quarkonium state is "compact," it is possible to power count each term in a Fock state expansion of the hadronic state. For example, a J/ψ meson, which in a potential quark model can be described as a $^3S_1$ state of charm quark and an antiquark, has a Fock state expansion,

$$|J/\psi\rangle = A_{c\bar{c}}|c\bar{c}(^3S_1(1))\rangle + A_{c\bar{c}g}|c\bar{c}(^3P_J(8))g\rangle$$
$$+ A_{c\bar{c}gg}|c\bar{c}(^3S_1(8))gg\rangle + B_{c\bar{c}gg}|c\bar{c}(^3S_1(1))gg\rangle$$
$$+ C_{c\bar{c}gg}|c\bar{c}(^3D_J(8))gg\rangle + D_{c\bar{c}gg}|c\bar{c}(^3D_1(1))gg\rangle$$
$$+ B_{c\bar{c}g}|c\bar{c}(^1S_0(8))g\rangle + \ldots . \qquad (6.19)$$

Here g denotes the *dynamical* i.e. non-potential ultrasoft gluon. Similar expansions can be done for other quarkonium states. In Eq. (6.19) we used spectroscopic notations to denote angular momentum-spin state of $Q\bar{Q}$ (for example, $^3S_1$ states that the $Q\bar{Q}$ is in the relative s-wave spin-triplet state), and 1(8) to denote singlet (octet) color state of $Q\bar{Q}$. While for the light quark states (and also for a heavy-light states) this expansion is not very useful, the fact that $Q\bar{Q}$ systems sit at the bottom of the QCD Coulomb potential makes the distance between the heavy quarks small compared to the distance scale set by soft QCD effects. Thus, a multipole expansion makes it possible to assign v-scaling to various coefficients of the Fock state expansion, which in turn affects power counting of NRQCD matrix elements in a major way.

Let us use a simple argument [Fleming and Maksymyk (1996)] to power count the coefficients A, B, C and D in Eq. (6.19). For the J/ψ state in our example, the "baseline" $Q\bar{Q}$ state has $c\bar{c}$ in the $^3S_1$ state. Thus, we expect that $A_{c\bar{c}} \sim v^0$. Radiating a non-potential gluon can be done in two different ways. First, one can radiate an *electric-type* gluon, via chromoelectric dipole operator, that would change angular momentum $L \to L'$ by one unit without changing the relative spin state of the $c\bar{c}$ pair, i.e. $L' = L \pm 1$ and $S' = S$. This process scales with a single power of the heavy quark's momentum, so $A_{c\bar{c}g} \sim v$. Also, $c\bar{c}$ is now in a relative color octet state to maintain the overall color-neutrality of the bound state. Emission of another electric-type gluon would further change angular momentum by another unit, in either direction, and might or might not change the color state of $c\bar{c}$. The corresponding coefficients receive another power of velocity suppression, i.e. $A_{c\bar{c}gg}, B_{c\bar{c}gg}, C_{c\bar{c}gg}, D_{c\bar{c}gg} \sim v^2$. Alternatively, a *magnetic-type* gluon can be emitted from the $^3S_1(1)$ state. Emission of such gluon leads to a spin flip of one of the heavy quarks. Since this process is mediated by a chromo magnetic dipole transition with $L' = L$ and $S' = S \pm 1$. This process would scale with the momentum of a gluon, i.e. $B_{c\bar{c}g} \sim v^2$.

Assembling together the power counting of operators and states gives us relative velocity scaling rules for the matrix elements, which enables a non relativistic expansion of matrix elements. Since in both quarkonium annihilation or production, only a finite number of matrix elements would contribute at each order in v, we have a predictive theory. Of course this is only so if a finite number of experiments is performed to fix those universal non relativistic matrix elements.

Additional simplification and reduction of the number of relevant matrix elements comes from heavy quark spin symmetry relations. Since heavy-quark spin-flipping interactions are governed by a chromo magnetic dipole transition, leading order matrix elements for the same angular momentum (but different spin) states would be the same. Moreover, one can even estimate how good the symmetry relation would be! This works for matrix elements of both color-singlet and color-octet operators. In particular,

$$\langle J/\psi|\mathcal{O}_1^{\eta_c}(^3S_1)|J/\psi\rangle = \langle \eta_c|\mathcal{O}_1^{\eta_c}(^1S_0)|\eta_c\rangle + \mathcal{O}(v^2),$$

$$\langle \chi_J|\mathcal{O}_1^{\chi_J}(^1P_J)|\chi_J\rangle = \langle h_0|\mathcal{O}_1^h(^1P_1)|h_0\rangle + \mathcal{O}(v^2), \quad J = 0,1,2, \quad (6.20)$$

$$\langle \chi_J|\mathcal{O}_8^{\chi_J}(^3S_1)|\chi_J\rangle = \langle h_0|\mathcal{O}_8^h(^1S_0)|h_0\rangle + \mathcal{O}(v^2), \quad J = 0,1,2.$$

Similar relations can be found for other states as well.

To show the pros of NRQCD, let us consider three examples of inclusive annihilation of s and p-wave quarkonium states. We shall concentrate on the decays of the charmonium states, where velocity power counting works (as $v^2 \simeq 0.3$), but $m_c v$ is not large compared to Λ_{QCD}.

6.3.1 Electromagnetic decays of the η_c

Let us first consider the simplest case of $\eta_c \to \gamma\gamma$. We argue that the decay amplitude can be factorized into a convolution of two processes: the scattering of two charm quarks into a final state $\gamma\gamma$ (hard process), and an amplitude for finding those two quarks in the same point inside the η_c (soft process).

It is possible to argue for this order-by-order in perturbative QCD. However, it would be sufficient for our purposes to simply illustrate this fact by Fig. 6.3. Indeed, while the average momentum of the charm quarks in the η_c bound state is of order $m_c v$, the momentum flowing through the propagator scales with m_c, so in the limit $m_c \to \infty$ (and $v \to 0$) soft charm

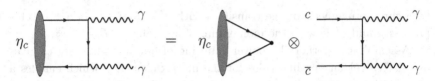

Fig. 6.3 Factorization of one of the two possible diagrams for the electromagnetic decay $\eta_c \to \gamma\gamma$. The second graph involves crossing the final state photon lines.

quarks in the bound state cannot resolve the details of the virtual c-quark in the propagator. Thus, it can be shrunk to a point, and the *decay amplitude* can be factorized,

$$\mathcal{A}(\eta_c \to \gamma\gamma) = \langle 0|\chi^\dagger \psi(0)|\eta_c\rangle \times \mathcal{A}(\bar{c}c \to \gamma\gamma). \qquad (6.21)$$

Here the first part on the right hand side represents a convolution of the *soft part* representing physics of the $\bar{c}c$ bound state, and the *hard part* representing hard scattering. In our example the convolution is simply a product of two amplitudes. We shall see later that *amplitude* factorization is not always possible in NRQCD.

Writing the Feynman diagrams in Fig. 6.3, expanding the 4-component spinors into the 2-component ones, and retaining the leading part in p/m_c yields

$$\Gamma(\eta_c \to \gamma\gamma) = \frac{32\pi\alpha^2}{81 m_c^2} \left|\langle 0|\chi^\dagger \psi(0)|\eta_c\rangle\right|^2. \qquad (6.22)$$

In the quark model, the matrix elements in Eq. (6.22) can be related to the values of quarkonium wave functions at the origin via

$$\left|\langle 0|\chi^\dagger \psi(0)|\eta_c\rangle\right|^2 = \frac{3}{2\pi} |R(0)|^2, \qquad (6.23)$$

where $R(r)$ is quarkonium's radial wave function.

6.3.2 Inclusive decays of the η_c into light hadrons

In order to calculate the inclusive rate of η_c annihilation into the light hadrons we can employ quark-hadron duality. That is to say, the sum over all possible light hadron states would be equivalent to the sum over quark and gluon states. In the limit $m_c \to \infty$ the sum over quark and gluon states is dominated by a rather simple two gluon state. Yet, even though Feynman graphs for electromagnetic annihilation and inclusive decay look similar, there is a huge difference between them: hadronic decay diagrams do not factorize at the amplitude level, as soft gluons emitted from the initial state charmed quarks can interact with the final state hadrons.

This stumbling block can be overcome with help from the Kinoshita-Lee-Nauenberg (KLN) theorem that states that infrared divergences from the coupling of soft gluons to jets cancel when summed over all possible final states, i.e. soft gluons decouple from jets. This means that factorization can be expected at the decay *rate* level, i.e. at the level of the squared amplitudes.

With this realization, the annihilation rate calculation can be done directly in NRQCD with the use of the optical theorem. In order to calculate the inclusive annihilation width for heavy quarkonia in NRQCD we can use the fact that the total decay rate is -2 times the imaginary part of the energy of the state, as follows from unitarity. Since $\mathcal{H}_{\text{eff}} = -\mathcal{L}_{\text{eff}}$ we can calculate the energy of the η_c state,

$$E - \frac{i}{2}\Gamma(H_c \to X) = -\langle H_c|\mathcal{L}_{\text{eff}}|H_c\rangle. \tag{6.24}$$

A non-zero contribution to the annihilation partial decay width comes from the 4-fermion part of the effective NRQCD Lagrangian in Eq. (6.10), i.e.

$$\Gamma(H_c \to X) = 2\text{Im}\langle H_c|\mathcal{L}_{4F}|H_c\rangle. \tag{6.25}$$

Let us choose a non-relativistic normalization of states, $\langle H_c(\vec{P}')|H_c(\vec{P})\rangle = (2\pi)^3\delta^3(\vec{P}' - \vec{P})$, and go to the rest frame of the decaying charmonium state, so $|H_c\rangle = |H_c(\vec{P}=0)\rangle$.

A particular case of Eq. (6.25) for the inclusive annihilation of the η_c is

$$\Gamma(\eta_c \to X) = \frac{2\text{Im} f_1(^1S_0)}{m_c^2}\langle \eta_c|O_1(^1S_0)|\eta_c\rangle, \tag{6.26}$$

where, as we recall from Eq. (6.11), $O_1(^1S_0) = \psi^\dagger\chi\chi^\dagger\psi$. Also, from Eq. (6.12)

$$\text{Im} f_1(^1S_0) = \alpha_s^2 C_F \left(\frac{C_A}{2} - C_F\right)\pi = \frac{2\pi\alpha_s^2}{9}. \tag{6.27}$$

Finally, as we already know from Eq. (6.17), to leading order in v we can factorize

$$\langle \eta_c|O_1^{\eta_c}(^1S_0)|\eta_c\rangle = \left|\langle 0|\chi^\dagger\psi(0)|\eta_c\rangle\right|^2 \tag{6.28}$$

and thus relate, if we choose, $\langle \eta_c|O_1(^1S_0)|\eta_c\rangle$ to the value of η_c's wave function at origin. This is usually not needed, as the matrix elements work perfectly fine as independent universal parameters. Thus, at leading order in heavy quark velocity,

$$\Gamma(\eta_c \to X) = \frac{4\pi\alpha_s^2}{9m_c^2}\langle\eta_c|\mathcal{O}_1^{\eta_c}(^1S_0)|\eta_c\rangle. \qquad (6.29)$$

This result has the standard form of a factorization theorem, with the long-distance physics collected in the unknown matrix element. Thus, if we would like to improve the accuracy of our predictions, there are two types of corrections to this formula. One type results from the $\mathcal{O}(\alpha_s^n)$ corrections to the short-distance coefficients f_i and can be calculated in QCD perturbation theory. The other type results from the long distance physics. Those corrections are suppressed by powers of heavy quark velocity v. For example, including all long-distance corrections up to order $\mathcal{O}(v^5)$, the inclusive annihilation rate is

$$\begin{aligned}\Gamma(\eta_c \to X) = &\frac{2\mathrm{Im}f_1(^1S_0)}{m_c^2}\langle\eta_c|\mathcal{O}_1(^1S_0)|\eta_c\rangle + \frac{2\mathrm{Im}f_8(^1S_0)}{m_c^2}\langle\eta_c|\mathcal{O}_8(^1S_0)|\eta_c\rangle \\ &+ \frac{2\mathrm{Im}f_8(^3S_1)}{m_c^2}\langle\eta_c|\mathcal{O}_8(^3S_1)|\eta_c\rangle \qquad (6.30) \\ &+ \frac{2\mathrm{Im}g_1(^1S_0)}{m_c^4}\langle\eta_c|\mathcal{P}_1(^1S_0)|\eta_c\rangle + \frac{2\mathrm{Im}f_8(^1P_1)}{m_c^4}\langle\eta_c|\mathcal{O}_8(^1P_1)|\eta_c\rangle.\end{aligned}$$

Eq. (6.30) contains more non-perturbative parameters that need to be determined elsewhere.

6.3.3 Inclusive decays of the χ_{cJ} into light hadrons

One of the best illustrations of the power of NRQCD factorization techniques involve decays of the p-wave quarkonium states, such as χ_{cJ} for $J = 0, 1, 2$. χ_{cJ} is a bound state of two charmed quarks in a relative p-wave, or ^3P_J state [Bodwin et al. (1992)]. If we are to calculate the inclusive decay rates for those states in the quark model, they will be proportional to the value of the *derivative* of the $c\bar{c}$ wave function at the origin, rather than the value of the wave function itself. Taking, for example, the χ_{c0} state,

$$\Gamma(\chi_{c0} \to X) = \left[6\alpha_s^2 + \left(\frac{8n_f}{9\pi}\log\frac{m_c}{\mu} + ...\right)\alpha_s^3\right]\frac{|R'(0)|^2}{m_c^4}, \qquad (6.31)$$

where $R'(0)$ is the derivative of the radial wave function of the $c\bar{c}$ state. Eq. (6.31) has a problem: the computed rate contains infrared divergence at $\mathcal{O}(\alpha_s^3)$! This divergence is cut off by imposing an infrared scale μ.

A solution can be found in the framework of NRQCD. Since

$$R'(0) \sim \langle 0|\chi^\dagger \vec{\sigma} \cdot \overleftrightarrow{D} \psi|\chi_0\rangle, \tag{6.32}$$

one can forgo calculation of $R'(0)$, working instead directly with matrix elements of NRQCD operators. In particular,

$$|R'(0)|^2 \sim \langle \chi_{cJ}|\mathcal{O}_1^{\chi_{cJ}}(^1P_J)|\chi_{cJ}\rangle \tag{6.33}$$

Power counting rules immediately tell us that $\langle \chi_J|\mathcal{O}_1^{\chi_J}(^1P_J)|\chi_J\rangle \sim \mathcal{O}(v^5)$, as each field scales as $v^{3/2}$ and each derivative scales as v, so $4 \times (3/2) + 2 \times 1 - 3 = 5$.

Now it turns out that $\langle \chi_{cJ}|\mathcal{O}_8^{\chi_{cJ}}(^3S_1)|\chi_{cJ}\rangle$ would also scale as $\mathcal{O}(v^5)$! This follows from the fact that while a color-octet s-wave state of $c\bar{c}$ quarks only contributes to $|\chi_{c0}\rangle$ via higher Fock state $|c\bar{c}(^3S_1(8))g\rangle$, i.e. suppressed by v^2 compared to the "baseline," the operator for $c\bar{c}(^3S_1(8))$ does *not* contain any derivatives. Thus, it is effectively enhanced by two powers of v. This offers an elegant solution to the problem of the infrared divergence of $\Gamma(\chi_{c0} \to H)$ that we discovered in the quark model. In NRQCD both operators must be taken into account. Moreover, the infrared divergence of the α_s^3 term can be absorbed into the matrix element of the $\mathcal{O}_8^{\chi_J}(^3S_1)$ operator! Thus, in NRQCD,

$$\Gamma(\chi_{c0} \to X) = \frac{2\mathrm{Im} f_1(^3P_J)}{m_c^4}\langle \chi_{c0}|\mathcal{O}_1(^3P_J)|\chi_{c0}\rangle$$

$$+ \frac{2\mathrm{Im} f_8(^3S_1)}{m_c^2}\langle \chi_{c0}|\mathcal{O}_8(^3S_1)|\chi_{c0}\rangle. \tag{6.34}$$

Notice that, contrary to HQET, it is the powers of heavy quark *velocity* that determine the power counting. In particular, both terms in Eq. (6.34) scale the same in the velocity power counting even though the first term has additional suppression by m_c^2.

Now, reading off the coefficients f_i from Eq. (6.12) we obtain,

$$\Gamma(\chi_{c0} \to X) = \frac{4\pi\alpha_s^2(m_c)}{3m_c^4}\langle \chi_{c0}|\mathcal{O}_1^{\chi_{c0}}(^3P_J)|\chi_{c0}\rangle$$

$$+ \frac{n_f \pi \alpha_s^2(m_c)}{6m_c^2}\langle \chi_{c0}|\mathcal{O}_8^{\chi_{c0}}(^3S_1)|\chi_{c0}\rangle, \tag{6.35}$$

which now contains two unknown parameters, but no infrared divergences! The unknown matrix elements can be either calculated with some non-perturbative methods (such as lattice QCD), or fitted from the data. In the later method one can rely on heavy quark spin symmetry relations that relates matrix elements of χ_{c0} to the matrix elements of other p-wave states χ_{cJ}. Global fit yields

$$\langle \chi_{cJ} | \mathcal{O}_1^{\chi_{cJ}}(^3P_J) | \chi_{cJ} \rangle = (7.2 \pm 0.9) \times 10^{-2} \text{ GeV}^5,$$

$$\langle \chi_{cJ} | \mathcal{O}_8^{\chi_{cJ}}(^3S_1) | \chi_{cJ} \rangle = (4.3 \pm 0.9) \times 10^{-2} \text{ GeV}^3. \qquad (6.36)$$

NRQCD factorization theorems allow for a consistent description of not only heavy quarkonium decays, but also production, both in pp and $p\bar{p}$ collisions, as well as in heavy ion collisions. While some problems in the description of p_T distributions and polarization parameters still exist at the time of the writing of this book, NRQCD has proved to be a powerful technique for calculating processes with heavy quarkonium states.

6.4 Going down: perturbative scales $m_Q v$ and $m_Q v^2$. pNRQCD

For large m_Q the physical quarkonium scales $m_Q v$ and $m_Q v^2$ can still be perturbative. In other words, there might still exist a perturbative matching scale μ such that $m_Q v \ll \mu \ll m_Q v^2, \Lambda_{QCD}$. If this is the case, the soft scale $m_Q v$ can also be integrated out. The emerging theory is then called *potential NRQCD* or pNRQCD [Pineda and Soto (1998); Brambilla et al. (2000)]. This is a theory of $Q\overline{Q}$ bound states and ultrasoft gluons with energies about $m_Q v^2$.

The typical size of a $Q\overline{Q}$ bound state is $r \sim 1/m_Q v$. This distance scale is small compared to the typical wave lengths of ultrasoft gluons, so we can systematically expand the gluon fields in r in a fashion similar to the standard multipole expansion in electrodynamics. We introduce new interpolating fields for the $Q\overline{Q}$ states, which we decompose in terms of a *singlet* **S** and an *octet* **O** fields. A typical energy associated with S and O is mv^2. Finally, we denote by $\vec{R}$ the coordinate of the center of mass of the $Q\overline{Q}$ state and by $\vec{r}$ their relative coordinate. Upon performing a multipole expansion we can write a pNRQCD effective Lagrangian in terms of **S**, **O**, and ultrasoft gluons. All non-analytic terms in r will be encoded in the matching coefficients of the pNRQCD Lagrangian, which can be interpreted as the quark-anti-quark potential.

At the leading order the most general effective Lagrangian for pNRQCD can be written as

$$\mathcal{L}_{\text{pNRQCD}} = \text{Tr}\left[\mathbf{S}^\dagger\left(i\partial_0 - \frac{\vec{p}^2}{m_Q} - V_s(r) + \cdots\right)\mathbf{S}\right]$$
$$+ \text{Tr}\left[\mathbf{O}^\dagger\left(iD_0 - \frac{\vec{p}^2}{m_Q} - V_o(r) + \cdots\right)\mathbf{O}\right] \quad (6.37)$$
$$+ gV_A(r)\text{Tr}\left[\mathbf{O}^\dagger\vec{r}\cdot\vec{E}\mathbf{S} + \mathbf{S}^\dagger\vec{r}\cdot\vec{E}\mathbf{O}\right]$$
$$+ \frac{g}{2}V_B(r)\text{Tr}\left[\mathbf{O}^\dagger\vec{r}\cdot\vec{E}\mathbf{O} + \mathbf{O}^\dagger\mathbf{O}\,\vec{r}\cdot\vec{E}\right],$$

where the Wilson coefficients $V_s(r)$ and $V_o(r)$ are the singlet and octet $Q\overline{Q}$ static potentials, respectively. We shall see how to calculate $V_s(r)$ in the next section. Before we do so, however, let us define our expectations for the form of $V_s(r)$ and $V_o(r)$,

$$V_s(r) = -C_F\frac{\alpha_{V_s}}{r}, \quad V_o(r) = \left(\frac{C_A}{2} - C_F\right)\frac{\alpha_{V_o}}{r}, \quad (6.38)$$

where, at the lowest order in α_s we should expect that $\alpha_{V_s} = \alpha_{V_o} = \alpha_s$, and $V_A = V_B = 1$.

Let us note that in practical calculations it might be convenient to redefine singlet and octet fields of pNRQCD so that they have a proper normalization in the color space [Pineda and Soto (1998); Brambilla et al. (2000)],

$$\mathbf{S} = \frac{1_c}{\sqrt{N_c}}S, \quad \mathbf{O} = \frac{T^a}{\sqrt{T_F}}O^a, \quad (6.39)$$

where $T_F = 1/2$, as defined in Appendix A. The Feynman rules for those fields are available in [Brambilla et al. (2000)].

6.4.1 Example: heavy quarkonium potential

We denoted the pNRQCD matching coefficients $V_s(r)$ and $V_o(r)$ as potentials in the singlet and octet channels. It can be seen that this is indeed so! Applying the Euler-Lagrange equations to Eq. (6.37) for the singlet field S we get the

$$i\partial_0 S = \left(\frac{\vec{p}^2}{m_Q} + V_s(r)\right)S \quad (6.40)$$

as the equation of motion for S. Interpreting this as a Schrödinder equation for S, we can indeed identify the Wilson coefficient $V_s(r)$ as a static potential.

The pNRQCD Lagrangian should be matched to the four-fermion part of the NRQCD Lagrangian, so we would need to define how we calculate the appropriate Green's functions. So far we glossed over the coordinate dependence of the NRQCD and pNRQCD operators such as $\mathbf{S}(\vec{R},\vec{r})$ and $\mathbf{O}(\vec{R},\vec{r})$, but now we need to keep it explicit, as we need to match the four-body sector of NRQCD to the two-body sector of pNRQCD.

Let us concentrate on the calculation of the potential in the singlet channel and make the following identification,

$$\chi^\dagger(x_2,t)U(x_2,x_1,t)\psi(x_1,t) = Z_s^{1/2} S(\vec{R},\vec{r},t). \qquad (6.41)$$

Here we made the coordinate dependence of the fields explicit and defined a Wilson line

$$U(x_2,x_1,t) = P \exp\left(ig \int_0^1 ds\, (\vec{x}_2 - \vec{x}_1) \cdot \vec{A}(\vec{x}_1 - s(\vec{x}_1 - \vec{x}_2),t)\right) \qquad (6.42)$$

to make sure that we are dealing with gauge-invariant objects. As usual, P denotes path-ordering over the length segment s. Also, $\vec{R} = \vec{x}_2 + \vec{x}_1$, while $\vec{r} = \vec{x}_2 - \vec{x}_1$. We will match while simultaneously expanding in r/R, which is equivalent to the multipole expansion in electrodynamics. Also, we shall suppress explicit time reference in what follows.

Recall that when performing a matching calculation, it is irrelevant what external states are used for the procedure, as long as the quantum numbers of the corresponding operators match[5]. Calculating singlet Green's functions $G^{(s)} = \langle S|S\rangle$ in NRQCD and pNRQCD (with identification provided by Eq. (6.41)), we can obtain both $V_s(r)$ and Z_s. In NRQCD,

$$\begin{aligned}G^{(s)}_{\text{NRQCD}} &= \langle 0|\chi^\dagger(x_2)U(x_2,x_1)\psi(x_1)\psi^\dagger(y_1)U(y_1,y_2)\chi(y_2)|0\rangle \\ &= \delta^3(\vec{x}_1 - \vec{y}_1)\delta^3(\vec{x}_2 - \vec{y}_2)\langle W_\square\rangle,\end{aligned} \qquad (6.43)$$

where $\langle W_\square\rangle$ is the rectangular Wilson loop averaged over gauge fields and light quarks, and we only retained the leading, $\mathcal{O}(1/m_Q^0)$ term. The coordinates of the angles of the box are $x_1 = (T/2,\vec{r}/2)$, $x_2 = (T/2,-\vec{r}/2)$, $y_1 = (-T/2,\vec{r}/2)$, and $y_2 = (-T/2,-\vec{r}/2)$. In accordance with [Fischler (1977); Appelquist et al. (1977); Eichten and Feinberg (1981)], we will need a large-time asymptotic $T \to \infty$ in order to extract the potential.

In pNRQCD we obtain [Pineda and Soto (1998); Brambilla et al. (2000)]

$$G^{(s)}_{\text{pNRQCD}} = Z_s^{1/2} \delta^3(\vec{x}_1 - \vec{y}_1)\delta^3(\vec{x}_2 - \vec{y}_2) \exp(-iTV_s(r)), \qquad (6.44)$$

The matching is represented by Fig. 6.4. We only need the first term on the pNRQCD side for the leading-order matching.

---

[5] An analogue can be found in lattice formulation of QCD, where it is sufficient for lattice interpolating operators to "have some overlap" with the corresponding external states.

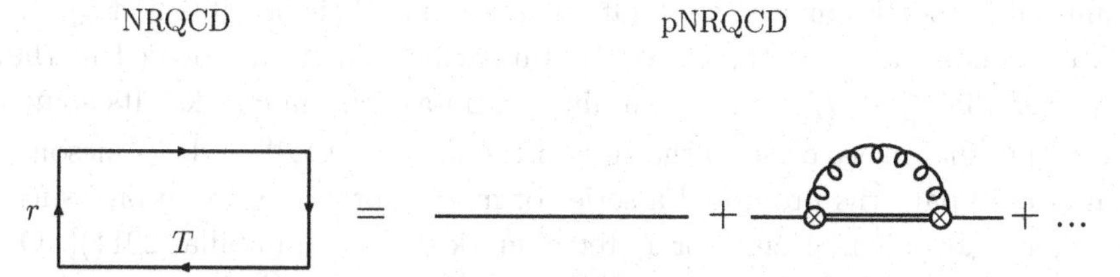

Fig. 6.4 Matching of NRQCD (left) and pNRQCD (right) Green's function at next-to-leading order in multipole expansion.

Matching Eqs. (6.43) and (6.44) in the limit $T \to \infty$ by taking the logarithm of both $G^{(s)}_{\text{NRQCD}}$ and $G^{(s)}_{\text{pNRQCD}}$ and expanding in $1/T$,

$$\frac{i}{T} \log \langle W_\Box \rangle = u_o(r) + \frac{iu_1(r)}{T} + \cdots \qquad (6.45)$$

we obtain

$$V_s(r) = u_0(r), \quad \log Z_s(r) = u_1(r). \qquad (6.46)$$

Calculating the leading order contribution in α_s we arrive at the results,

$$V_s(r) = -C_F \frac{\alpha_s}{r}, \quad Z_s(r) = N_c, \qquad (6.47)$$

which is consistent with our expectations announced in Eq. (6.38). Calculations of the octet potential, as well as next-to-leading order corrections are not as straightforward as the calculation done above. They can be found in [Pineda and Soto (1998); Brambilla et al. (2000)].

6.5 Different degrees of freedom: hadronic molecules

The first part of this chapter was concerned with the physics of $Q\bar{Q}$ bound states. This is probably the most common application of non-relativistic theories as applied to the bound states of two heavy objects. Yet, the list of non-$Q\bar{Q}$ bound states is rather long. For example, any nucleus is a bound state of several nucleons. In particular, deuterium, a very loosely bound state of a proton and a neutron, is a nuclear system where EFTs analogous to the ones described in the first part of this chapter can be applied. Indeed, effective theory techniques for such nuclei have been developed [Kaplan et al. (1996)].

Here we shall consider a simpler example, the bound state of two bosons. This example is quite interesting, as experimental data from the heavy flavor experiments BaBar and Belle revealed a host of new hadrons whose unusual properties invited speculations regarding their possible non-$q\bar{q}$ nature. Among these is the $X(3872)$ state which, being discovered in the decay $X(3872) \to J/\psi \pi^+\pi^-$, contains charm-anticharm quarks. Its somewhat unusual mass, close to the sum of the masses of D^{*0} and $\overline{D}^0$ mesons, and decay patterns prompted a series of more exotic interpretations, such as of a "hadronic molecule" or a "tetra quark state" [Brambilla (2011)]. Of course, states of different "nature" can mix if they have the same quantum numbers, further complicating the correct interpretation of experimental data. Since the mass of the $X(3872)$ state lies tantalizingly close to the $D^{*0}\overline{D}^0$ threshold of 3871.3 MeV, it is tempting to assume that $X(3872)$ could be a $D^{*0}\overline{D}^0$ $J^{PC} = 1^{++}$ molecular state, so we shall explore this to discuss possible venues for theoretical analyses of heavy-meson molecular states. In particular we shall set up the formalism by tasking ourselves with extracting the bound state energy of the $D^{*0}\overline{D}^0$ molecule.

The lowest-energy bound state of D and $\overline{D}^*$ is an eigenstate of charge conjugation. The two eigenstates of charge conjugation will be given by

$$|X_\pm\rangle = \frac{1}{\sqrt{2}} \left[|D^*\overline{D}\rangle \pm |D\overline{D}^*\rangle \right]. \qquad (6.48)$$

To find the bound-state energy of $X(3872)$ with $J^{PC} = 1^{++}$, we shall look for a pole of the transition amplitude $T_{++} = \langle X_+|T|X_+\rangle$ [AlFiky et al. (2006)].

This study is possible due to the multitude of scales present in QCD. The tiny binding energy of this molecular state introduces an energy scale which is much smaller than the mass of the lightest particle, the pion, whose exchange can provide binding. Thus, for a suitable description of this state in the framework of effective field theory, the pion, along with other particles providing possible binding contributions (i.e. the ρ-meson and other higher mass resonances), must be integrated out. The resulting Lagrangian should contain only heavy-meson degrees of freedom with interactions approximated by local four-boson terms constrained only by the symmetries of the theory.

In order to describe the molecular states of heavy mesons we need to supplement the two-body effective Lagrangian discussed in Chapter 5 with the one containing four-particle interaction terms. In order to write an effective Lagrangian describing the properties of $X(3872)$, we need to couple the fields $H_a^{(Q)}$ and $H^{(\overline{Q})a}$ so that the resulting Lagrangian respects the heavy-quark spin and chiral symmetries.

The general effective 4-body Lagrangian consistent with heavy-quark spin and chiral symmetries can be written as

$$\mathcal{L}_4 = -\frac{C_1}{4}\operatorname{Tr}\left[\overline{H}^{(Q)} H^{(Q)} \gamma_\mu\right] \operatorname{Tr}\left[H^{(\overline{Q})}\overline{H}^{(\overline{Q})} \gamma^\mu\right]$$
$$-\frac{C_2}{4}\operatorname{Tr}\left[\overline{H}^{(Q)} H^{(Q)} \gamma_\mu\gamma_5\right] \operatorname{Tr}\left[H^{(\overline{Q})}\overline{H}^{(\overline{Q})} \gamma^\mu\gamma_5\right]. \quad (6.49)$$

This Lagrangian describes the scattering of D and D^* mesons at energies above m_π. Integrating out the pion degrees of freedom at tree level corresponds to a modification $C_2' \to C_2 + (2/3)(g/f)^2$. This Lagrangian can be used for the calculation of the bound state properties of $X(3872)$.

Evaluating the traces yields for the $D\overline{D}^*$ sector

$$\mathcal{L}_{4,DD^*} = -C_1 D^{(Q)\dagger} D^{(Q)} D_\mu^{*(\overline{Q})\dagger} D^{*(\overline{Q})\mu} - C_1 D_\mu^{*(Q)\dagger} D^{*(Q)\mu} D^{(\overline{Q})\dagger} D^{(\overline{Q})}$$
$$+ C_2 D^{(Q)\dagger} D_\mu^{*(Q)} D^{*(\overline{Q})\dagger\mu} D^{(\overline{Q})} + C_2 D_\mu^{*(Q)\dagger} D^{(Q)} D^{(\overline{Q})\dagger} D^{*(\overline{Q})\mu}. \quad (6.50)$$

Similarly, one obtains the component Lagrangian governing the interactions of D and $\overline{D}$,

$$\mathcal{L}_{4,DD} = C_1 D^{(Q)\dagger} D^{(Q)} D^{(\overline{Q})\dagger} D^{(\overline{Q})}. \quad (6.51)$$

To extract the bound state energy E_b, we shall consider the scattering amplitude of all four possible constituents of the $|X_+\rangle$ state, D^*, $\overline{D}$, D, and $\overline{D}^*$. We define the following transition amplitudes,

$$T_{11} = \langle D^*\overline{D}|T|D^*\overline{D}\rangle, \quad T_{12} = \langle D^*\overline{D}|T|D\overline{D}^*\rangle,$$
$$T_{21} = \langle D\overline{D}^*|T|D^*\overline{D}\rangle, \quad T_{22} = \langle D\overline{D}^*|T|D\overline{D}^*\rangle, \quad (6.52)$$

which correspond to the scattering of D and D^* mesons. At tree level, $T_{ii} \sim C_1$ and $T_{ij,\,i\neq j} \sim C_2$, since we consider only contact interactions. As we discussed in the introduction to Chapter 5, bound states cannot be described by any fixed-order calculation in perturbation theory, so we also have to include a resummation of loop contributions to complete the leading order [Weinberg (1990)]. These transition amplitudes satisfy a system of Lippmann-Schwinger equations,

$$iT_{11} = -iC_1 + \int \frac{d^4q}{(2\pi)^4} T_{11} G_{PP^*} \cdot C_1 - \int \frac{d^4q}{(2\pi)^4} T_{12} G_{PP^*} \cdot C_2,$$
$$iT_{12} = iC_2 - \int \frac{d^4q}{(2\pi)^4} T_{11} G_{PP^*} \cdot C_2 + \int \frac{d^4q}{(2\pi)^4} T_{12} G_{PP^*} \cdot C_1,$$
$$iT_{21} = iC_2 + \int \frac{d^4q}{(2\pi)^4} T_{21} G_{PP^*} \cdot C_1 - \int \frac{d^4q}{(2\pi)^4} T_{22} G_{PP^*} \cdot C_2, \quad (6.53)$$
$$iT_{22} = -iC_1 - \int \frac{d^4q}{(2\pi)^4} T_{21} G_{PP^*} \cdot C_2 + \int \frac{d^4q}{(2\pi)^4} T_{22} G_{PP^*} \cdot C_1,$$

where G_{PP*} are the Green's functions for two bosons in the momentum space,

$$G_{PP*} = \frac{1}{4} \frac{1}{\left(\vec{p}^{\,2}/2m_{D*} + q_0 - \Delta - \vec{q}^{\,2}/2m_{D*} + i\epsilon\right)}$$
$$\times \frac{1}{\left(\vec{p}^{\,2}/2m_D - q_0 - \vec{q}^{\,2}/2m_D + i\epsilon\right)}, \quad (6.54)$$

$\vec{p}$ is the momentum of one of the mesons in the center-of-mass system, and we canceled out factors of $m_D m_{D*}$ appearing on both sides of Eq. (6.53). Note that in Eq. (6.54) the vector propagator includes the mass difference Δ [AlFiky et al. (2006)], which is equivalent to measuring bound state energies with respect to the "constituent mass" of the system, which is in our case twice the pseudoscalar mass. A different choice of phase will give different propagators, but also a different "constituent mass".

Since we are interested in the pole of the amplitude T_{++}, we must diagonalize this system of equations rewritten in an algebraic matrix form,

$$\begin{pmatrix} T_{11} \\ T_{12} \\ T_{21} \\ T_{22} \end{pmatrix} = \begin{pmatrix} -C_1 \\ C_2 \\ C_2 \\ -C_1 \end{pmatrix} + i\tilde{A} \begin{pmatrix} -C_1 & C_2 & 0 & 0 \\ C_2 & -C_1 & 0 & 0 \\ 0 & 0 & -C_1 & C_2 \\ 0 & 0 & C_2 & -C_1 \end{pmatrix} \begin{pmatrix} T_{11} \\ T_{12} \\ T_{21} \\ T_{22} \end{pmatrix}. \quad (6.55)$$

Notice that the matrix is in the block-diagonal form, which allows us to solve Eq. (6.55) in two steps working only with 2×2 matrices. The solution of Eq. (6.55) produces the T_{++} amplitude,

$$T_{++} = \frac{1}{2}(T_{11} + T_{12} + T_{21} + T_{22}) = \frac{\lambda}{1 - i\lambda\tilde{A}}, \quad (6.56)$$

with $\lambda = -C_1 + C_2$, and $\tilde{A}$ is a (divergent) integral

$$\tilde{A} = \frac{i}{4} 2\mu_{DD*} \int \frac{d^3q}{(2\pi)^3} \frac{1}{\vec{q}^{\,2} - 2\mu_{DD*}(E - \Delta) - i\epsilon}, \quad (6.57)$$

where $E = \vec{p}^{\,2}/2\mu_{DD*}$, μ_{DD*} is the reduced mass of the $DD*$ system, and we have used the residue theorem to evaluate the integral over q^0. The divergence of the integral of Eq. (6.57) is removed by renormalization, as usual. We choose to define a renormalized λ_R within the $\overline{MS}$ subtraction scheme in dimensional regularization. In this scheme the integral $\tilde{A}$ is finite, which corresponds to an implicit subtraction of power divergences in Eq. (6.57). Computing the integral in Eq. (6.57) by analytically continuing to $d - 1$ dimensions yields

$$\tilde{A} = -\frac{1}{8\pi}\mu_{DD*}|\vec{p}|\sqrt{1 - \frac{2\mu_{DD*}\Delta}{\vec{p}^{\,2}}}. \quad (6.58)$$

This implies for the transition amplitude

$$T_{++} = \frac{\lambda_R}{1 + (i/8\pi)\lambda_R \mu_{DD^*}|\vec{p}|\sqrt{1 - 2\mu_{DD^*}\Delta/\vec{p}^2}}. \quad (6.59)$$

The position of the pole of the molecular state on the energy scale can be read off Eq. (6.59),

$$E_{\text{Pole}} = \frac{32\pi^2}{\lambda_R^2 \mu_{DD^*}^3} - \Delta. \quad (6.60)$$

This is the amount of energy we must subtract from the "constituent mass" of the system, determined above as $2m_D$, in order to calculate the mass of the molecular X-state

$$M_X = 2m_D - E_{\text{Pole}} = 2m_D + \Delta - \frac{32\pi^2}{\lambda_R^2 \mu_{DD^*}^3}. \quad (6.61)$$

The binding energy can be obtained remembering that that $m_{D^*} = m_D + \Delta$, so from Eq. (6.61)

$$E_b = \frac{32\pi^2}{\lambda_R^2 \mu_{DD^*}^3}. \quad (6.62)$$

If $E_b = 0.5$ MeV we obtain for the four-boson coupling

$$\lambda_R \simeq 8.4 \times 10^{-4} \text{ MeV}^{-2}. \quad (6.63)$$

The small binding energy of the $X(3872)$ state implies that the scattering length a_D is large and can be written as

$$a_D = \sqrt{(2\mu_{DD^*} E_b)^{-1}} = \frac{\lambda_R \mu_{DD^*}}{8\pi}, \quad (6.64)$$

yielding a numerical value $a_D = 6.3$ fm. This is large, as a typical size of an ordinary hadron is about 1 fm! Since the scattering length is large, universality implies that the leading-order wave function of $X(3872)$ is known,

$$\psi_{DD^*}(r) = \frac{e^{-r/a_D}}{\sqrt{2\pi a_D r}}. \quad (6.65)$$

This can be used to predict the production and decay properties of $X(3872)$. For this purpose, pion fields must be included in our effective theory. Non relativistic effective field theories for explicitly containing pion fields we discussed in [Fleming et al. (2007)].

6.6 Notes for further reading

The theory of non relativistic bound states of two heavy quarks is a rather mature subfield of effective theories. A good introduction to NRQCD and NRQED can be found in [Caswell and Lepage (1986); Bodwin et al. (1995)]. Power counting in NRQCD and NRQED is also discussed in [Luke and Manohar (1997)]. Several reviews of NRQCD and quarkonium production in general [Braaten et al. (1996); Brambilla et al. (2005); Brambilla (2011)] are also recommended.

Applications of EFT techniques to nuclear physics have been attempted since the early nineties [Weinberg (1990)]. Methods of effective theories describing the deuteron that use the fact that the scattering length in the two-nucleon system is large, thus generating another scale that affects the power counting were discussed in [Kaplan et al. (1996, 1998a,b)]. Universality in systems with large scattering lengths is discussed in the review [Braaten and Hammer (2006)], which contains applications to both particle and condensed matter physics.

Finally, lattice QCD methods enjoy applications of effective field theories [Kronfeld (2002)]. We invite interested reader to read them to gain deeper understanding of effective field theories for the systems with two heavy particles.

Problems for Chapter 6

(1) **Positronium decays**

Compute the amplitude for e^+e^- annihilation into 2 photons in the extreme nonrelativistic limit (i.e. keep only the term proportional to zero powers of the electron and positron momentum). Use this result, together with the formalism for fermion-antifermion bound states to compute the rate of annihilation of the 1S states of positronium into 2 photons. You should find that the spin-1 states of positronium do not annihilate into 2 photons. For the spin-0 state of positronium, you should find a result proportional to the square of the 1S wavefunction at the origin. Inserting the value of this wavefunction from nonrelativistic quantum mechanics, you should find

$$\frac{1}{\tau} = \Gamma = \frac{\alpha^5 m_e}{2} \approx 8.03 \times 10^9 sec^{-1}. \quad (6.66)$$

Compare it to the experimental result. Comment on what would change in the case of 1S charmonium decay into 2 gluons and estimate the rate.

(2) **Field redefinitions and NRQCD Lagrangian**

A strategy in writing any effective Lagrangian includes writing out all possible terms consistent with symmetries of the theory arranged according to the chosen power counting scheme. Applying this idea to the NRQCD Lagrangian of Eq. (6.3), one can immediately notice that some of the possible terms that are consistent with symmetries and power counting are missing. Let us investigate what is going on.

(a) Show that the term

$$\delta \mathcal{L} = \frac{c}{m_Q} \psi^\dagger \left(iD^0\right)^2 \psi$$

should be admitted by symmetry considerations, including discrete symmetries, such as parity.

(b) Show that the term in (a) can be eliminated by the following field redefinition

$$\psi \to \psi - \frac{c}{2m_Q} iD^0 \psi,$$

at least up to the order $1/m_Q^2$. What does this imply for the higher-order terms $\psi^\dagger (iD^0)^n \psi$?

Chapter 7

Effective Field Theories of Type-III. Fast Particles in Effective Theories

7.1 Infrared divergences

Consider the process $b \to s\gamma$: a heavy b quark is undergoing a two-body decay to a (massless) s quark and a photon. If we stick to a simple tree-level, quark-level analysis, the process is easy to understand. Conservation of energy and momentum in the b quark's rest frame lead us to conclude that the photon and s quark each carry $m_b/2$ of energy and an equal and opposite amount of momentum. The decay width can be calculated by a straightforward QFT calculation. The problem seems simple enough.

Unfortunately, the answer you get from such an analysis does not have much to do with what you see in the real world. The problem is that this decay happens within a miasma of QCD, and virtual quarks and gluons being exchanged through non-perturbative interactions makes matching the above calculation to any observable quantity very difficult. But the good news is that, while the QCD haze tends to bury the details of the underlying process, it occurs at a radically different scale: the decay process involves energy-momentum transfers near m_b, while the QCD background lives at Λ_{QCD}, which is an order of magnitude lower. This problem is crying out to be done in an EFT framework, and it is precisely what HQET was invented to solve! So you can do an HQET analysis of the problem, prove that (at least to leading order in Λ_{QCD}/m_b) the QCD effects factorize from the underlying event into some universal matrix elements and form factors, and you are saved from the challenge of needing to account for non-perturbative hadronization effects.

But the story does not end there. To see what else can go wrong, let us parametrize our phase space as

$$E_\gamma = xm_b/2, \qquad E_s = (2-x)m_b/2 , \qquad (7.1)$$

where $0 < x < 1$ is the fraction of energy that the photon carries away from the decay, and E_s represents the energy carried away by the outgoing jet. Conserving momentum and keeping the photon on-shell means that both the photon and the jet carry and equal and opposite momentum of magnitude $xm_b/2$. Then when calculating the momentum transfer in the jet, sometimes called the "off-shellness," we find

$$Q_s^2 = m_b^2(1-x) \,. \tag{7.2}$$

This is a very interesting result: when we start to compute loops in HQET, we expect to find terms that go as

$$\alpha_s \log^2 \left(\frac{Q_s^2}{m_b^2} \right) = \alpha_s \log^2(1-x) \,, \tag{7.3}$$

and these logarithms become large as $x \to 1$. This "endpoint region" of phase space is where the event is looking predominantly 2-body; that is, all of the jet fragments from the s quark are collinear with the original quark. This is an infrared (collinear) divergence: HQET has no mechanism for handling such divergences.

Now is a good time to talk about infrared and ultraviolet divergences. When a UV divergence is found in a calculation, we know precisely what that means. Our theory is not able to give us meaningful predictions of physics at the smallest scales, so we are not surprised by these divergences. We absorb them into unspecified coefficients, and then either rely on experiment to fix these coefficients, or (if available) use a UV-completion of our theory to match these coefficients to something finite. In short, UV divergences carry no real information other than the already-known fact that our theory cannot be used to predict physics at all scales.

However, IR divergences are much more serious. All physical observables *must* be IR-safe. When you compute an IR divergence in your final answer, it means that you have done something wrong. The most famous example of this is in QED bremsstrahlung, which has a soft divergence: the cross section for the emission of a single photon with vanishing energy is infinite. But if you think about it, that is not a surprise at all: how can you even think to *ask* such a question as, "What is the probability of emitting a *single* photon at zero energy?" At low energy, the electron is surrounded by a sea of virtual photons that can live for longer and longer times by the uncertainty principle. You cannot devise an experiment that would test your prediction, and QED was kind enough to let you know that you asked a bad question!

To make it a good question again, you have a few options: you can be completely pragmatic about it and say, "I can never build an experiment to measure zero-energy photons, but I will always have a cutoff E_c; so rather than compute the total cross section, let me restrict my phase space to photons with energy *at least as large as* E_c." Then you can expect to have an answer that diverges as E_c goes to zero, but remains finite at any real value of E_c. It is very important to realize that this is *not* the same thing as momentum-cutoff regularization, even on a conceptual level. The usual momentum cutoff is not a real scale but an arbitrary scale that is used to cutoff troublesome *virtual* corrections; it has no physical meaning and cannot affect any physical observable. However, E_c is a real scale set, not by QED, but by your detector. You would expect that as you make your detector better and better so that E_c gets lower, your cross section would increase since you can start to measure more and more photons. But you will never actually measure all the way to zero energy, so your result will always be finite.

The problem with this approach, as you might have seen by the last paragraph, is that as you make E_c smaller, you increase the cross section without bound, and sooner or later you are going to loose perturbativity. While unlikely to be a serious problem for QED for reasonable choices of E_c, it is a very real concern when QCD is involved. Instead, theorists would prefer to somehow absorb these divergent terms into something finite. The idea was originally shown to work for QED-bremsstrahlung by Weinberg (see [Peskin and Schroeder (1995)] for example), who showed that if you do a calculation to all orders in α, but leading order in E_γ/m_e, the divergent terms can be resummed into an exponential, called a "Sudakov-factor". This is remarkably like renormalization-group running, only something is very different, since these are not UV-divergences. One of the greatest powers of EFT is that it converts IR-divergences to UV-divergences, and allows this kind of resummation to occur by using the RG-machine!

Now let us return to our $b \to s\gamma$ example. When $x < 1$, soft divergences can easily be resummed by the machinery of HQET and all of our results are fine. But what about when $x \to 1$? Then we saw collinear divergences appear. What is going on? The answer can be seen by looking at Eq. (7.2). From this we see that the strange quark jet is becoming massless. That means that we cannot integrate out this degree of freedom from QCD. In HQET, we integrate out all of the hard degrees of freedom, leaving behind only soft QCD and non-dynamical gluon sources represented by Wilson lines. All other "hard" QCD effects can be absorbed in Wilson coefficients, like any other EFT. But in this limit of hard+collinear jets, we cannot get away with this: we have contributions from a hard scale that are still dynamical, and can thus still produce IR divergences. What we need is an effective theory that describes not only the soft degrees of freedom we usually want, but these hybrid modes as well. With these additional degrees of freedom, we can cancel off the extra IR-collinear divergences, and resum the $\log(1-x)$ terms.

For this, we need to turn to Soft-Collinear Effective Theory (SCET).

7.2 Soft-Collinear Effective Theory

7.2.1 *Quantum mechanics of fast particles*

In previous chapters we considered a situation when part of the solution of the field equation was known. Indeed, expanding around the static solution for a system with one or two heavy particles allowed us to build effective theories of bound states containing those particles. There are other situations when an approximate solution is known. An important example is fast-moving particles. As a warm up to what is coming, let us recall how to dealt with those in nonrelativistic quantum mechanics.

Imagine a free particle moving along the z-axis. A solution of Schrödinger's equation for such a particle is a plane wave,

$$\psi(\vec{r})_{\text{free}} = e^{ikz}. \tag{7.4}$$

If we now consider the case that our interaction potential $U(r)$ has a finite range and is not very strong, and our particle is fast[1], the wave function should not change much from the free solution of Eq. (7.4),

$$\psi(\vec{r}) = e^{ikz} + e^{ikz}\phi_1(\vec{r}) + e^{ikz}\phi_2(\vec{r}) + \ldots = e^{ikz}\Phi(\vec{r}), \tag{7.5}$$

where small changes of the wave function $\phi_i(\vec{r})$ obtained after each iteration of the Lippmann-Schwinger equation

$$\psi(\vec{r}) = e^{ikz} - \frac{m}{2\pi\hbar^2}\frac{e^{ikr}}{r}\int d^3r' e^{-i\vec{k}\cdot\vec{r}'}U(r')\psi(\vec{r}') \tag{7.6}$$

resum into $\Phi(\vec{r})$. This is very similar to what we have observed in Chapter 5: a solution of the wave equation was written by separating out a large

---

[1] In principle, all of those qualifiers need to be quantified, which would give us a range of applicability of our perturbative expansion. This can be found in almost any standard quantum mechanics text.

mechanical part that describes the free particle. Now, all we have to do is to plug this solution back into the Schrödingier equation,

$$\left(\nabla^2 + \vec{k}^2\right) \psi(\vec{r}) = \frac{2m}{\hbar^2} U(r) \psi(\vec{r}), \quad (7.7)$$

to see that the equation for $\Phi(\vec{r})$ is a first order differential equation,

$$\frac{\partial \Phi(\vec{r})}{\partial z} + \frac{im}{k\hbar^2} U(r) \Phi(\vec{r}) = 0. \quad (7.8)$$

Here we used some simplifying assumptions, such as the fact that most changes in $\Phi(\vec{r})$ would still happen along the z-axis. The solution of equation Eq. (7.8), consistent with the boundary condition that the wave function $\psi(\vec{r})$ must be given by Eq. (7.4) for $z \to -\infty$ and $U(r) \to 0$ is

$$\psi(\vec{r}) = \exp\left(ikz - \frac{im}{k\hbar^2} \int_{-\infty}^{z} U(r) dz\right). \quad (7.9)$$

This solution can be used to derive an "improved Born" (also known as Glauber or eikonal) approximation for the scattering amplitude.

Let us use Eq. (7.9) in Eq. (7.6) to extract the scattering amplitude as the coefficient of the spherical wave e^{ikr}/r (extracted from the second term in Eq. (7.6) in the limit $r \to \infty$). Introducing a transverse coordinate vector $\vec{\rho}$ one can show that the nonrelativistic scattering amplitude $f(\theta)$ is

$$f(\theta) = -\frac{ik}{2\pi} \int d^2\rho\, e^{i\vec{q}\cdot\vec{\rho}} \left(e^{2i\delta(\vec{\rho})} - 1\right), \text{ where}$$

$$\delta(\vec{\rho}) = -\frac{m}{2\hbar^2 k} \int_{-\infty}^{\infty} U(r) dz. \quad (7.10)$$

A differential or total cross section can be found by taking absolute value squared of Eq. (7.10) and integrating over relevant angular variables.

We shall do something similar as a first step in constructing the Soft-Collinear Effective Theory in this chapter. Not surprisingly, a field-theoretical description of high energy scattering is immensely more difficult. Yet, application of SCET makes it still easier to separate scales in problems with fast moving particles. A solution written in the scale-separated form is called a *factorization theorem*, and our goal will be to prove these theorems for various processes.

7.2.2 SCET power counting

7.2.2.1 Momentum scaling

We wish to construct an EFT that describes light, fast-moving particles. These particles satisfy $E \gg m$, and so have 4-momenta that are very nearly pointed along the light-cone. Therefore, it makes sense to use light-cone coordinates to describe these particles. Let us define the vectors:

$$n^\mu = (1,0,0,1), \qquad \bar{n}^\mu = (1,0,0,-1). \tag{7.11}$$

These satisfy $n^2 = \bar{n}^2 = 0$, $n \cdot \bar{n} = 2$. In terms of these vectors, any vector can be broken up as:

$$p^\mu = \frac{n^\mu}{2}(\bar{n} \cdot p) + \frac{\bar{n}^\mu}{2}(n \cdot p) + p_\perp^\mu . \tag{7.12}$$

This momentum vector has a component $p^- = \bar{n} \cdot p$ along the "forward edge" of the light cone, and $p^+ = n \cdot p$ along the "back edge" of the light cone. For fast-moving forward particles, we expect $p^- = E + p$ to be large, while $p^+ = E - p$ should be small ($p_\perp$ should be whatever is needed to keep the particle on-shell). In what follows, p^- scales like the "hard" momentum, while p^+ scales like the "soft" momentum. In terms of light-cone coordinates, we write $p^\mu = (p^+, p^-, p_\perp)$, and

$$p^2 = p^+ p^- - p_\perp^2 . \tag{7.13}$$

If our hard scale is Q and our soft scale is Λ,[2] then collinear momenta should scale like

$$p_c \sim (\Lambda, Q, \sqrt{Q\Lambda}) = Q(\lambda^2, 1, \lambda), \tag{7.14}$$

where $\lambda = \sqrt{\Lambda/Q}$ is our small power-counting parameter. When we talk about the "offshellness" of a particle, we are referring to how much p^2 can safely differ from zero while still being consistent with zero within the power-counting error; this is the typical value of momentum transfer that occurs in any process involving these particles. This is given by

$$p^2 \simeq p^+ p^- \simeq p_\perp^2 , \tag{7.15}$$

and for the collinear particles it is $p_c^2 \simeq Q\Lambda$. If $\Lambda \ll Q$, then the collinear offshellness can be large and QCD can be perturbative.[3] Thus separating this scale is useful and important if we want to have a theory that does not have any large logarithms that might destroy perturbation theory.

---

[2] $\Lambda = \Lambda_{\text{QCD}}$ in most contexts, but we often drop the "QCD" label to make the notation simpler. Also note that Λ is always an IR scale in this chapter, and never a UV cutoff.

[3] For this reason, these modes are sometimes called "hard-collinear" in the SCET literature.

7.2.2.2 Field scaling

Now that we have identified our small power counting parameter, we need to decide how to power count fields. Like before, the easiest way to do this is to determine how the field must scale so that it is consistent with the propagator. Let us start with an ordinary fermion field that is meant to describe a collinear object (such as a quark), denoted by $\xi(x)$:

$$\langle \bar{\xi}(x)\xi(y)\rangle = \int \frac{d^4k}{(2\pi)^4} e^{ik(x-y)} \frac{i\slashed{k}}{k^2+i\varepsilon}. \tag{7.16}$$

If this field is only to have support over the collinear region on phase space, the momentum k should scale like Eq. (7.14). Using the fact that $\slashed{k} = \frac{\slashed{n}}{2}k^- + \cdots \sim \mathcal{O}(1)$, and $k_c^2 \sim \lambda^2$ we have

$$\langle \bar{\xi}(x)\xi(y)\rangle \sim (\lambda^4)\frac{(1)}{(\lambda^2)} \sim \lambda^2. \tag{7.17}$$

Therefore $\xi \sim \lambda$ in the power counting.

We can perform the same trick with gauge fields. Although not absolutely required, determining power counting rules for these fields is most straightforward in the renormalization-gauges[4]

$$\langle A^\mu(x)A^\nu(y)\rangle = \int \frac{d^4k}{(2\pi)^4} e^{ik(x-y)} \frac{-i(g^{\mu\nu} - (1-\alpha)\frac{k^\mu k^\nu}{k^2})}{k^2+i\varepsilon}, \tag{7.18}$$

where we suppress any gauge indices for simplicity. Noting that in light-cone coordinates, $g^{++} = g^{--} = 0$ (can you prove this?), you can show that $A_c^\mu \sim (\lambda^2, 1, \lambda)$, which is the same scaling as p_c^μ. This is not a surprise, since it means that the covariant derivatives will have a homogeneous scaling.

We can also determine the power counting for ghost fields as well. This is left as an exercise.

We can also derive scaling rules for fields which annihilate "soft" momentum states. We call these *ultrasoft* fields.[5] Such fields have momentum that scale like $p_{us}^\mu \simeq (\Lambda, \Lambda, \Lambda) = Q(\lambda^2, \lambda^2, \lambda^2)$. Thus $p_{us}^2 \sim \lambda^4$ and using the same propagators we find:

$$\langle \bar{q}_{us}(x) q_{us}(y)\rangle \sim (\lambda^8)\frac{(\lambda^2)}{(\lambda^4)} \sim \lambda^6. \tag{7.19}$$

$$\langle A_{us}^\mu(x) A_{us}^\nu(y)\rangle \sim (\lambda^8)\frac{(1)}{(\lambda^4)} \sim \lambda^4. \tag{7.20}$$

Thus, $q_{us} \sim \lambda^3$ and $A_{us}^\mu \sim \lambda^2 \sim p_{us}^\mu$.

---

[4]Often referred to as "R_ξ" gauge in [Peskin and Schroeder (1995)]; here we use α to avoid confusion with the collinear quarks.

[5]There are also "soft" fields, but we will save them for the exercises.

7.2.2.3 A word of warning

Notice that this is already looking very strange compared to other EFTs that we have encountered. There are two places that might have caused you to take pause. First of all, we are choosing to work in a reference frame where the energetic particle is picking out a given direction in space-time, yet QCD is a relativistically covariant theory, and therefore it should not matter what reference frame we work in! Why not boost into a reference frame where the energetic particle is not so fast-moving? Of course we can do this, and there have been some efforts in that direction [Freedman and Luke (2012)]. However, the SCET choice of frame is a particularly good one for these problems, usually coming from the decay of a heavy object in its rest frame, or jet production from a hard central collision. So while we might be able to work in a frame where all of the hard momenta go away, the effects will always be present; it is just a matter of where they appear.

Given that we have chosen to work in this reference frame, there is still another problem: it seems that these particles have "hard" momenta in one direction, and so according to the rules of EFT, they should be integrated out; and yet they have "soft" momenta in another direction, so by the same rules, we have to leave them in! As strange as HQET was compared to Fermi's model of weak interaction, this is much worse! At least in HQET, we integrated out the heavy quark dynamics in every direction, leaving behind a heavy, static source of soft gluons. Now we are going to have something similar, but with a part of the field still dynamical! We will integrate out the parts of the field that can create fluctuations in some directions *but not all directions*.

These questions are still being discussed by the users of SCET, and time will tell if we ever come up with a satisfactory answer to them. For now, however, we will skirt over the tricky subtleties and focus on getting a handle on what SCET can do in practice.

7.2.3 SCET action

The next step is to write down the leading order action for the fields in SCET. There are two ways of achieving this: a position-space formalism described in [Beneke *et al.* (2002); Beneke and Feldmann (2003)], or a momentum-space approach described in [Bauer *et al.* (2001a, 2002b)]. Of course, both theories are equivalent, but they each have their own advantages and disadvantages, just like in ordinary field theory. The position space formalism makes it very clear how the nonlocalities of the problem appear, and is also more inline with the notations of work done earlier in QCD. The momentum space approach hides the nonlocality of the theory somewhat, but it might be more natural for doing actual calculations. We will be following the momentum-space approach here, and leave the position-space formalism to the literature for any interested readers.

For the ultrasoft fields, matching from QCD is very simple: these fields are remnants from the full QCD, and therefore their actions are the same as ordinary QCD. Since all of the gauge fields and momentum components scale the same way, nothing is power-suppressed relative to the leading terms that define the propagator. The collinear fields require more work, however.

Like in HQET, only the top two components of the spinor that annihilate the quark state are large enough to contribute. So let us choose some projection operators that will pick out the larger components:

$$\Pi_+ = \frac{\not{n}\not{\bar{n}}}{4}, \qquad \Pi_- = \frac{\not{\bar{n}}\not{n}}{4}. \qquad (7.21)$$

These satisfy the usual relations required for projection operators:

$$\Pi_+ + \Pi_- = 1, \qquad \Pi_\pm^2 = \Pi_\pm. \qquad (7.22)$$

The quark fields will be broken up into a large part ($\Pi_+ \xi = \xi$) and a small part ($\Pi_- \Xi = \Xi$). Notice that this means $\Pi_- \xi = 0 \Rightarrow \not{n}\xi = 0$. This will be useful in what follows.

Our next step is to extract the large momentum components. As in HQET, we can break the momentum up into a "label" momentum and a "residual momentum"

$$p = \tilde{p} + k, \qquad (7.23)$$

with $\tilde{p} = \frac{n^\mu}{2} p^- + p_\perp^\mu$ (note how there is no $\tilde{p}^+$ component), and k^μ scales like an ultrasoft momentum. As we described in Section 7.2.1, we extract an eikonal factor from the spinors that carry the dependence on the label momentum components:

$$q_c(x) = \sum_{\tilde{p}} e^{i\tilde{p}\cdot x} \left[\xi_{\tilde{p}}(x) + \Xi_{\tilde{p}}(x) \right]. \qquad (7.24)$$

The fields $\xi_{\tilde{p}}$, $\Xi_{\tilde{p}}$ also depend on label momenta as subscript indices; we often leave those out for notational convenience, but you should always remember that each large momentum mode is getting its *own* fields, just like what happens in HQET with the velocity (h_v).

Now that all the large momentum has been factored out of the fields, derivatives will only pull out residual momenta that scale like Λ. Therefore, if we need to refer to label momenta, it is convenient to define label-momentum operators that explicitly pull out these large components ($\mathcal{P}^-$, $\mathcal{P}_\perp$). These operators pull down the label momentum of everything *to the right* of the operator; $\mathcal{P}^\dagger$ acts to the left.

By plugging our expansion of Eq. (7.24) into the quark action and using the projection properties of the two fields, we can then integrate out the small components of the collinear spinor:

$$\Xi_{\tilde{p}} = \frac{1}{\mathcal{P}^- - gA_c^-} \left(\slashed{\mathcal{P}}_\perp + g\slashed{A}_{c\perp} \right) \frac{\slashed{\bar{n}}}{2} \xi_{\tilde{p}}. \qquad (7.25)$$

We leave the proof of this to the exercises. Notice that from this solution, it can be seen that $\Xi \sim \lambda^2$, so these fields do represent power-suppressed collinear contributions and can (should!) be integrated out.

Looking at this equation, we can see something very confusing happening: the denominator is *highly* nonlocal. In previous situations like HQET, the $\mathcal{P}^-$ was replaced by a mass, and it made sense to Taylor expand the denominator as a series of power-suppressed operators. Our introduction of the label momenta, which scale like the hard scale Q, gives us hope that we can do something similar, but unfortunately it will not work. Any attempt to expand this denominator would give us an expansion in powers of $A_c^-/\mathcal{P}^- \sim \mathcal{O}(1)$ in the power counting. So even though this series of operators is suppressed by the coupling constant, it is not power suppressed, and is therefore not an appropriate expansion to do in the effective action. Indeed, SCET often deals with nonperturbative calculations where an expansion in g would not work. We must find another way to make sense of this denominator.

The solution is to introduce a "collinear Wilson line" which allows for the possibility of a collinear quark to bremsstrahlung an arbitrary number of collinear gluons without penalty.[6] Let us define the path-ordered exponential:

$$W(x) = \mathbf{P} \exp \left(ig \int_{-\infty}^1 ds \, \bar{n} \cdot A_c(sx) \right). \qquad (7.26)$$

This Wilson line is chosen to obey the equation

$$i\bar{n} \cdot DW = 0 \implies W^\dagger i\bar{n} \cdot DW = \mathcal{P}^- W. \qquad (7.27)$$

---
[6] Of course, we could always choose to work in light-cone gauge which turns off this polarization of gluon ($\bar{n} \cdot A = 0$) but we will continue to work in general gauge.

This will eliminate the gauge field from the denominator of Eq. (7.25) at the expense of introducing a nonlocal Wilson line in the action.

Putting this all together and making all collinear fields explicit, we can write down the final Lagrangian for the collinear quarks:

$$\mathcal{L}_c = \sum_{\tilde{p}} \bar{\xi}_{\tilde{p}} \left[in \cdot D + g A_c^+ + (\mathcal{P}_\perp + g\mathcal{A}_{c\perp}) W \frac{1}{\mathcal{P}^-} W^\dagger (\mathcal{P}_\perp + g\mathcal{A}_{c\perp}) \right] \frac{\bar{n}}{2} \xi_{\tilde{p}}, \tag{7.28}$$

where $n \cdot D$ is an ultrasoft covariant derivative; the collinear gluon fields have been written out explicitly. Each term scales as λ^4, corresponding to an action that scales as λ^0.

With the exception of the first term, there are no ultrasoft gluons coupling to the collinear quarks. Although we could use the action as it is, it is often convenient to do a similar Wilson-line trick to eliminate the ultrasoft gluons from this Lagrangian, thus proving soft-collinear factorization at the dynamical level. To do that, we need to introduce another Wilson line

$$Y(x) = \mathbf{P} \exp\left(ig \int_{-\infty}^{1} ds\, \bar{n} \cdot A_{\text{us}}(sx) \right), \tag{7.29}$$

$$Y^\dagger in \cdot D Y = in \cdot \partial. \tag{7.30}$$

Now we do a field-redefinition on the quark fields: $\xi_{\tilde{p}} = Y \xi_{\tilde{p}}^{(0)}$, and $W = Y^\dagger W^{(0)} Y$. By inserting several $Y^\dagger Y$ into Equation (7.28) we can convert the covariant derivative to an ordinary derivative without altering any other term. Our final result is

$$\mathcal{L}_c = \sum_{\tilde{p}} \bar{\xi}_{\tilde{p}}^{(0)} \left[in \cdot \partial + g A_c^+ \right.$$

$$\left. + (\mathcal{P}_\perp + g\mathcal{A}_{c\perp}) W^{(0)} \frac{1}{\mathcal{P}^-} W^{(0)\dagger} (\mathcal{P}_\perp + g\mathcal{A}_{c\perp}) \right] \frac{\bar{n}}{2} \xi_{\tilde{p}}^{(0)}. \tag{7.31}$$

Now there are no ultrasoft fields in the collinear Lagrangian: we have proved factorization to leading order in λ! However, when you make this field redefinition, you must also remember to replace the fields in the currents, and thus collinear-ultrasoft couplings reappear as contact vertices. In what follows, we will omit the zero-superscript on the collinear fields for notational simplicity, and assume that the field redefinition has been done. We can also do a similar analysis for the collinear gauge fields themselves.

When looking at power-suppressed interactions, it becomes convenient to define collinear and mixed collinear-ultrasoft gluon field strengths

$$ig B_c^{\perp \mu} = [i\bar{n} \cdot D_c, D_c^{\perp \mu}] \tag{7.32}$$

$$ig \mathcal{M} = [i\bar{n} \cdot D_c, i\mathcal{D}_{\text{us}} + \frac{\bar{n}}{2} g n \cdot A_c] \tag{7.33}$$

At this order in the power counting, (ultra)soft and collinear modes no longer decouple at the Lagrangian level, making proofs of factorization a bit more interesting. There are three subleading operators that are consistent with the symmetries of SCET (see the next section):

$$\mathcal{L}_{\xi q}^{(1)} = \bar{\xi} W \frac{1}{\mathcal{P}^-} W^\dagger ig\slashed{B}_c^\perp W q_{\text{us}} \tag{7.34}$$

$$\mathcal{L}_{\xi q}^{(2a)} = \bar{\xi} W \frac{1}{\mathcal{P}^-} W^\dagger ig\slashed{M} W q_{\text{us}} \tag{7.35}$$

$$\mathcal{L}_{\xi q}^{(2b)} = \bar{\xi} \frac{\slashed{\bar{n}}}{2} i\slashed{D}_c^\perp W \frac{1}{(\mathcal{P}^-)^2} W^\dagger ig\slashed{B}_c^\perp W q_{\text{us}} \tag{7.36}$$

These vertices must be included as a time-ordered product with soft-collinear currents in any Feynman diagram. Notice that all of these operators require the emission of at least one collinear gluon. The details of deriving these terms, as well as showing how they show up as time-ordered products with subleading currents, is done very well in [Pirjol and Stewart (2003)]. Also included in this paper are the Feynman rules for SCET, including these subleading terms. We direct the reader to this paper for the details.

7.3 Symmetries of SCET

Now we will turn to two important symmetries in SCET that will help us determine the allowed operators at a given order in the power counting.

7.3.1 Gauge invariance

By separating collinear and ultrasoft fields at the Lagrangian level, we have managed to create an enhanced symmetry for the theory. QCD gauge invariance also splits up into separate gauge invariances for collinear and ultrasoft fields. Having this extra symmetry gives us another very powerful tool for determining the forms of operators and currents in the effective theory.

The idea is straightforward enough: since the gauge fields break up into collinear and ultrasoft gluon fields, and there are no couplings between collinear and ultrasoft fields in the leading-order Lagrangian, it follows that the gauge transformations can be broken up into a collinear-gauge transformation that only involves A_c^μ and an ultrasoft gauge transformation that only involves A_{us}^μ. Upon studying the full gauge symmetries of all of the operators in the theory, we can construct a table of how all of the fields transform; this is given in Table 7.1.

Table 7.1 SCET Fields and their collinear and ultrasoft gauge transformations and scaling dimensions.

	Field	Scaling	U_c	U_{us}
collinear	ξ_n	λ	$U_c \xi_n$	$U_{us} \xi_n$
	A_n^μ	$(\lambda^2, 1, \lambda)$	$U_c A_n^\mu U_c^\dagger + \frac{i}{g} U_c \left[iD_c^\mu, U_c^\dagger \right]$	$U_{us} A_n^\mu U_{us}^\dagger$
	W_n	1	$U_c W_n$	$U_{us} W_n U_{us}^\dagger$
ultrasoft	q	λ^3	q	$U_{us} q$
	A_{us}^μ	λ^2	A_{us}^μ	$U_{us} \left(A_{us}^\mu + \frac{i}{g} \partial^\mu \right) U_{us}^\dagger$
	Y	1	Y	$U_{us} Y$

Since each field has its own separate transformation rule under collinear and ultrasoft gauge transformations, we must demand all operators be invariant under both tranformations. This puts strong constraints on how operators can be formed. For example, whenever a collinear field is present in a current, it must come in the combination $W^\dagger \xi$ in order to be invariant under collinear gauge invariance.

If this splitting of gauge invariance into two separate symmetries gives you pause, then you are to be commended! There is something very subtle going on here, that gives us yet another reason to be wary of Type-III EFTs. It is not immediately obvious that you can factorize gauge transformations this way without loosing some important physics. Naively, it certainly makes sense: you do not want to allow symmetry transformations that spoil your power counting. But the full QCD has these transformations, and by not allowing them you are taking an awful risk. For example, you might generate spurious new terms that should not be present. Such terms can presumably be removed by a field redefinition similar to the ones in Eqs. (7.29-7.31), but it is not always clear how that would work in general.

Another example where this factorized gauge invariance may fail is in the renormalization group. In effect, there are now two gauge couplings: g_c and g_{us}. What is the relationship between them? Do they have the same beta function? Why should they, when they include different fields?! Furthermore, are we supposed to run each of these coupling constants separately, or are we supposed to do some sort of "correlated running" that sums both collinear and ultrasoft logarithms at the same time?[7] If a correlated-

---

[7] In NRQCD, it is known that this approach to the RG is needed in some circumstances. It goes under the name of the *velocity renormalization group* and we refer the reader to external sources such as [Luke et al. (2000)].

running approach is required, then it is impossible to maintain a factorized gauge invariance at every scale, and RG effects would spoil factorization. Some have claimed that this is proof that we cannot do a correlated running scheme in SCET, while others worry that this means factorized gauge invariance is doomed to fail at higher order.

The short answer is that we are still unsure what this means. We can add it to the list of concerns over Type-III EFT. In the meantime, insisting on this factorized gauge invariance has led to great success, so until we explicitly find a contradiction, we will continue to use it.

7.3.2 Reparametrization invariance

In Section 5.2.4, we found that HQET has a symmetry we called reparametrization invariance (RPI) that came from shifting the heavy quark velocity and residual momentum by a small number. This was just a manifestation of a hidden Lorentz invariance coming from the arbitrary split of a heavy quark momentum into "large" and "small" pieces. Since SCET also involves splitting a collinear momentum into a "label" momentum and a "residual" momentum, there is an RPI symmetry here as well, but it is more complex. It turns out that there are actually three forms of RPI in SCET, first described in [Manohar et al. (2002)]. We can describe these symmetries in terms of how they affect the collinear direction n^μ:

$$\text{Type I}: \quad n^\mu \to n^\mu + \Delta_\perp^\mu, \quad \bar{n}^\mu \text{ unchanged}, \tag{7.37}$$

$$\text{Type II}: \quad \bar{n}^\mu \to \bar{n}^\mu + \epsilon_\perp^\mu, \quad n^\mu \text{ unchanged}, \tag{7.38}$$

$$\text{Type III}: \quad n^\mu \to e^\alpha n^\mu, \quad \bar{n}^\mu \to e^{-\alpha} \bar{n}^\mu, \tag{7.39}$$

where $\Delta_\perp \sim \lambda$ and $\alpha, \epsilon_\perp \sim \lambda^0$. The parameters of the RPI transformation are perpendicular vectors to enforce the conditions $n^2 = \bar{n}^2 = 0$, $n \cdot \bar{n} = 2$. All three of these transformations leave the vector $V^\mu = \frac{\bar{n}^\mu}{2}(n \cdot V) + \frac{n^\mu}{2}(\bar{n} \cdot V) + V_\perp^\mu$ invariant.

Type-III RPI in Eq. (7.39) implies that any operator with an n^μ must either also have a $\bar{n}^\mu$, or else have an n^μ in the denominator. We can see that this is true for the Lagrangian in Eq. (7.31). Notice that this symmetry also forbids potentially dangerous operators such as $\bar{\xi}(i\bar{n} \cdot D)\xi$, which scale like λ^{-2} in the action, since they must come with a factor of $\not{n}$, and $\not{n}\xi = 0$ as we saw below Eq. (7.22).

Under Type-I RPI in Eq. (7.37), we have the following behavior of the

SCET fields:

$$n \cdot D \to n \cdot D + \Delta_\perp \cdot D_\perp, \tag{7.40}$$

$$D_\perp^\mu \to D_\perp^\mu - \frac{\Delta_\perp^\mu}{2} \bar{n} \cdot D - \frac{\bar{n}^\mu}{2} \Delta_\perp \cdot D_\perp, \tag{7.41}$$

$$\xi_n \to \left(1 + \frac{\slashed{\Delta}_\perp \slashed{\bar{n}}}{4}\right) \xi_n, \tag{7.42}$$

while $\bar{n} \cdot D$ and W are invariant. Under Type-II RPI in Eq. (7.38), we have the following behavior of the SCET fields:

$$\bar{n} \cdot D \to \bar{n} \cdot D + \epsilon_\perp \cdot D_\perp, \tag{7.43}$$

$$D_\perp^\mu \to D_\perp^\mu - \frac{\epsilon_\perp^\mu}{2} n \cdot D - \frac{n_\mu}{2} \epsilon_\perp \cdot D_\perp, \tag{7.44}$$

$$\xi_n \to \left(1 + \frac{\slashed{\epsilon}_\perp}{2} \frac{1}{i\bar{n} \cdot D} i\slashed{D}_\perp\right) \xi_n, \tag{7.45}$$

$$W_n \to \left(1 - \frac{1}{i\bar{n} \cdot D} i\epsilon_\perp \cdot D_\perp\right) W_n, \tag{7.46}$$

while $n \cdot D$ is invariant. We leave it as an exercise to show that Eq. (7.31) is invariant under all of these symmetries.

7.4 Examples

Many of these ideas from this chapter are especially complex and abstract. To see how we put them into practice, let us consider a collection of problems that have been studied with SCET, to varying degrees of success.

7.4.1 $B \to X_s \gamma$

The first problem solved with SCET was computing the decay distribution of a B meson decaying to a photon and a jet containing a strange quark in the endpoint region [Bauer et al. (2000)]; this was the process mentioned at the beginning of the chapter. As long as the jet was made of soft particles, this problem could be completely worked out using HQET. However, as we saw earlier, when the jet becomes strongly columnated, there is an additional degree of freedom that becomes massless, namely the jet itself. The corresponding infrared divergence is a collinear divergence, and we must include fields that can reproduce these divergences in our low-energy matrix elements in order to get an answer that makes sense.

The original proof that this extra field was needed came from studying a *factorization theorem* that said, to leading order in power-counting, contributions from different regions in phase space cannot affect divergences coming from other regions in phase space. As a result, the decay rate formula can be written as a product[8] of functions that can be calculated by considering only hard and soft (and now collinear) interactions. Without the collinear interactions, the factorization theorem failed, and this was a signal that something was missing.

In this section we will sketch how the factorized result for $B \to X_s \gamma$ comes about, but we will leave out most of the details. These will be relegated to the literature.

In the standard model of particle physics, this process occurs through a magnetic-dipole "penguin" operator:

$$\mathcal{O}_7 = \frac{e}{16\pi^2} m_b \bar{s} \sigma_{\mu\nu} \mathcal{P}_R b F^{\mu\nu}. \tag{7.47}$$

Normally, we can use the optical theorem to write decay widths as a product of currents separated in spacetime; these currents can be identified by matching operators like Eq. (7.47) onto EFT operators that are allowed by the symmetries. We can then use the OPE to compute this nonlocal product in terms of a series of local operators. When we calculate QCD corrections to the matrix element for this process, we find terms that go like $\alpha_s \log(1-x)$, where $x = 2E_\gamma/m_b$. As $x \to 1$ these logarithms overcome the already large value of α_s and perturbativity is lost. Therefore, a proper analysis in the endpoint region requires resummation, which in turn is calling out for an effective theory such as SCET.

We match the strange quark to a collinear field ξ, and the heavy bottom quark to an HQET field h_v. Then we must match the operator in Eq. (7.47) to a SCET current that couples to a hard photon. The details of this were worked out in the literature. The operator that satisfies all of the constraints of RPI and gauge invariance, to leading order in the power counting, is

$$\mathcal{J}^\mu = (\bar{\xi}W)\gamma_\perp^\mu \mathcal{P}_R(Y^\dagger h_v). \tag{7.48}$$

The final result of using this current is that the decay width takes the form

$$\frac{1}{\Gamma}\frac{d\Gamma}{dx} = H(m_b, x, \mu) \times T(x, \mu), \tag{7.49}$$

where H is a hard function containing the electroweak constants as well as matching coefficients that only depend on the large scales such as m_b, and

---

[8] In this case, the product is a convolution.

$T(x)$ is the forward scattering amplitude

$$T = i \int d^4y \, e^{i[m_b(\bar{n}/2)-q]\cdot y} \langle \bar{B}_v | T\left\{ \mathcal{J}_\mu^\dagger(y) \mathcal{J}^\mu(0) \right\} | \bar{B}_v \rangle. \tag{7.50}$$

with q^μ the 4-momentum of the photon. Plugging in the expression (7.48) into Eq. (7.50), and rearranging terms inside the matrix element, we find

$$T = i \int d^4y \, e^{i[m_b(\bar{n}/2)-q]\cdot y} \langle \bar{B}_v | T\left\{ \left[\bar{h}_v Y \right](y) P_L \gamma_{\perp \mu} \right.$$
$$\left. \times \left[\tilde{J}(y,0) \right] \gamma_\perp^\mu P_R \left[Y^\dagger h_v \right](0) \right\} | \bar{B}_v \rangle, \tag{7.51}$$

where we have combined collinear fields into the "jet function":

$$\left[\tilde{J}(y,0) \right] = \langle 0 | \left[W^\dagger \xi \right](y) \left[\bar{\xi} W \right](0) | 0 \rangle$$
$$\equiv i \int \frac{d^4k}{(2\pi)^4} e^{-ik\cdot y} J(k) \frac{\not{n}}{2}. \tag{7.52}$$

The jet function is basically the collinear quark propagator, and since (at this order in the power counting) the only relevant interactions of these fields involves $n \cdot \partial$ which brings down a k^+, it follows that the jet function only depends on this variable.[9] Thus the integrals over k^-, $k_\perp$ are simply delta functions. When we plug this back into Eq. (7.51), integrate over the delta functions and perform the spinor algebra, we get

$$T = \frac{1}{8\pi} \int dk^+ J(k^+) \int dy^- e^{-i(2E_\gamma - m_b - k^+)y^-/2}$$
$$\times \langle \bar{B}_v | T\left\{ \left[\bar{h}_v Y \right]((y^-/2)n^\mu) \left[Y^\dagger h_v \right](0) \right\} | \bar{B}_v \rangle. \tag{7.53}$$

The final matrix element involves only ultrasoft particles (heavy quarks and gluons) and no hard or collinear objects. Thanks to our previous results of soft-collinear factorization in the Lagrangian, this means that this last matrix element only receives contributions from soft, nonperturbative (but universal) QCD. This could, in principle, be calculated on the lattice, or it could be made to cancel between other processes such as $B \to X_u l \bar{\nu}$, since this term is flavor blind. The jet and hard functions are both computable in perturbation theory at the scale $Q_c^2 = m_b \Lambda$ and $Q_h^2 = m_b^2$, respectively.

Thus the problem is, at least in principle, solved to this order in our power counting. The beauty of SCET is that it also allows us to derive expressions for the subleading terms in our power-counting, which is very difficult to do in the full QCD. Subleading terms can be written as sum of factorized terms. For more information on that calculation, see the references at the end of the chapter.

---

[9]This is specific to this process. In general, the jet function is a function of the full k^2.

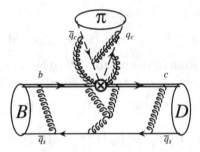

Fig. 7.1 The decay process $B \to D\pi$. This is a "Type-I" process, meaning that the spectator stays soft. All collinear gluons (springs with lines) stay within the π, while the spectator and heavy quark interact only through soft QCD. The $\otimes$ symbolizes an insertion of the decay operator $\mathcal{Q}_1^1$ described in the text.

7.4.2 $B \to D\pi$

Exclusive processes tend to be much harder to compute than inclusive processes, since we are imposing extra constraints on the phase space by forcing the spectator quarks and gluons to go the right way. But there are a few exclusive processes that are relatively simple and can be analyzed with little additional work. One such process is the decay of a B meson to a D meson and a pion [Bauer et al. (2001b)], shown schematically in Fig. 7.1.

This is an example of a "Type-I" process, meaning that the soft spectator from the B meson goes into the D meson and thus remains soft. This is the dominant contribution to this decay. There are other decays, such as $B \to \bar{D}K^0$ where this cannot happen, and it must be that the spectator ends up in the light meson (in this case, the K^0). This is called a "Type-II" process, and is much more difficult, because we are now forced to include soft-collinear interactions that could give a kick to the spectator and make it collinear. This requires the use of a new theory, called SCET$_{\text{II}}$, and is beyond the scope of this book. We discuss this possibility briefly in the exercises at the end of the chapter.

Once again, we wish to prove a factorization theorem for the decay width. The first step is to consider the operators that could generate the process in the full theory (we will treat Fermi's theory as the "full theory" in this context). After taking all of the symmetries of the problem into account, such as gauge invariance and parity, there is only one operator that can contribute:

$$\mathcal{H} = \frac{4G_F}{\sqrt{2}} V_{ud}^* V_{cb} \, (\bar{c}\gamma^\mu P_L b)(\bar{d}\gamma_\mu P_L u). \tag{7.54}$$

When matching this onto SCET, we want to match both the heavy c and b quarks to *ultrasoft* fields c_v, b_v[10] and the light u, d quarks that will end up in the pion will be matched to collinear fields ξ; we assume isospin invariance and make no reference to quark flavor to reduce complications. Color octet operators will not contribute to this order in power counting since our matrix elements will always be between color-singlet states. Also, parity requires that no γ^5 appears between the heavy quark fields. Therefore, the matched current in SCET is

$$\mathcal{Q}_1^1 = \sum_{\tilde{p},\tilde{p}'} \left[\bar{c}_{v'} \frac{\not{n}}{2} b_v \right] \left[(\bar{\xi}W)_{\tilde{p}'} \, C(\bar{\mathcal{P}} + \bar{\mathcal{P}}^\dagger) \frac{\not{n}}{4}(1-\gamma^5)(W^\dagger \xi)_{\tilde{p}} \right], \quad (7.55)$$

where we have explicitly included label-momenta on the collinear fields. The matching coefficient C depends on the label momenta, but since these are included in the fields, order of operations matters. We also took advantage of the fact that while the coefficient should depend on both $\bar{\mathcal{P}}^\dagger$ and $\bar{\mathcal{P}}$, the combination $\bar{\mathcal{P}} - \bar{\mathcal{P}}^\dagger = 2E_\pi$ is just a constant. This method of making the matching coefficients dependent on the label-momentum operators was described in [Bauer and Stewart (2001)]; note that when doing this, the ordering of operators (where we put the matching coefficient) matters.

Now when we match the matrix element for $\mathcal{H}$ to the matrix element for $\mathcal{Q}_1^1$, we note again that ultrasoft and collinear particles do not interact at this order in the power counting, so we can separate the matrix elements

$$\langle D\pi | \mathcal{H} | B \rangle \to \langle D_{v'} \pi_n | \mathcal{Q}_1^1 | B_v \rangle$$
$$= \langle D_{v'} | \bar{c}_{v'} \frac{\not{n}}{2} b_v | B_v \rangle$$
$$\times \sum_{\tilde{p},\tilde{p}'} \langle \pi_n | (\bar{\xi}W)_{\tilde{p}'} \frac{\not{n}}{4}(1-\gamma^5) C(\bar{\mathcal{P}} + \bar{\mathcal{P}}^\dagger)(W^\dagger \xi)_{\tilde{p}} | 0 \rangle. \quad (7.56)$$

The first matrix element is the $B - D$ form factor $m_B F_{B \to D}(0)$ from heavy quark physics; it is well known in the literature. To better understand the second matrix element, it is useful to pull out the matching coefficient with the help of a Dirac delta function:

$$\int d\omega \, C(\omega) \sum_{\tilde{p},\tilde{p}'} \langle \pi_n | (\bar{\xi}W)_{\tilde{p}'} \frac{\not{n}}{4}(1-\gamma^5) \delta(\omega - (\bar{\mathcal{P}} + \bar{\mathcal{P}}^\dagger))(W^\dagger \xi)_{\tilde{p}} | 0 \rangle$$
$$\equiv \int d\omega \, C(\omega) M_c(\omega). \quad (7.57)$$

---

[10] We will use this notation rather than the more cumbersome $h_v^{(b)}$ for b_v, etc.

Now we parametrize the large "parton" label momenta in this matrix element in terms of the pion momentum and the fraction $0 < x < 1$:

$$\bar{n} \cdot \tilde{p} = x\bar{n} \cdot p_\pi = 2E_\pi x, \qquad (7.58)$$

$$\bar{n} \cdot \tilde{p}' = (1-x)\bar{n} \cdot p_\pi = 2E_\pi(1-x); \qquad (7.59)$$

then the sum over label momenta is equivalent to an integral over x:

$$M_c(\omega) = \frac{1}{4}\int_0^1 dx \, \langle \pi_n | (\bar{\xi}W)_{1-x} \not{\bar{n}} \gamma^5 \delta(\omega - 2E_\pi(2x-1))(W^\dagger \xi)_x | 0 \rangle. \quad (7.60)$$

Now plug this back into Eq (7.57) to get

$$\frac{1}{4}\int_0^1 dx \, C\Big[(2x-1)2E_\pi\Big] \langle \pi_n | (\bar{\xi}W)_{1-x} \not{\bar{n}} \gamma^5 (W^\dagger \xi)_x | 0 \rangle. \qquad (7.61)$$

This matrix element is very well known. If we work in the gauge where $W = 1$ and ignore the x dependence in the matching coefficient, then this is nothing more than ($\bar{n}$ times) the axial current matrix element. The answer would then be

$$if_\pi \bar{n} \cdot p_\pi = 2iE_\pi f_\pi. \qquad (7.62)$$

Because of the x-dependence, however, the matrix element can be written as

$$2iE_\pi f_\pi \phi_\pi(x). \qquad (7.63)$$

Here, $\phi_\pi(x)$ is the "light-cone pion wavefunction," giving the probability that there is a parton in the pion with light-cone momentum $\tilde{p}^- = xE_\pi$. It has been normalized so that the integral of this function over the range of x is unity.

Finally, we can make a few definitions to simplify the notation:

$$N = \frac{i}{2} m_B f_\pi E_\pi,$$

$$T(x) = C\Big[(2x-1)2E_\pi\Big],$$

and write down the completely factorized formula for $B \to D\pi$ matrix element in SCET:

$$\langle D_{v'} \pi_n | \mathcal{H} | B_v \rangle = N F_{B \to D}(0) \int_0^1 dx \, T(x) \phi_\pi(x). \qquad (7.64)$$

Once again, the final result gives an answer in terms of a product of a hard function T, a soft function F, and collinear function ϕ_π, each independent of the others. This factorization is independent of perturbation theory, and is therefore true to all orders in α_s, although only to leading order in Λ/m_b.

7.4.3 Deep Inelastic Scattering

Deep Inelastic Scattering (DIS) provides one of the nicest examples of how SCET can be used to solve problems outside of precision flavor physics. There are many similarities between it and the analysis for $B \to X_s\gamma$, but there are also several key differences. There are many great sources that explain DIS from the full QCD point of view; you can start with [Peskin and Schroeder (1995)]. Our focus is on how to apply EFT techniques to derive expressions for DIS form factors in the "endpoint region" (defined below); we will closely follow the paper by [Manohar (2003)].

The process involves the scattering of a lepton (usually an electron) on a nucleon. From the theory point of view, the lepton is merely the source of a virtual photon with momentum transfer squared $q^2 = -Q^2$. The process is shown schematically in Fig. 7.2a. It was discovered in the late 1960s that the resulting form factors in the cross section only depend on a single kinematic variable, known as the Bjorken scaling variable [Bjorken (1969)]:

$$x = \frac{Q^2}{2P \cdot q}, \tag{7.65}$$

and not directly on Q^2 at all; this is known as *Bjorken scaling*. This variable can be thought of as the fraction of the proton momentum that is carried by a parton such as a quark or gluon. It was later learned after the discovery of QCD that Q^2 dependence does enter the structure functions in the form of logarithmic corrections, known today as *Dokshitzer-Gribov-Lipatov-Altarelli-Parisi (DGLAP) Evolution* [Gribov and Lipatov (1972); Dokshitzer (1977); Altarelli and Parisi (1977)]. We often like to describe the process in the *Breit frame*, where the proton is moving at very high speed and then recoils hard in the opposite direction. For this reason it is also called the infinite momentum frame.

For arbitrary values of x, the cross section is described quite well in terms of perturbative QCD, but when we take the limit that $x \to 1$, we begin to encounter problems. Large corrections of the form $\alpha_s \log(1-x)$ can spoil the perturbation expansion and we must resum these corrections to get a reasonable answer. Just like in $B \to X_s\gamma$ this is calling for an EFT approach.

The reason for these large logarithms can be understood in terms of what is happening to the jet of particles created after the nucleon has shattered. If we compute the energy-momentum of the resulting jet, we find that it has a "mass"

$$p_X^2 = \frac{Q^2}{x}(1-x), \tag{7.66}$$

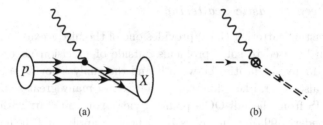

Fig. 7.2 Deep Inelastic Scattering. (a) A schematic view of the process, where a hard, virtual photon scatters off of a parton inside a nucleon. (b) The process can be matched onto a SCET current in the Breit frame, with a collinear quark (dashed line) and an "anti-collinear" quark (double-dashed line). The $\otimes$ represents an insertion of the current in Eq. (7.67).

which goes to zero as $x \to 1$. This is analogous to what happens in $B \to X_s \gamma$ where the resulting s quark jet goes on shell. Following the same analogy there, we can identify the "endpoint region" of DIS when $1 - x \simeq \Lambda_{\rm QCD}/Q$, which is usually taken to be pretty small, since Q^2 is typically at least several GeV^2 in any DIS experiment. We therefore find $p_X^2 \simeq Q\Lambda$ and we have another collinear scale in the problem. To properly handle this new scale, we cannot simply do an OPE at the hard scale Q since we cannot integrate out the jet yet. This is the source of the large logarithms.

Following the $B \to X_s \gamma$ analogy further, it seems that the best solution is to match our QCD theory onto SCET at the hard scale Q, run the theory down to the jet scale $\sqrt{Q\Lambda}$, integrate out the jet to get a nonlocal product and *then* do an OPE. If we wish, we can then continue to run the theory down to some hadronic scale and match the result to a parton distribution function.

When working in the target frame, which has a stationary nucleon struck by a hard, virtual photon, DIS and $B \to X_s \gamma$ are remarkably similar, with the b quark's role taken on by a parton in the nucleon, and the s quark's role taken on by the scattered parton. In the Breit frame, however, the incoming parton is moving very fast, and is therefore properly described by a n-collinear quark ξ_n in SCET, while the post-collision parton is a fast-moving particle moving in the *opposite direction* and is therefore described by a $\bar{n}$-collinear quark $\xi_{\bar{n}}$. The QED current $\bar{q}\gamma^\mu q$ is therefore matched onto a new SCET current at leading order in the power counting:

$$J = (\bar{\xi}_n W_n)\gamma^\mu (W_{\bar{n}}^\dagger \xi_{\bar{n}}). \tag{7.67}$$

This current is shown (at tree level) in Fig. 7.2b. Here we have our first example of a problem that has more than one collinear direction.

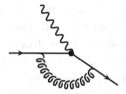

Fig. 7.3 The full QCD graph representing the renormalization of the QED vertex inside the nucleon.

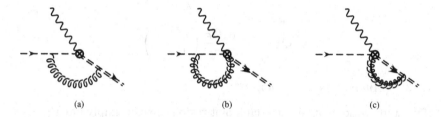

Fig. 7.4 SCET one-loop matching calculation. In dimensional regularization, all three of these diagrams are exactly zero, but they can still contribute to the anomalous dimension. (a) Soft gluon contribution. (b) Collinear quark - collinear gluon (emitted from the W_n). (c) Anti-collinear quark - anti-collinear gluon (emitted from the $W_{\bar{n}}$). Not shown are the self-energy diagrams (gluons emitted and absorbed on the same quark line). Notice that the collinear and anti-collinear sectors are decoupled, giving us factorization at the operator level.

7.4.3.1 *Matching coefficient*

We need to match QCD onto SCET and compute the matching coefficient. The procedure is standard by now: we compute matrix elements of the full QED-QCD current and the QED-SCET current, to one loop order in α_s. In QCD, the diagram that needs to be calculated is given in Fig. 7.3, and the resulting matrix element is

$$\langle j^\mu \rangle = \gamma^\mu \left[1 + \frac{\alpha_s C_F}{4\pi} \left(-\frac{2}{\epsilon^2} - \frac{1}{\epsilon} \left(2\log\left(\frac{\mu^2}{Q^2}\right) + 3 \right) \right. \right.$$
$$\left. \left. - \log^2\left(\frac{\mu^2}{Q^2}\right) - 3\log\left(\frac{\mu^2}{Q^2}\right) - 8 + \frac{\pi^2}{6} \right) \right]. \quad (7.68)$$

Notice that all the poles here are IR-divergences; there are no counterterms in QCD besides wavefunction renormalization since this is a conserved current.

Now in SCET, the corresponding matrix element involves three diagrams, with the exchange of an ultrasoft gluon, a n-collinear gluon from the Wilson line, and an $\bar{n}$-collinear gluon from the other Wilson line; these are shown in Fig. 7.4. However, all of these diagrams vanish in dimensional regularization, since they are all scalelss integrals (see Appendix C). Therefore, the matching coefficient is simply the finite terms in Eq. (7.68). The IR-divergent terms are reproduced by the matrix elements in both theories (cancelled by the UV-divergent terms – more on this below). To avoid any troublesome logarithms, we should match at the scale $\mu^2 = Q^2$, and we get

$$C(Q) = 1 + \frac{\alpha_s C_F}{4\pi}\left(\frac{\pi^2}{6} - 8\right). \tag{7.69}$$

7.4.3.2 *Anomalous dimension*

If we were working outside the endpoint region, we can simply integrate out the (massive) jet at this stage, and perform an OPE to get an expression for the cross section. But when $1-x \simeq \Lambda/Q$ we can no longer do this, since the jet can generate collinear divergences that will create large logarithms which destroy the validity of perturbation theory. We must therefore continue to treat the jet as a massless object and run the theory down to the jet scale. This means we must compute the anomalous dimension of the current operator in Eq (7.67).

There are no couplings in the action between n- and $\bar{n}$-collinear fields, since such couplings would require a field to go offshell. This means that the only graphs that contribute to the anomalous dimensions are the same ones that contributed to the matching. We mentioned above that these diagrams all vanish in dimensional regularization since they are scaleless. However, that happens because there is an effective cancelation between UV and IR divergences. That is fine for a matching calculation, but the anomalous dimension is only sensitive to the UV behavior of the theory. Therefore we need to find the UV divergent part and ignore the IR divergent part when trying to compute the anomalous dimension.

One way to extract the nonzero result is to regulate the IR divergences with some different regulator. A very nice choice is to assume that the quarks are off-shell by a small amount $p^2 \neq 0$. This will safely regulate all the IR divergences, meaning that whatever poles are left over from dimensional regularization must be UV divergences. This is worked out very nicely and carefully in Manohar's paper.

However, there is a brilliant shortcut to doing more loop integrals! Since this cancellation occurs between the UV and IR divergences in SCET, it must be that *the UV divergent terms are exactly equal to (minus) the IR divergent terms in Eq (7.68)*. This is true because SCET must give exactly the same IR divergences as QCD, otherwise the theories would not match. Thus the SCET IR divergences are the same as the QCD IR divergences. But if the SCET UV divergences are to cancel the SCET IR divergences, then we have our result. This means we can simply read off the counter term in SCET from the matching calculation without doing any more work:

$$Z_{\text{SCET}} = 1 - \frac{\alpha_s C_F}{4\pi}\left(\frac{2}{\epsilon^2} + \frac{1}{\epsilon}\left(2\log\left(\frac{\mu^2}{Q^2}\right) + 3\right)\right). \quad (7.70)$$

This result is truly amazing – SCET has converted an IR divergence in the full theory (QCD) into a UV divergence in the EFT. And UV divergences mean that we can bring the full machinery of the renormalization group to bear! You cannot renormalize IR divergences: there are no counterterms. Previously, physicists have worked very hard to discover how to resum IR logarithms like the Sudakov double-log, but now we know what to do: simply convert these IR divergences into UV ones, and use the RG. The price is to give up on using the full QCD and to focus instead on an EFT approach: a very small price to pay!

Now that we have the counter term we can read off the anomalous dimension from (two times) the coefficient of the $1/\epsilon$ term:

$$\gamma = -\frac{\alpha_s C_F}{\pi}\left[\log\left(\frac{\mu^2}{Q^2}\right) + \frac{3}{2}\right]. \quad (7.71)$$

This anomalous dimension has an explicit $\log(\mu)$ in it, which is much different than most results. This additional logarithm is the Sudakov double-log that appears when there is a combined soft and collinear divergence. Notice that the Q^2 dependence is not suggesting the appearance of derivatives in the anomalous dimension, which would not make sense: as far as SCET is concerned, Q is simply a label (like v in HQET), and it can appear anywhere.

This anomalous dimension has the form

$$\gamma = \Gamma \log(\mu^2/Q^2) + B. \quad (7.72)$$

In this context, Γ is known as the *cusp anomalous dimension*, because in QCD it comes from evaluating products of Wilson lines that meet at sharp points. These terms are common when multiple jets are produced in a process.

7.4.3.3 Running

Now that we have an expression for the anomalous dimension, we can perform the RG running down to the jet scale $p_X^2 = Q^2(1-x)$ to resum the large $\alpha_s \log(1-x)$ terms. We need to solve

$$\mu \frac{d}{d\mu} C(\mu) = \gamma(\mu) C(\mu), \qquad (7.73)$$

subject to the matching condition in Eq. (7.69). There is scale dependence in two places: explicitly from the logarithm, and implicitly from the strong coupling constant, which at one loop is

$$\alpha_s(\mu) = \frac{2\pi}{b_0 \log(\mu/\Lambda)}. \qquad (7.74)$$

The solution to the running matching coefficient at the jet scale is

$$C_{\text{jet}} = \left[1 + \frac{\alpha_s(Q) C_F}{4\pi}\left(\frac{\pi^2}{6} - 8\right)\right]$$
$$\times \exp\left\{\left(\frac{8\pi C_F}{b_0^2 \alpha_s(Q)}\right)[1 - \zeta + \log(\zeta)] + 3 C_F \log(\zeta)\right\}, \qquad (7.75)$$

where

$$\zeta \equiv \frac{\alpha_s(Q)}{\alpha_s(p_X)}. \qquad (7.76)$$

7.4.3.4 OPE

Now that we have integrated out the physics at the intermediate jet scale, we can safely integrate out the hard-collinear jet without worrying about large $\log(1-x)$ terms that might spoil perturbation theory. At this point, we match the cross section by performing an operator product expansion. A very nice description of this is given in [Manohar and Wise (2000)], for example.

For the case at hand, we consider matrix elements of the product of SCET currents

$$W^{\mu\nu} = \frac{1}{2\pi} \int d^d y \, e^{iq\cdot y} \mathcal{J}^\mu(y) \mathcal{J}^\nu(0) \longrightarrow C_{\text{jet}}^2 \mathcal{O}. \qquad (7.77)$$

From the optical theorem, the total cross section is related to the imaginary part of this tensor. $\mathcal{O}$ is a nonlocal operator that we have matched $W^{\mu\nu}$ onto, that only contains ultrasoft and n-collinear fields, but no $\bar{n}$-collinear fields:

$$\mathcal{O} = \frac{1}{4\pi} \int dz \, e^{izr^+} \left\{(\bar{\xi} W)_n(nz) \cdot \slashed{n} \mathcal{Y}_n(z,0)(W^\dagger \xi)_n(0)\right\}; \qquad (7.78)$$

$\mathcal{Y}$ is a product of ultrasoft Wilson lines that are required to maintain ultrasoft gauge invariance in the operator. This operator was first written down in QCD by [Collins and Soper (1982)].

At this point, we can follow the usual results from QCD, where it was shown that upon performing a Mellin-transform on the cross section the final result can be factorized into a hard function and a parton-distribution function. We can also continue to apply RG-techniques and run the corresponding matrix elements of $\mathcal{O}$ down to a lower hadronic scale, resumming potentially large DGLAP-logarithms, but we will stop at this point.

7.4.4 Jet production

The previous example of DIS was useful for two reasons: first, because DIS is an important experiment in its own right; and second, because it is a relatively simple example where (in the Breit frame) there are *two* collinear directions to worry about! By crossing symmetry, this same current gives us processes where two hard quarks are created in a collision and fly off back-to-back. This is the dominant underlying process that describes *jet physics* at collider experiments.

Jets are a real challenge to the effective field theorist, because like exclusive decays, they necessarily involve hard cuts on phase space, and these cuts always lead to inherent IR divergences that need to be handled very carefully. For example, the simplest kind of jet, the *cone jet*, is defined so that all of the particles emitted are contained within a cone of opening angle δ. Such cuts on the phase space of particles will necessarily involve contributions to the cross section that go like $\log \delta$, and for small enough cone angle, this logarithm will threaten perturbation theory. At first, this might sound like a perfect chance for SCET to do its thing, resumming these logarithms through a RG-type calculation. Unfortunately, it is not obvious how to resum these logarithms, because they are not coming from any *dynamical* cutoff but rather are enforced by hand. It is therefore not clear which Feynman diagrams you are resumming when you try to run your theory down to a lower scale.

Nevertheless, there have been many attempts to try and cast the question in terms of an EFT approach, by trying to chose $\delta \simeq \Lambda/Q$, with Q representing the center-of-mass energy of the collision. There has been limited success with this approach to the problem; see [Trott (2007); Cheung et al. (2009)].

Another tactic that was suggested was to study weighted cross sections;

see [Bauer et al. (2004)] for example. In this case, an observable e is constructed in terms of the kinematics of the problem, and the differential cross section is augmented by a delta function:

$$\frac{d\sigma}{de} = \int_{\text{P.S.}} d\sigma \, \delta(e - f_e(\{p\})), \qquad (7.79)$$

where $f_e(\{p\})$ is a function of phase space. Upon integrating over all of phase space (without any applied cuts), the result is called an *Event Shape Distribution*. Although there are no explicit cuts in the integrals, certain limiting values of e can often be interpreted as special configurations of phase space. The most famous example is the *thrust* distribution, where thrust is defined as

$$f_T(\{\mathbf{p}_i\}) = \frac{1}{\sqrt{s}} \max_{\hat{\mathbf{t}}} \sum_i |\hat{\mathbf{t}} \cdot \mathbf{p}_i|. \qquad (7.80)$$

In the limit that $T \to 1$, the phase space configuration is that of two back-to-back jets; you can see this is true by looking at Eq (7.80) and noting that for $T = 1$ all of the momenta must be collinear or anti-collinear to a single direction $\hat{t}$, known as the *thrust axis*. It is not a surprise that the corresponding distribution has contributions that go like $\log(1 - T)$; these terms, however, *can* be resummed in an unambiguous way. Other event shape distributions have similar behavior.

Yet another approach to jet physics in SCET is to directly compute production cross sections with a given number of final particles each going in a different lightlike direction [Bauer and Schwartz (2007)]. This is possible because in the twist expansion of QCD, the dominant contributions to the cross section involve strongly-ordered jets, where $Q \gg p_T^{(1)} \gg p_T^{(2)} \gg \cdots$; therefore we can apply an SCET description between each hard splitting. The process can be thought of as starting from a two-quark production current such as the one given in Eq (7.67). The theory can then be run down from the hard scale $Q = \sqrt{s}$ to the first hard emission scale given by $p_T^{(1)}$. At this scale, SCET is matched onto a new theory which contains *two* collinear directions n_1, n_2, and the new operators now involve three fields. One can then run the theory down further until the next splitting occurs and match onto yet another SCET with three lightlike directions, and so on until one reaches a scale close to Λ_{QCD} where you can no longer make a hard splitting. The resulting combination of matching-and-running can be shown to reproduce the usual parton-shower result, but it is also hoped that by considering higher order terms in the anomalous dimensions and matching coefficients we can incorporate subleading terms into the Monte-Carlo generators that produce jets. This is an ongoing branch of research.

7.5 Notes for further reading

SCET was first used to study Sudakov logs in [Bauer et al. (2000)], and was worked out in more detail and generality in [Bauer et al. (2001a, 2002b,a)]. All of these references are excellent sources for students to see how to derive and apply the techniques of this chapter. The methods of using SCET$_{II}$ are described first in [Bauer et al. (2003)], and all of the Feynman rules and matching calculations to NLO in power counting are given in a great summary paper by [Pirjol and Stewart (2003)].

For a detailed description of computing processes like $b \to s\gamma$ at subleading order in the power counting, the reader should look at [Paz (2009); Benzke et al. (2010)] and their references.

Our treatment of deep inelastic scattering follows [Manohar (2003)] pretty closely. Another nice paper on using SCET in DIS is [Becher et al. (2007)]. Although he does not use SCET directly, the reader can also see [Paz (2010)] to see how DIS can be treated from an EFT point of view.

The somewhat controversial method of "Zero-Bin Subtraction" was introduced in [Manohar and Stewart (2007)]. It brings up the very subtle issue of double-counting the soft region when computing collinear loops. We did not have space to talk about this topic here, but those who wish to work in SCET would do well to read this paper carefully.

For more information on alternative formulations of SCET that do not require choosing a special frame, and thus avoid some of the criticisms mentioned in the text, the reader should take a look at [Freedman and Luke (2012)].

A very detailed review article focussed specifically on SCET was written by [Becher et al. (2014)].

Problems for Chapter 7

(1) **Filling in gaps**

(a) Plug the expansion in Eq. (7.24) into the QCD-quark action, and derive Eq. (7.25).

(b) Derive the collinear-gluon action for $SU(3)$ in a general-covariant gauge like R_α gauge. Don't forget to include collinear ghosts!

(c) Prove the RPI equations (7.40-7.46). Use these transformations on the operators in the SCET collinear Lagrangian to show that the theory is invariant under all of the RPI transformations.

(2) **Soft Fields and SCET$_{II}$**

Another region that appears in many multi-scale QCD problems is the *soft* region, where light-like momenta scale like $Q(\lambda, \lambda, \lambda)$. This region of phase space is often relevant when handling *exclusive* decays. In that case, we dispense with ultrasoft fields (which are contained in the soft fields) but collinear fields whose momenta scale like $Q(\lambda^2, 1, \lambda)$ can still be present.

(a) Show that you cannot couple soft and collinear fields together in the Lagrangian (to leading order in λ) in an effective theory. *HINT:* Remember that our theory can only contain on-shell modes. This result shows that all soft-collinear interactions factorize automatically to leading order in λ, and all soft-collinear interactions must reside in currents.

(b) When the soft region is relevant, it is often convenient to first match onto an intermediate theory at $\mu = Q$ with ultrasoft fields and *hard-collinear fields* that scale like $Q(\eta^2, \eta^2, \eta^2)$ and $Q(\eta^2, 1, \eta)$ respectively, where $\eta = \sqrt{\lambda}$. This is called SCET$_I$; it is identical to the theory we have been working with before. Then at the intermediate "hard-collinear scale" $\mu = Q\eta$ we integrate out the hard collinear modes. We match the ultrasoft modes directly to soft modes (since $\eta^2 = \lambda$), and we introduce new *collinear* modes that scale properly in terms of the true power-counting parameter. This final EFT is called SCET$_{II}$. As a relatively simple example, consider the decay $B^0 \to \bar{D}^0 K^0$, first discussed in [Mantry et al. (2003)].

    i. Write down the leading operators that mediate the decay in QCD, and match them onto currents in SCET$_I$.

    ii. Now compute the matrix element of these currents at tree level. Note that you will need to convert hard-collinear quarks from the current into ultrasoft quarks so they can play the role of spectators. To do this, you can use an insertion of $\mathcal{L}^{(1)}_{\xi q}$.

    iii. Next, match the corresponding matrix elements to new operators in SCET$_{II}$. This comes from shrinking the hard-collinear gluon propagators from the SCET$_I$ diagrams to a point. You should find three diagrams: one contact term where all six quarks come from a single vertex, as well as two diagrams where the hard-collinear quarks are matched onto collinear quarks. These two diagrams require the insertion of a subleading SCET$_{II}$ Lagrangian $\mathcal{L}^{(1)}_{\xi\xi qq}$. Work out the form of this operator.

iv. Rewrite the final result in terms of a factorized amplitude. What is the interpretation of each factor?

(3) **SCET Event Generators**

When trying to simulate a jet event, we can use currents similar to Eq. (7.67), with more general Lorentz-Dirac, color or flavor structure; but instead of $n, \bar{n}$ to represent the (anti-)collinear direction, we use n_q^μ, $n_{\bar{q}}^\mu$. A gluon can also be bremm'ed off a (anti-)quark line with the insertion of $B_\mu^\perp$, descibed in Eq. (7.32).

(a) Write down the leading order operator responsible for the emission of a collinear gluon in a third direction n_g^μ, assuming a general Lorentz-Dirac structure Γ.

(b) Compute the matching coefficients and anomalous dimensions for 2-jet operators, matched at the hard scale Q. Note that this is similar to the DIS result, but changed by a crossing symmetry, so that the logarithm actually creates an imaginary matching coefficient. What does the imaginary part mean?

(c) Now repeat the calculation for the 3-jet operator. Begin by matching from QCD at the hard scale Q. Then run the scale down to the transverse momentum of the emitted gluon $p_T^{(1)}$, and match the result to a new EFT with three collinear directions. Notice that the 2-jet operator has a nonzero matrix element to produce a 3-jet event at scales above $p_T^{(1)}$, but not below. This makes the matching a bit nontrivial.

(d) Compare your result from the previous question to the usual Parton Shower result used in programs like PYTHIA.

Chapter 8

Standard Model as an Effective Field Theory

8.1 Introduction

Our previous discussions in this book dealt with various infrared limits of the standard model (SM). The applications of EFT techniques were useful there because of the presence of several well separated scales. In all cases the standard model was a nice, renormalizable field theory to which we could often do a meaningful if not perturbative matching. Besides, it survived years of intense experimental and theoretical attempts to find any significant deviations between predictions and measurements. Yet, there are reasons to believe that the SM is not a complete theory. For example, it does not contain candidates for dark matter particles and has difficulties in describing baryogenesis [Peskin and Schroeder (1995); Donoghue et al. (1992)]. Could it be that the standard model is a low-energy approximation to some other, more complete and fundamental theory?

If the standard model is indeed a leading order approximation to some more fundamental theory in the EFT expansion it would be truly unique. No other effective theory that we considered so far had the leading order term that by itself represented a perturbatively renormalizable quantum field theory. In other words, the leading order term has no encoded information about the scale at which it stops being a meaningful representation of a complete theory. Technically, the SM is ok all the way to the Planck scale at which the quantum gravity effects become important. Most theoretical excursions into the realm of new physics were chiefly motivated by aesthetic considerations such as requirement of gauge coupling unification or the absence of fine-tuning of the Higgs mass.

There is one clear indication of physics beyond the standard model: the neutrino oscillations. Their observation clearly indicates the need for massive neutrinos that are not present in the standard model. There are, then, several possibilities for how new physics scale(s) Λ to enter

(1) The scale Λ associated with new physics could be close to the electroweak scale $v = 246$ GeV. This is, of course, the most exciting possibility, as new physics particles can be directly observed in high energy experiments, such as the ones that are being carried out at the Large Hadron Collider (LHC). This situation, however, might not be the best for the EFT description of new particles, as the relevant expansion parameters, v/Λ could be rather large.
(2) The new physics scale could be significantly larger that the SM electroweak scale, $\Lambda \gg v$. This is exactly the situation that EFT is designed to deal with. If the new physics scale is too large for any current or future collider to probe directly, indirect searches for quantum effects of new particles might be the only way to study Beyond Standard Model (BSM) physics.
(3) There is no new physics beyond that currently discovered (plus sterile right-handed neutrinos), or there are no new scales all the way to the Planck scale. This possibility stems from the fact that the SM is a perturbatively renormalizable theory. Adding the right-handed neutrino as a dark matter candidate removes the need for any BSM physics if one accepts all fine tunings present in the standard model. Finally, conformal symmetry might be somehow realized for some window of energies below Planck mass.

In what follows we shall assume that the separation of scales between the standard model and new physics is sufficiently large, so meaningful description of the SM as effective field theory is possible [Appelquist and Carazzone (1975)]. If this is so, the effective Lagrangian can be written as

$$\mathcal{L} = \mathcal{L}_{\text{SM}} + \mathcal{L}^{(5)} + \mathcal{L}^{(6)} + \cdots$$
$$= \mathcal{L}_{\text{SM}} + \frac{1}{\Lambda} \sum_k C_k^{(5)} Q_k^{(5)} + \frac{1}{\Lambda^2} \sum_k C_k^{(6)} Q_k^{(6)} + \mathcal{O}\left(\frac{1}{\Lambda^3}\right). \quad (8.1)$$

If we are to treat the standard model as an EFT, we need to follow the usual set of rules. That is to say, the effective field theory that we are to write should (i) be invariant under the same set of symmetries as the SM, that is, $SU(3)_C \times SU(2)_L \times U(1)_Y$, and (2) should contain only the SM degrees of freedom. This assumes that whatever the full theory is, the SM degrees of freedom are incorporated into it as fundamental or composite fields. We are going to assume in what follows that the electroweak symmetry breaking is realized linearly.

8.2 Standard model as the leading term in the EFT expansion

The leading term in this effective field theory is the standard model, which is described in details in a number of textbooks [Peskin and Schroeder (1995); Donoghue et al. (1992); Itzykson and Zuber (1980)]. We shall not review it here. However, since the EFT that we are about to build is required to be invariant under the large class of symmetries, it might be useful to set up short-hand notations for the quantities that we will be working with in this chapter. The SM Lagrangian can be written as

$$\mathcal{L}_{\rm SM} = \mathcal{L}_{\rm gauge} + \mathcal{L}_{\rm kinetic} + \mathcal{L}_{\rm Higgs} + \mathcal{L}_{\rm Yukawa}, \qquad (8.2)$$

where $\mathcal{L}_{\rm gauge}$ and $\mathcal{L}_{\rm kinetic}$ contains kinetic terms of the gauge and fermion fields, respectively, $\mathcal{L}_{\rm Higgs}$ contains Higgs kinetic and potential terms, and $\mathcal{L}_{\rm Yukawa}$ contains Yukawa sector of the standard model. In terms of our short-hand notation, the gauge kinetic terms are

$$\mathcal{L}_{\rm gauge} = -\frac{1}{4}G^A_{\mu\nu}G^{A\mu\nu} - \frac{1}{4}W^I_{\mu\nu}W^{I\mu\nu} - \frac{1}{4}B_{\mu\nu}B^{\mu\nu}, \qquad (8.3)$$

where the gauge field strength tensors are defined in the usual way (see, e.g. Appendix B) The fermion kinetic terms are given by

$$\mathcal{L}_{\rm kinetic} = \bar{l}i\slashed{D}l + \bar{e}i\slashed{D}e + \bar{q}i\slashed{D}q + \bar{d}i\slashed{D}d. \qquad (8.4)$$

Note that the matter fields in Eq. (8.4) bear weak isospin $j = 1, 2$, color $\alpha = 1, 2, 3$, and generation indices $p = 1, 2, 3$. The matter content includes left-handed lepton weak doublet fields l^j_{Lp} with weak hypercharge $Y = -1/2$, right-handed weak singlet lepton fields e_{Rp} with $Y = -1$, left-handed quark weak doublet fields $q^{\alpha j}_{Lp}$ with $Y = 1/6$, and right-handed weak singlet fields up and down quark fields u^α_{Rp} and d^α_{Rp} with weak hypercharge $Y = 2/3$ and $-1/3$, respectively. In this chapter we shall suppress the L or R chirality indices, as they can be immediately restored by looking at the Dirac structure of operators.

The covariant derivatives in Eq. (8.4) are defined as

$$(D_\mu q)^{\alpha j} = \left(\partial_\mu + ig_s T^A_{\alpha\beta} G^A_\mu + ig S^I_{jk} W^I_\mu + ig' Y_q B_\mu\right) q^{\beta k}, \qquad (8.5)$$

where $T^A = \lambda^A/2$ with Gell-Mann matrices λ^A represent the generators of color $SU(3)$, while $S^I = \tau^I/2$ with Pauli matrices τ^I represent the $SU(2)$

generators of weak interactions. Y_q represents the hypercharge of the quark of type q as defined above.

The Higgs-related part of the SM Lagrangian is defined as

$$\mathcal{L}_{\text{Higgs}} = (D_\mu H)^\dagger (D^\mu H) + m^2 H^\dagger H - \frac{\lambda}{2} \left(H^\dagger H\right)^2. \tag{8.6}$$

It will be useful to define a Hermitian derivative acting on the Higgs doublet field H,

$$H^\dagger i \overleftrightarrow{D}_\mu H = H^\dagger i \left(D_\mu - \overleftarrow{D}_\mu\right) H,$$

$$H^\dagger i \overleftrightarrow{D}_\mu^I H = H^\dagger i \left(\tau^I D_\mu - \overleftarrow{D}_\mu \tau^I\right) H. \tag{8.7}$$

Finally, the Yukawa part of the SM Lagrangian is

$$-\mathcal{L}_{\text{Yukawa}} = \Gamma_e \bar{l} e H + \Gamma_u \bar{q} u \widetilde{H} + \Gamma_d \bar{q} d H + \text{h.c.}, \tag{8.8}$$

where Γ_i are the matrices of Yukawa coupling constants. Here $\widetilde{H} \equiv \widetilde{H}^j = \epsilon_{jk} \left(H^k\right)^*$ is a conjugated Higgs field. We define the totally antisymmetric tensor ϵ_{jk} with $\epsilon_{12} = +1$.

Quark and charged lepton masses appear after spontaneous symmetry breaking when the Higgs field acquires a vacuum expectation value. In unitary gauge [Peskin and Schroeder (1995); Donoghue et al. (1992)],

$$H = \frac{1}{\sqrt{2}} \begin{pmatrix} 0 \\ v + h(x) \end{pmatrix}. \tag{8.9}$$

Setting $h(x) \to 0$, all quark and charged lepton masses are proportional to a product $\Gamma_p v$. It is rather unsettling that a large hierarchy of quark and lepton masses is simply parameterized by an unnaturally large hierarchy of the values of Yukawa couplings. While technically natural, this situation is aesthetically not pleasing and goes under the name of the standard model *flavor problem*[1]. Its solution is often addressed in various BSM scenarios which, in principle, could be described by higher order terms in EFT description of the BSM physics which we now turn to.

8.2.1 Dimension-5 operators: fermion number violation

The discussion of dimension-five operators happens to be extremely interesting. As it turns out, if we impose the standard model gauge symmetry

---

[1]It might well be that the solution of the SM flavor problem is of the "just so" type. After all, there are other hierarchies in physics that do not require additional symmetries for explanation. For example, a hierarchy of masses of planets in our Solar System is simply accepted.

constraints on $\mathcal{L}^{(5)}$ of Eq. (8.1) we are left with only a single operator,

$$Q^{(5)} = \epsilon_{jk}\epsilon_{mn}H^j H^m \left(l_p^k\right)^T C l_r^n \equiv \left(\tilde{H}^\dagger l_p\right)^T C \left(\tilde{H}^\dagger l_r\right). \tag{8.10}$$

This curious fact was noticed by S. Weinberg in [Weinberg (1979a)], so the contribution of Eq. (8.10) is sometimes referred to as "Weinberg operator". Here C is a charge-conjugation operator, $C = i\gamma^2\gamma^0$ in the Dirac representation of gamma-matrices.

This operator violates lepton number. There are no terms in the standard model Lagrangian that do this. Therefore, experimental searches for lepton-number violation could boost the idea that SM is indeed an effective theory for some more fundamental one.

What would be the experimental consequences of the Weinberg operator? The most important one is the generation of (Majorana) neutrino masses. As the reader might remember, the minimal standard model only contains left-handed neutrino fields. Thus, no mass term of the type $m_D \bar{\nu}_L \nu_R +$ h.c. is possible as there are no right-handed neutrino fields. If we simply postulate the existence of (non-interacting) right-handed neutrino fields, we could extend the usual Higgs mechanism to neutrinos to generate "Dirac masses".

Alternatively, if neutrinos are their own antiparticles, i.e. they are the Majorana fermions, it is possible to write a mass term with purely left-handed fields, which would violate the lepton number

$$\mathcal{L}_\nu = M\bar{\nu}_L^c \nu_L + \text{h.c.} \tag{8.11}$$

where ν_L^c stands for the charge-conjugated neutrino field. After spontaneous symmetry breaking, the Weinberg operator of Eq. (8.10) leads to Majorana neutrino masses,

$$(M)_{pq} \sim \frac{1}{2} C_{pq}^{(5)} \frac{v^2}{\Lambda}. \tag{8.12}$$

This relation immediately implies $\Lambda > 10^{14}$ GeV to generate neutrino masses $M_\nu < 1$ eV, if we set $C^{(5)} \sim 1$. This is a rather high scale that is unlikely to be probed in experiments at particle accelerators. This scale can be somewhat lowered by making the Wilson coefficients smaller. Alternatively, a requirement of lepton number conservation can be placed on the effective operators, assuming that neutrino masses are generated by some other mechanism. In this case, the first new physics correction would be parameterized by the operators of dimension six.

8.2.2 Dimension-6 operators: parameterizing new physics

The set of operators of dimension six contains fermion-number conserving as well as fermion-number violating operators, which we break into several classes. These operators parameterize contributions from BSM physics. While historically it was the paper [Buchmuller and Wyler (1986)] that first attempted to address a complete set of those effective operators, there are, in fact, several ways to choose a complete set of non-redundant dimension-6 operators [Grzadkowski et al. (2010); Giudice et al. (2007); Duehrssen-Debling et al. (2015)]. All of those sets (or *bases*) are equivalent up to operator redefinitions, Fierz transformations, equations of motion and other operator identities. In what follows we shall use the so-called "Warsaw basis" [Grzadkowski et al. (2010)]. This set contains 59 baryon-number conserving and 5 baryon-number violating operators, which are organized into twelve subsets (see Tables 8.1-8.4). Among those subsets there are five that contain various four-fermion operators (Tables 8.3 and 8.4). Spelling out different flavors of quarks and leptons gives a total of 2499 baryon-number conserving operators. As one can see, the complete basis of only dimension-6 operators is rather large.

The SM effective Lagrangian can be built the same way as any effective Lagrangian discussed in previous chapters: by combining relevant fields (degrees of freedom) consistent with the symmetries of the theory and then using equations of motion, field redefinitions, or other operator identities to eliminate redundant operators to find the minimal set. For example, the use of equations of motion is really simplified for the derivation of the set of dimension six operators. Since any operators suppressed as $1/\Lambda^3$ can be dropped, only EOMs obtained from the operators of dimension four (i.e. $\mathcal{L}_{\text{SM}}$ of Eq. (8.2)) would be needed. They are

$$\begin{aligned}
(D^\mu D_\mu H)^j &= m^2 H^j - \lambda \left(H^\dagger H\right) H^j - \bar{e}\Gamma_e^\dagger l^J + \epsilon_{jk}\bar{q}^k\Gamma_u u - \bar{d}\Gamma_d^\dagger q^j, \\
(D^\alpha G_{\alpha\mu})^A &= g_s \left(\bar{q}\gamma_\mu T^A q + \bar{u}\gamma_\mu T^A u + \bar{d}\gamma_\mu T^A d\right), \\
(D^\alpha W_{\alpha\mu})^I &= \frac{g}{2}\left(H^\dagger i \overleftrightarrow{D}_\mu^I H + \bar{l}\gamma_\mu \tau^I l + \bar{q}\gamma_\mu \tau^I q\right), \\
\partial^\alpha B_{\alpha\mu} &= g' Y_H H^\dagger i \overleftrightarrow{D}_\mu H + g' \sum_{f=l,e,q,u,d} Y_f \bar{f}\gamma_\mu f.
\end{aligned} \qquad (8.13)$$

Notice that all bosonic operators – that is X^3, $X^2 H^2$, H^4, and $H^2 D^2$ sets – are automatically Hermitian.

Table 8.1 Operators with H^n, sets X^3, H^6, H^4D^2, and ψ^2H^3. See text for the operator name conventions.

X^3		H^6 and H^4D^2		ψ^2H^3+ h.c.	
Q_G	$f^{ABC}G_\mu^{A\nu}G_\nu^{B\rho}G_\rho^{C\mu}$	Q_H	$(H^\dagger H)^3$	Q_{eH}	$(H^\dagger H)(\bar{l}_p e_r H)$
$Q_{\widetilde{G}}$	$f^{ABC}\widetilde{G}_\mu^{A\nu}G_\nu^{B\rho}G_\rho^{C\mu}$	$Q_{H\Box}$	$(H^\dagger H)\Box(H^\dagger H)$	Q_{uH}	$(H^\dagger H)(\bar{q}_p u_r \widetilde{H})$
Q_W	$\epsilon^{IJK}W_\mu^{I\nu}W_\nu^{J\rho}W_\rho^{K\mu}$	Q_{HD}	$(H^\dagger D^\mu H)^*(H^\dagger D_\mu H)$	Q_{dH}	$(H^\dagger H)(\bar{q}_p d_r H)$
$Q_{\widetilde{W}}$	$\epsilon^{IJK}\widetilde{W}_\mu^{I\nu}W_\nu^{J\rho}W_\rho^{K\mu}$				

Table 8.2 Operators with H^n, sets X^2H^2, ψ^2XH, and ψ^2H^2D. See text for the operator name conventions.

X^2H^2		ψ^2XH+ h.c.		ψ^2H^2D	
Q_{HG}	$H^\dagger H G_{\mu\nu}^A G^{A\mu\nu}$	Q_{eW}	$(\bar{l}_p \sigma^{\mu\nu} e_r)\tau^I H W_{\mu\nu}^I$	$Q_{Hl}^{(1)}$	$(H^\dagger i\overleftrightarrow{D}_\mu H)(\bar{l}_p \gamma^\mu l_r)$
$Q_{H\widetilde{G}}$	$H^\dagger H \widetilde{G}_{\mu\nu}^A G^{A\mu\nu}$	Q_{eB}	$(\bar{l}_p \sigma^{\mu\nu} e_r) H B_{\mu\nu}$	$Q_{Hl}^{(3)}$	$(H^\dagger i\overleftrightarrow{D}_\mu^I H)(\bar{l}_p \tau^I \gamma^\mu l_r)$
Q_{HW}	$H^\dagger H W_{\mu\nu}^I W^{I\mu\nu}$	Q_{uG}	$(\bar{q}_p \sigma^{\mu\nu} T^A u_r)\widetilde{H} G_{\mu\nu}^A$	Q_{He}	$(H^\dagger i\overleftrightarrow{D}_\mu H)(\bar{e}_p \gamma^\mu e_r)$
$Q_{H\widetilde{W}}$	$H^\dagger H \widetilde{W}_{\mu\nu}^I W^{I\mu\nu}$	Q_{uW}	$(\bar{q}_p \sigma^{\mu\nu} u_r)\tau^I \widetilde{H} W_{\mu\nu}^I$	$Q_{Hq}^{(1)}$	$(H^\dagger i\overleftrightarrow{D}_\mu H)(\bar{q}_p \gamma^\mu q_r)$
Q_{HB}	$H^\dagger H B_{\mu\nu} B^{\mu\nu}$	Q_{uB}	$(\bar{q}_p \sigma^{\mu\nu} u_r)\widetilde{H} B_{\mu\nu}$	$Q_{Hq}^{(3)}$	$(H^\dagger i\overleftrightarrow{D}_\mu^I H)(\bar{q}_p \tau^I \gamma^\mu q_r)$
$Q_{H\widetilde{B}}$	$H^\dagger H \widetilde{B}_{\mu\nu} B^{\mu\nu}$	Q_{dG}	$(\bar{q}_p \sigma^{\mu\nu} T^A d_r) H G_{\mu\nu}^A$	Q_{Hu}	$(H^\dagger i\overleftrightarrow{D}_\mu H)(\bar{u}_p \gamma^\mu u_r)$
Q_{HWB}	$H^\dagger \tau^I H W_{\mu\nu}^I B^{\mu\nu}$	Q_{dW}	$(\bar{q}_p \sigma^{\mu\nu} d_r)\tau^I H W_{\mu\nu}^I$	Q_{Hd}	$(H^\dagger i\overleftrightarrow{D}_\mu H)(\bar{d}_p \gamma^\mu d_r)$
$Q_{H\widetilde{W}B}$	$H^\dagger \tau^I H \widetilde{W}_{\mu\nu}^I B^{\mu\nu}$	Q_{dB}	$(\bar{q}_p \sigma^{\mu\nu} d_r) H B_{\mu\nu}$	Q_{Hud}	$i(\widetilde{H}^\dagger D_\mu H)(\bar{u}_p \gamma^\mu d_r)$

8.2.3 Experimental tests and observables

The presence of dimension six operators would modify several experimental observables and low-energy constants. Most of those modifications were discussed in the original paper [Buchmuller and Wyler (1986)]. For example, the Fermi constant is modified

$$G_F = \left[\frac{1}{4\sqrt{2}}\frac{g^2}{m_W^2}\right]\left(1+\mathcal{O}\left(\frac{v^2}{\Lambda^2}\right)\right), \tag{8.14}$$

where terms $\mathcal{O}(v^2/\Lambda^2)$ parameterize terms coming from the operators listed in Tables 8.1-8.4, including their Wilson coefficients $C_{ll}^{(3)}$, $C_{Hl}^{(3)}$, C_{HW}, etc. Here we assumed that all matrix elements of dimension six operators scale as v^2. The term in square brackets in the equation above is often denoted as $G_F^{(0)}$ [Buchmuller and Wyler (1986)].

Here we briefly discuss how dimension six operators affect weak decays of the top quark [Zhang and Willenbrock (2011)]. There are only two of

Table 8.3 Four-fermion operators, sets $(\bar{L}L)(\bar{L}L)$, $(\bar{R}R)(\bar{R}R)$, and $(\bar{L}L)(\bar{R}R)$. See text for the operator name conventions.

$(\bar{L}L)(\bar{L}L)$		$(\bar{R}R)(\bar{R}R)$		$(\bar{L}L)(\bar{R}R)$	
Q_{ll}	$\left(\bar{l}_p\gamma^\mu l_r\right)\left(\bar{l}_s\gamma^\mu l_t\right)$	Q_{ee}	$(\bar{e}_p\gamma^\mu e_r)(\bar{e}_s\gamma^\mu e_t)$	Q_{le}	$\left(\bar{l}_p\gamma^\mu l_r\right)(\bar{e}_s\gamma^\mu e_t)$
$Q_{qq}^{(1)}$	$\left(\bar{q}_p\gamma^\mu q_r\right)\left(\bar{q}_s\gamma^\mu q_t\right)$	Q_{uu}	$(\bar{u}_p\gamma^\mu u_r)(\bar{u}_s\gamma^\mu u_t)$	Q_{lu}	$\left(\bar{l}_p\gamma^\mu l_r\right)(\bar{u}_s\gamma^\mu u_t)$
$Q_{qq}^{(3)}$	$\left(\bar{q}_p\gamma^\mu \tau^I q_r\right)\left(\bar{q}_s\gamma^\mu \tau^I q_t\right)$	Q_{dd}	$(\bar{d}_p\gamma^\mu d_r)(\bar{d}_s\gamma^\mu d_t)$	Q_{ld}	$\left(\bar{l}_p\gamma^\mu l_r\right)(\bar{d}_s\gamma^\mu d_t)$
$Q_{lq}^{(1)}$	$\left(\bar{l}_p\gamma^\mu l_r\right)\left(\bar{q}_s\gamma^\mu q_t\right)$	Q_{eu}	$(\bar{e}_p\gamma^\mu e_r)(\bar{u}_s\gamma^\mu u_t)$	Q_{qe}	$\left(\bar{q}_p\gamma^\mu q_r\right)(\bar{e}_s\gamma^\mu e_t)$
$Q_{lq}^{(3)}$	$\left(\bar{l}_p\gamma^\mu \tau^I l_r\right)\left(\bar{q}_s\gamma^\mu \tau^I q_t\right)$	Q_{ed}	$(\bar{e}_p\gamma^\mu e_r)(\bar{d}_s\gamma^\mu d_t)$	$Q_{qu}^{(1)}$	$\left(\bar{q}_p\gamma^\mu q_r\right)(\bar{u}_s\gamma^\mu u_t)$
		$Q_{ud}^{(1)}$	$(\bar{u}_p\gamma^\mu u_r)(\bar{d}_s\gamma^\mu d_t)$	$Q_{qu}^{(8)}$	$\left(\bar{q}_p\gamma^\mu T^A q_r\right)(\bar{u}_s\gamma^\mu T^A u_t)$
		$Q_{ud}^{(8)}$	$(\bar{u}_p\gamma^\mu T^A u_r)(\bar{d}_s\gamma^\mu T^A d_t)$	$Q_{qd}^{(1)}$	$\left(\bar{q}_p\gamma^\mu q_r\right)(\bar{d}_s\gamma^\mu d_t)$
				$Q_{qd}^{(8)}$	$\left(\bar{q}_p\gamma^\mu T^A q_r\right)(\bar{d}_s\gamma^\mu T^A d_t)$

Table 8.4 Four-fermion operators, sets $(\bar{L}R)(\bar{R}L)$, and B (baryon-number violating). See text for the operator name conventions.

$(\bar{L}R)(\bar{R}L)$		B-violating	
Q_{ledq}	$\left(\bar{l}_p^j e_r\right)\left(\bar{d}_s q_t^j\right)$	Q_{duq}	$\epsilon^{\alpha\beta\gamma}\epsilon_{jk}\left[(d_p^\alpha)^T C u_r^\beta\right]\left[(q_s^{\gamma j})^T C l_t^k\right]$
$Q_{quqd}^{(1)}$	$\left((\bar{q}_p^j u_r)\epsilon_{jk}(\bar{q}_s^k d_t)\right)$	Q_{qqu}	$\epsilon^{\alpha\beta\gamma}\epsilon_{jk}\left[(q_p^{\alpha j})^T C q_r^{\beta k}\right]\left[(u_s^\gamma)^T C e_t\right]$
$Q_{quqd}^{(8)}$	$\left((\bar{q}_p^j T^A u_r)\epsilon_{jk}(\bar{q}_s^k T^A d_t)\right)$	$Q_{qqq}^{(1)}$	$\epsilon^{\alpha\beta\gamma}\epsilon_{jk}\epsilon_{mn}\left[(q_p^{\alpha j})^T C q_r^{\beta k}\right]\left[(q_s^{\gamma m})^T C l_t^n\right]$
$Q_{lequ}^{(1)}$	$\left((\bar{l}_p^j e_r)\epsilon_{jk}(\bar{q}_s^k u_t)\right)$	$Q_{qqq}^{(3)}$	$\epsilon^{\alpha\beta\gamma}\left(\tau^I\epsilon\right)_{jk}\left(\tau^I\epsilon\right)_{mn}\left[(q_p^{\alpha j})^T C q_r^{\beta k}\right]\left[(q_s^{\gamma m})^T C l_t^n\right]$
$Q_{lequ}^{(3)}$	$\left((\bar{l}_p^j \sigma_{\mu\nu} e_r)\epsilon_{jk}(\bar{q}_s^k \sigma^{\mu\nu} u_t)\right)$	Q_{duu}	$\epsilon^{\alpha\beta\gamma}\left[(d_p^\alpha)^T C u_r^\beta\right]\left[(u_s^\gamma)^T C e_t\right]$

those operators that would contribute to top decay,

$$Q_{Hq}^{(3)} = i\left(H^\dagger \tau^I D_\mu H\right)\left(\bar{q}_p \gamma^\mu \tau^I q_r\right),$$
$$Q_{tW} = \left(\bar{q}_p \gamma^\mu \tau^I q_r\right)\widetilde{H} W_{\mu\nu}^I. \tag{8.15}$$

These operators could be found in Table 8.2. After spontaneous symmetry breaking those operators would generate dimension 6 terms in the effective Lagrangian affecting the semileptonic top quark decay,

$$\mathcal{L}_{tb} = \frac{C_{Hq}^{(3)}}{\Lambda^2}\frac{gv^2}{\sqrt{2}}\bar{b}_L \gamma^\mu t_L - 2\frac{C_{tW}}{\Lambda^2} v\, \bar{b}_L \sigma^{\mu\nu} t_R\, \partial_\nu W_\mu^- + \text{h.c.}, \tag{8.16}$$

where the first term comes from the $Q_{Hq}^{(3)}$ and the second one comes form the Q_{tW} operators respectively.

We could see that the first operator in Eq. (8.16) has exactly the same structure as the leading, dimension four operator. Thus, the only effect of this operator would amount to rescaling the SM Wtb vertex by a factor

$$\left(1 + \frac{C_{Hq}^{(3)}}{\Lambda^2} \frac{v^2}{V_{bt}}\right), \tag{8.17}$$

and not to affect any dynamical distributions. Here V_{tb} is the CKM matrix element responsible for the $t \to b$ transition. The second term could affect some dynamical distributions, like the electron energy spectrum in $t \to be\bar{\nu}$ decay.

Both of those operators, however, would affect the partial width of semileptonic top decay. Neglecting all but the top quark mass [Zhang and Willenbrock (2011)],

$$\Gamma(t \to be\bar{\nu}) = V_{tb}^2 \left(1 + \frac{2C_{Hq}^{(3)}}{\Lambda^2} \frac{v^2}{V_{bt}}\right) \frac{g^4}{\pi^2 \Gamma_W} \frac{m_t^2}{3072\sqrt{x_W}} \left(1 - 3x_W^2 + 2x_W^3\right)$$

$$+ \frac{\text{Re}\,[C_{tw}]}{\Lambda^2} \frac{V_{tb} g^2}{\pi^2 \Gamma_W} \frac{m_t^4}{64\sqrt{2}} x_W \left(1 - x_W\right)^2. \tag{8.18}$$

where Γ_W is the SM tree-level decay width of the W-boson, $\Gamma_W = (3\alpha_W/4)M_W$, and $x_W = M_W^2/m_t^2$. The factor of two in the first bracket signifies that the effect comes from the interference of the SM and new physics contributions. By the same token, the production cross section of a top quark could also be affected and probed experimentally, provided that Λ is not that large compared to v [Zhang and Willenbrock (2011)].

8.3 BSM particles in EFT

As mentioned in section 8.1, new physics degrees of freedom could lie just above the energy scale associated with electroweak symmetry breaking. In this case, it would be appropriate to include them explicitly in the effective theory. This effective theory can be built based on the same principles as all EFTs that we have discussed so far: by writing a complete set of non-redundant operators consistent with the chosen symmetry group that should include the SM symmetries as a subset.

There are many possible extensions of the standard model that can be used to build different effective theories. Instead of discussing them here, we choose to provide two simple examples, which include particles that transform as singlets under the SM symmetry groups. Their discussion is

Table 8.5 Dimension 6 and 7 effective operators describing interactions of Dirac fermion (D) WIMPs χ with quarks and gluons. Operator classification is from [Goodman et al. (2010)].

Name	Operator
D1	$(m_q/\Lambda^2)\,\overline{\chi}\chi\overline{q}q$
D2	$(im_q/\Lambda^2)\,\overline{\chi}\gamma_5\chi\overline{q}q$
D3	$(im_q/\Lambda^2)\,\overline{\chi}\chi\overline{q}\gamma_5 q$
D4	$(m_q/\Lambda^2)\,\overline{\chi}\gamma_5\chi\overline{q}\gamma_5 q$
D5	$(1/\Lambda^2)\,\overline{\chi}\gamma^\mu\chi\overline{q}\gamma_\mu q$
D6	$(1/\Lambda^2)\,\overline{\chi}\gamma^\mu\gamma_5\chi\overline{q}\gamma_\mu q$
D7	$(1/\Lambda^2)\,\overline{\chi}\gamma^\mu\chi\overline{q}\gamma_\mu\gamma_5 q$
D8	$(1/\Lambda^2)\,\overline{\chi}\gamma^\mu\gamma_5\chi\overline{q}\gamma_\mu\gamma_5 q$
D9	$(1/\Lambda^2)\,\overline{\chi}\sigma^{\mu\mu}\chi\overline{q}\sigma_{\mu\nu} q$
D10	$(i/\Lambda^2)\,\overline{\chi}\sigma^{\mu\mu}\gamma_5\chi\overline{q}\sigma_{\mu\nu} q$
D11	$(1/4\Lambda^3)\,\overline{\chi}\chi\alpha_s G^A_{\mu\nu} G^{A\mu\nu}$
D12	$(i/4\Lambda^3)\,\overline{\chi}\gamma_5\chi\alpha_s G^A_{\mu\nu} G^{A\mu\nu}$
D13	$(i/4\Lambda^3)\,\overline{\chi}\chi\alpha_s \widetilde{G}^A_{\mu\nu} G^{A\mu\nu}$
D14	$(1/4\Lambda^3)\,\overline{\chi}\gamma_5\chi\alpha_s \widetilde{G}^A_{\mu\nu} G^{A\mu\nu}$

quite interesting phenomenologically, as they can serve as candidates for dark matter. Thus, their studies at the high energy colliders such as the LHC could shed some light onto the problem of dark matter and even lead to experimental discovery of dark matter particles.

8.3.1 Dark matter at colliders: effective operators

Weakly-interacting massive particles (WIMPs) [Bertone et al. (2005)], stable neutral states which exist in many extensions of the standard model, provide a good solution to the cosmological dark matter (DM) problem. The most widely-used models, which also provide solutions to other well-known shortcomings of the standard model, include supersymmetry, which naturally gives rise to a WIMP candidate, provided that R-parity is conserved. Other natural candidates include Kaluza-Klein excitations of the SM particles, where stability can be realized thanks to momentum conservation in the compactified extra dimension(s). Correct identification of the nature of dark matter will lend support to one or another extension of the standard model. It is thus not surprising that much of the recent efforts in both high energy and astrophysical experiments has been directed towards searches for those states.

According to the WIMP hypothesis, DM particles could have masses not much larger than that of the SM particles. We shall build an effective field theory that includes DM degrees of freedom. We consider all interactions of dark matter with the standard model that are permitted by a minimal set of assumptions: DM particles are (1) SM singlets, and (2) invariant under a Z_2 symmetry. The latter requirement insures that DM particles are stable and can only be produced in pairs. In recognition that the operators describing DM interactions could generically be introduced by some heavy particles that have been integrated out of the spectrum, operators of dimensions higher than four must be included as well. Their masses collectively indicate the scale Λ at which this EFT breaks down. Depending on the WIMP's spin, two sets of operators can be introduced [Goodman et al. (2010)] (see Table 8.5 for spin-1/2 DM particles and Table 8.6 for scalar DM particles). Note that the set of quark-DM operators in Table 8.5 is complete, all operators of the type $\bar{\chi}\Gamma q \bar{q} \Gamma q \chi$ for any Dirac matrix Γ can be related to the operators D1-D14 by applying Fierz relations.

We shall not write out the complete basis set of EFT operators. Since we are interested in a direct production of WIMP particles at colliders (say, a pp collider such as the LHC), the chosen set must include DM, quark and gluon degrees of freedom. If we are only interested in such spectacular experimental signatures as jets of partons recoiling against "missing energy"[2], no other degrees of freedom need to be included in the effective operators. Indeed, if a particular experimental search includes other SM degrees of freedom produced in association with DM, they would have to be included in effective operators as explicit degrees of freedom as well. We shall provide an example of such a search in the next section, with the experimental signature of a Higgs boson recoiling against missing energy.

8.3.2 Mono-Higgs signatures of dark matter at LHC

What if other SM particles are produced in association with DM? Let us exemplify this by writing out a set of effective operators that describe associated production of DM and a single Higgs boson.

Let us write a set of effective operators that can possibly generate our experimental signature [Petrov and Shepherd (2014)]. In what follows we shall consider all possible operators suppressed by at most three powers of the new physics scale and study their implications for experimental signals.

---

[2]Since dark matter interacts with ordinary ("luminous") matter very weakly, the produced WIMPs will leave no trace in the detector.

Table 8.6 Dimension 5 and 6 effective operators describing interactions of complex (C) and real (R) scalar WIMPs χ with quarks and gluons. Operator classification is from [Goodman et al. (2010)].

Name	Operator
C1	$(m_q/\Lambda^2)\chi^\dagger\chi\bar{q}q$
C2	$(im_q/\Lambda^2)\chi^\dagger\chi\bar{q}\gamma_5 q$
C3	$(1/\Lambda^2)\chi^\dagger\partial_\mu\chi\bar{q}\gamma^\mu q$
C4	$(1/\Lambda^2)\chi^\dagger\partial_\mu\chi\bar{q}\gamma^\mu\gamma_5 q$
C5	$(1/4\Lambda^2)\chi^\dagger\chi\alpha_s G^A_{\mu\nu}G^{A\mu\nu}$
C6	$(i/4\Lambda^2)\chi^\dagger\chi\alpha_s \widetilde{G}^A_{\mu\nu}G^{A\mu\nu}$
R1	$(m_q/2\Lambda^2)\chi^2\bar{q}q$
R2	$(im_q/2\Lambda^2)\chi^2\bar{q}\gamma_5 q$
R3	$(1/8\Lambda^2)\chi^2\alpha_s G^A_{\mu\nu}G^{A\mu\nu}$
R4	$(i/8\Lambda^2)\chi^2\alpha_s \widetilde{G}^A_{\mu\nu}G^{A\mu\nu}$

For concreteness, let us assume that DM is a Dirac fermion and a singlet of the SM gauge group.

The lowest dimension at which the dark matter can interact with SM fields under these assumptions is five. These operators are

$$\mathcal{L}_5 = \frac{2C_1^{(5)}}{\Lambda}|H|^2\bar{\chi}\chi + \frac{2C_2^{(5)}}{\Lambda}|H|^2\bar{\chi}\gamma_5\chi. \tag{8.19}$$

Throughout, $C_i^{(n)}$ are the effective Wilson coefficients that characterize the strength of Higgs-DM interactions of dimension n in the effective theory.

Note that Eq. (8.19) does not contain any quark or gluon degrees of freedom. In order to contribute to the experimental signature of interest, the Higgs first needs to be produced with the SM processes, like $gg \to H$. This means that in terms of the physical Higgs field h we need to only consider the terms quadratic in the physical field. Thus, this operator can be written as

$$\mathcal{L}_5 = \frac{C_1^{(5)}}{\Lambda}h^2\bar{\chi}\chi + \frac{C_2^{(5)}}{\Lambda}h^2\bar{\chi}\gamma_5\chi. \tag{8.20}$$

We do not expect to have a strong bound on the scale Λ from those operators, for two reasons. First, Higgs production by itself is relatively rare at the LHC, with subsequent DM-Higgs interactions giving an additional suppression. Second, since the Higgs boson is in the s-channel, it has to be significantly off-shell in this process. We note that the differences between the scalar and pseudoscalar couplings to the DM pair, while very significant for dynamics in the low-velocity regime, are negligible at the LHC where all particles are produced relativistically.

Next, there are two operators of dimension six,

$$\mathcal{L}_6 = \frac{C_1^{(6)}}{\Lambda^2} H^\dagger \overleftrightarrow{D}_\mu H \, \overline{\chi} \gamma^\mu \chi + \frac{C_2^{(6)}}{\Lambda^2} H^\dagger \overleftrightarrow{D}_\mu H \, \overline{\chi} \gamma^\mu \gamma_5 \chi. \tag{8.21}$$

Once again, in unitary gauge and in terms of the physical Higgs field h, the operators that could generate the mono-Higgs signature can be derived from Eq. (8.21)

$$\mathcal{L}_6 = \frac{iC_1^{(6)} m_Z}{\Lambda^2} h Z_\mu \, \overline{\chi} \gamma^\mu \chi + \frac{iC_2^{(6)} m_Z}{\Lambda^2} h Z_\mu \, \overline{\chi} \gamma^\mu \gamma_5 \chi, \tag{8.22}$$

where we defined $Z_\mu = (g^2 + g'^2)^{-1/2}(gW_\mu^3 - g'B_\mu)$ and employed the relation $2m_Z = v\sqrt{g^2 + g'^2}$. With an off-shell Z boson in the s-channel this operator gives rise to the desired experimental signature.

At the next order (dimension seven) there are several classes of relevant operators. There are four operators that involve Higgs doublets and their derivatives,

$$\mathcal{L}_{7H} = \frac{C_1'^{(7)}}{\Lambda^3} \left(H^\dagger H\right)^2 \overline{\chi} \chi + \frac{C_2'^{(7)}}{\Lambda^3} \left(H^\dagger H\right)^2 \overline{\chi} \gamma_5 \chi,$$
$$+ \frac{C_3'^{(7)}}{\Lambda^3} |D_\mu H|^2 \, \overline{\chi} \chi + \frac{C_4'^{(7)}}{\Lambda^3} |D_\mu H|^2 \, \overline{\chi} \gamma_5 \chi. \tag{8.23}$$

The part of $\mathcal{L}_{7H}$ that generates the mono-Higgs signature at the LHC can be written as

$$\mathcal{L}_{7H} = \frac{3C_1'^{(7)}}{2} \frac{v^2}{\Lambda^3} h^2 \, \overline{\chi} \chi + \frac{3C_2'^{(7)}}{2} \frac{v^2}{\Lambda^3} h^2 \, \overline{\chi} \gamma_5 \chi,$$
$$+ \frac{C_3'^{(7)}}{2\Lambda^3} (\partial_\mu h)^2 \, \overline{\chi} \chi + \frac{C_4'^{(7)}}{2\Lambda^3} (\partial_\mu h)^2 \, \overline{\chi} \gamma_5 \chi. \tag{8.24}$$

We do not expect strong constraints on Λ from those operators, as they simply represent higher-order $1/\Lambda$ corrections to the operators discussed above (see Eq. (8.20)).

There are also four operators of dimension seven which describe coupling of dark matter to the SM fermions f,

$$\mathcal{L}_{7F} = \frac{2\sqrt{2}C_1^{(7)}}{\Lambda^3} \Gamma_d \, \overline{q}_L H d_R \, \overline{\chi} \chi + \frac{2\sqrt{2}C_1^{(7)}}{\Lambda^3} \Gamma_u \, \overline{q}_L \widetilde{H} u_R \, \overline{\chi} \chi \tag{8.25}$$
$$+ \frac{2\sqrt{2}C_2^{(7)}}{\Lambda^3} \Gamma_d \, \overline{q}_L H d_R \, \overline{\chi} \gamma^5 \chi + \frac{2\sqrt{2}C_2^{(7)}}{\Lambda^3} \Gamma_u \, \overline{q}_L \widetilde{H} u_R \, \overline{\chi} \gamma^5 \chi + h.c.$$

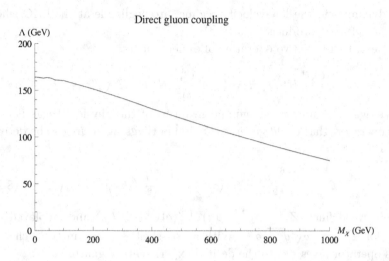

Fig. 8.1 Bound on the new physics scale Λ from DM and h couplings to gluons (from [Petrov and Shepherd (2014)]).

Here we scaled the Wilson coefficients to introduce Yukawa couplings Γ_p for each fermion flavor $f = u, d$ of up (u) or down (d) type. q_L is a standard electroweak doublet of left-handed fermions. This form of operators is invariant under electroweak $SU(2)_L$ group and also naturally suppresses DM couplings to the light fermions. We assume the couplings $C_i^{(7)}$ are flavor-blind, but permit them to be complex. In terms of the physical field h, Eq. (8.25) can be written as

$$\mathcal{L}_{7F} = \frac{Re\left(C_1^{(7)}\right)}{\Lambda^3} \Gamma_f \left(\bar{f}f\right) h \left(\bar{\chi}\chi\right) + \frac{Im\left(C_1^{(7)}\right)}{\Lambda^3} i\Gamma_f \left(\bar{f}\gamma_5 f\right) h \left(\bar{\chi}\chi\right) \quad (8.26)$$

$$+ \frac{Im\left(C_2^{(7)}\right)}{\Lambda^3} i\Gamma_f \left(\bar{f}f\right) h \left(\bar{\chi}\gamma_5\chi\right) + \frac{Re\left(C_2^{(7)}\right)}{\Lambda^3} \Gamma_f \left(\bar{f}\gamma_5 f\right) h \left(\bar{\chi}\gamma_5\chi\right)$$

Note that these operators are identical to those which have traditionally been known as D1-D4 discussed in section 8.3.1 and [Goodman et al. (2010)] in effective theories of DM scattering and production, with the sole difference being that the implied Higgs vev has been replaced by the dynamical Higgs field in these operators. We expect the strongest constraints to come from this and the next set of operators, even though they are operators of relatively high dimension.

There are also four operators that are formally of dimension 8 that describe DM couplings to the gluons and the physical Higgs,

$$\mathcal{L}_8 = \frac{C_1^{(8)}}{\Lambda^3 M_{EW}} (\bar{\chi}\chi) h\, G^{a\mu\nu} G_{\mu\nu}^a + \frac{C_2^{(8)}}{\Lambda^3 M_{EW}} (\bar{\chi}\gamma^5\chi) h\, G^{a\mu\nu} G_{\mu\nu}^a$$

$$+ \frac{C_3^{(8)}}{\Lambda^3 M_{EW}} (\bar{\chi}\chi) h\, G^{a\mu\nu} \widetilde{G}_{\mu\nu}^a + \frac{C_4^{(8)}}{\Lambda^3 M_{EW}} (\bar{\chi}\gamma^5\chi) h\, G^{a\mu\nu} \widetilde{G}_{\mu\nu}^a, \quad (8.27)$$

where we choose $M_{EW} = v$. Note that the presence of M_{EW} here makes these operators equivalent in power counting of the new physics scale to the dimension seven operators above. In fact, similarly to the operators in Eq. (8.26), these are equivalent to the well-known operators D11-D14 of [Goodman et al. (2010)] with a Higgs vev replaced by the dynamical field. The importance of these higher-order operators coupling directly to gluons is enhanced by the large gluonic luminosity of the LHC.

The contributions of operators described above can be constrained with the LHC data. Using the currently available datasets, constraints on the combinations $C^{(n)}/\Lambda^{n-4}$ can be placed [Petrov and Shepherd (2014)]. Since any measurement constrains the combination of Wilson coefficients and scales, we can assume that $C^{(n)} \sim 1$ to get constraints on the new physics scale, see, e.g. Fig. 8.1.

8.4 Notes for further reading

Effective Lagrangians for the standard model were first considered in [Buchmuller and Wyler (1986)]. Further refinement of the operator structure and elimination of some of the redundant operators lead to several equivalent operator bases, [Grzadkowski et al. (2010)], [Giudice et al. (2007)], and [Duehrssen-Debling et al. (2015)]. A good collection of references to EFT analyses of experimental observables can be found in [Cirigliano and Ramsey-Musolf (2013)]. One loop corrections and renormalization group analyses can be found in [Jenkins et al. (2013)]. There are many reviews of the concepts associated with dark matter studies. Among those, we recommend [Bertone et al. (2005)], but any other review would be useful as well.

Problems for Chapter 8

(1) **Operator identities**
Show that
$$\left(\bar{l}_p^j u_t^\alpha\right) \epsilon_{jk} \left(\bar{q}_s^k e_r\right) = \frac{1}{8} Q_{lequ}^{(3)} - \frac{1}{2} Q_{lequ}^{(1)}. \qquad (8.28)$$

(2) **Standard model equations of motion**
Using the standard model Lagrangian of Eq. (8.2), derive the equations of motion given in Eq. (8.13). Write out an explicit form of all covariant derivatives in Eq. (8.13).

Chapter 9

Effective Field Theories of Gravity

9.1 Introduction

In this chapter we shall discuss effective field theories of gravity, both classical and quantum. Admittedly, the applications of a quantum theory of gravity, whatever that theory might be, are at the moment quite limited by the currently available accuracy of experimental measurements. The classical theory of gravity, on the other hand, is irreplaceable in describing dynamics of galaxies, black holes and other astrophysical objects. This is given by General Relativity (GR), the theory of space-time geometry. It is possible to recast its geometrical language into a field-theoretic one. This field theory, however, is not like the ones that we encountered so far in this book. For one, a theory of gravity cannot be described with vector or scalar fields, like electrodynamics or Yukawa theory.

Why is that so? Let us imagine a theory of gravity where the interactions are described by a vector field, just like in electrodynamics. This would lead to a world that is very different from ours, since vector fields mediate both attractive and repulsive interactions of matter, something gravity does not do. Neither could the classical field theory of gravity be described by a scalar field: scalars cannot couple to energy in a Lorentz-invariant fashion as required by General Relativity. Thus, gravitational interactions have to be mediated by a tensor field, which, upon quantization, will describe exchanges of spin-2 particles called "gravitons." These particles would have to be massless to accommodate long-range gravitational interactions.

Quantum gravity is a theory that combines General Relativity and quantum mechanics. Indications are that it is not even described by a field theory – at least not a renormalizable one. A leading candidate for a *complete* quantum theory of gravity is string theory, which is known to heal many problems of the field-theoretic description of gravity, including the UV divergences. It does so by introducing a new *string scale*, which is related to string tension, that describes a scale at which particles are no longer point like objects. Here we shall not deal with string theory (or any other candidates for a quantum theory of gravity), referring the reader to the list of relevant literature at the end of this chapter. Instead we shall concentrate on the description of the effects of quantum gravity at low energies.

A low-energy description of quantum gravity is a well-defined and unambiguous theory. As was discussed earlier in this book, the exact description of the UV physics is not needed to correctly describe low-energy interactions, so long as the symmetries of the complete theory that survive in the low energy limit are known. The theories describing the high energy and low energy regimes do not even need to have the same degrees of freedom, such as χPT vs. QCD discussed in chapter 4. The low energy theory is still predictive provided there is enough experimental data to fix all of the relevant unknown constants.

Effective field theories of gravity describe quantum gravity at energy scales that are way below the Planck mass[1] $M_{\text{Pl}} \sim 10^{18}$ GeV. It was shown by J. Donoghue in 1994 [Donoghue (1994)] that it is possible to build an effective theory of gravity which, analogous to χPT, utilizes a derivative expansion to organize quantum corrections as a series of powers of $1/M_{\text{Pl}}$. Similarly to χPT, it is renormalizable order-by-order in this derivative expansion. This EFT is universal in its description of quantum corrections to GR. This is the effective theory that we shall discuss in the first part of this chapter. The second part of the chapter will discuss a fun application of EFT techniques to the classical theory of gravity (GR) under certain situations that are relevant for experiments such as LIGO. Before discussing either of these EFT approaches to gravity, however, we begin with a self-contained introduction to GR.

---

[1] We define $M_{\text{Pl}} \equiv (32\pi G_N)^{-1/2} = 1.22 \times 10^{18}$ GeV, in natural units $\hbar = c = 1$. This choice is not consistent throughout the literature, and factors of 2 and π can appear without warning between papers and textbooks. Let the reader beware!

9.2 Review of general relativity

9.2.1 Geodesics and affine connection

General Relativity is a theory built on the principle of general covariance. This principle states that physics is insensitive to the choice of coordinate system, i.e. the forms of physical laws are invariant under arbitrary differentiable coordinate transformations. This is a very powerful requirement, akin to gauge invariance in ordinary field theories.

Let us recall a couple of useful notions, which will be important for our discussions later in this chapter. In Einsten's theory, gravity changes the geometry of the space-time continuum. This means that objects, when not acted on by "outside forces," no longer move in straight lines. To reestablish Newton's First Law, this deviation from straight lines can be reinterpreted as coming from a new force: the force of gravity! So how do the basic properties of field theory – such as preservation of the scalar products – get modified in the presence of curved space, a.k.a. gravitational fields?

This question can be studied in a way that is similar to any gauge theory: by examining a parallel displacement of a vector along a path in the gravitational field. This leads to the notion of a covariant derivative [Peskin and Schroeder (1995)], which can be defined (suppressing the relevant space-time indices for now) as

$$DV = \lim_{s \to 0} \frac{1}{s} \left(V(s) - V^P(s) \right), \tag{9.1}$$

where s is called an "affine parameter" along the curve, $V(s)$ is some vector field that is restricted to the curve, and $V^P(s)$ is a parallel displacement of a vector $V(0)$ by distance s. As with gauge theories, the notion of a covariant derivative is very useful, as for any field χ the combination $(D\chi)$ transforms the same way as the field χ. This is so for any spin of χ.

We can derive an equation that describes the "shortest" path in curved space, called a *geodesic*. The geodesic equation is one of the first things derived in a GR course. It follows from the least action principle applied to the path length

$$S = -m \int ds, \tag{9.2}$$

where ds is the infinitesimal element of the path:

$$ds^2 = g_{\mu\nu} dx^\mu dx^\nu. \tag{9.3}$$

Requiring that

$$\delta S = 0 \tag{9.4}$$

results in the *geodesic equation*,

$$\frac{d^2 x^\mu}{ds^2} + \Gamma^\mu_{\alpha\beta} \frac{dx^\alpha}{ds} \frac{dx^\beta}{ds} = 0, \tag{9.5}$$

where we defined the *affine connection* or Christoffel symbols,

$$\Gamma^\lambda_{\alpha\beta} = \frac{g^{\lambda\sigma}}{2} \left(\partial_\alpha g_{\beta\sigma} + \partial_\beta g_{\alpha\sigma} - \partial_\sigma g_{\alpha\beta} \right). \tag{9.6}$$

The affine connection plays the role of a compensator for coordinate transformations in curved space-time. It is important to note that it is *not* a tensor. This can be seen by directly computing how it transforms under coordinate changes, as described in the next section, or by noting the definition of Christoffel symbol as a holonomic (coordinate) version of the more general Ricci rotation coefficients,

$$D_{e_\mu} e_\nu = \Gamma^\lambda_{\mu\nu} e_\lambda, \tag{9.7}$$

where D_{e_μ} denotes a covariant derivative acting on the basis vectors e_μ.

A combination of the affine connection and a derivative would in fact transform as a vector under general coordinate transformations. Thus, a covariant derivative's action on a vector field, for example, can be defined as

$$D_\mu A^\nu = \partial_\mu A^\nu + \Gamma^\nu_{\mu\alpha} A^\alpha. \tag{9.8}$$

This definition can be generalized to a tensor of any rank.

One useful fact that can be directly noted from the definition of the Christoffel symbols of Eq. (9.6) is that the connection involves one derivative acting on the metric. We shall use it later to build an effective Lagrangian for the gravitational field.

9.2.2 General relativity and the weak field limit

Let us now construct the Lagrangian of GR emphasizing its field-theoretic side. We by no means pretend to have a complete description of the GR-field theory and refer the reader to section 9.5 for an extensive, albeit incomplete, list of textbook suggestions.

It is important to remember that in describing gravity we are dealing with coordinate (Lorentz) transformations. This means that in writing the action we would need to make sure that we deal with the invariant integration measure. We can show that the integration measure $\int d^4 x \sqrt{-g}$, where $g = \det g_{\mu\nu}$ is a determinant of the metric tensor $g_{\mu\nu}$, is invariant under coordinate transformations

$$x'^\mu \to \Lambda^\mu_\nu x^\nu. \tag{9.9}$$

In other words, if $J = \det \Lambda$ is a Jacobian of the coordinate change of Eq. (9.9), then

$$\int d^4x' \sqrt{-g'} = \int d^4x \; JJ^{-1} \sqrt{-g} = \int d^4x \sqrt{-g}, \qquad (9.10)$$

This holds so long as $g'_{\mu\nu} = \overline{\Lambda}^\alpha_\mu \overline{\Lambda}^\beta_\nu \, g_{\alpha\beta}$, where $\overline{\Lambda}^\nu_\mu = [\Lambda^\mu_\nu]^{-1}$. We will see that this is true below.

General relativity can be cast in a form similar to a gauge theory under local coordinate transformations. This similarity was first noted by S. Gupta in [Gupta (1954)], who suggested that field-theoretic methods could be used to derive such quantities as the field strength tensor in GR. With this in mind, let us promote the global coordinate change of Eq. (9.9) to a *local* transformation with

$$\begin{aligned} x^\mu &\to x'^\mu = \Lambda^\mu_\nu(x) \, x^\nu, \\ dx^\mu &\to dx'^\mu = \Lambda^\mu_\nu(x) \, dx^\nu, \end{aligned} \qquad (9.11)$$

In the infinitesimal form this transformation becomes

$$x'^\mu = x^\mu + \epsilon^\mu(x), \qquad (9.12)$$

where $\epsilon^\mu(x)$ is an infinitesimal coordinate-dependent vector.

Because the interval ds^2 is, by definition, the same in all inertial frames of reference, it must be invariant under coordinate transformations

$$ds^2 = g'_{\mu\nu}(x') dx'^\mu dx'^\nu = g_{\mu\nu}(x) dx^\mu dx^\nu. \qquad (9.13)$$

This implies that the metric tensor $g_{\mu\nu}(x)$ should transform as a coordinate-dependent field

$$g'_{\mu\nu}(x') = \overline{\Lambda}^\alpha_\mu \; g_{\alpha\beta}(x) \; \overline{\Lambda}^\beta_\nu. \qquad (9.14)$$

This is precisely what we needed to show that Eq. (9.10) works. One can also check, recalling that $g^{\mu\nu} = (g_{\mu\nu})^{-1}$, that

$$g^{\mu\alpha}(x) \, g_{\alpha\nu}(x) = g'^{\mu\alpha}(x) \, g'_{\alpha\nu}(x) = \delta^\mu_\nu. \qquad (9.15)$$

Let us now build a Lagrangian describing gravitational interactions. Using the covariant derivative constructed in the previous section (see Eq. (9.8)), we can define the *Riemann-Christoffel curvature tensor*, which, just like $F^a_{\mu\nu}$ in gauge theories, can be determined by the commutator $[D_\mu, D_\nu] A_\alpha \equiv R^\beta_{\alpha\mu\nu} A_\beta$,

$$R^\beta_{\alpha\mu\nu} = \partial_\mu \Gamma^\beta_{\alpha\mu} - \partial_\nu \Gamma^\beta_{\alpha\mu} + \Gamma^\lambda_{\alpha\nu} \Gamma^\beta_{\lambda\mu} - \Gamma^\lambda_{\alpha\mu} \Gamma^\beta_{\lambda\nu}. \qquad (9.16)$$

As can be seen from its definition, $R^\beta_{\alpha\mu\nu} = -R^\beta_{\alpha\nu\mu}$. Notice that while $\Gamma^\lambda_{\mu\nu}$ need not vanish in flat spacetime (such as with spherical coordinates), the curvature tensor has all components equal to zero in flat space time in any coordinate system (Cartesian, polar, etc.).

Eq. (9.16) can be used to build other useful quantities. For example, the *Ricci tensor* can be obtained by tracing over two of its indices, i.e.

$$R_{\mu\nu} \equiv R^\alpha_{\mu\nu\alpha}, \qquad (9.17)$$

while the *Ricci scalar curvature* is defined as

$$R = g^{\mu\nu} R_{\mu\nu}. \qquad (9.18)$$

Subtracting off the Ricci traces, the trace-free curvature tensor is called the *Weyl curvature tensor*

$$C^\alpha_{\mu\nu\rho} \equiv R^\alpha_{\mu\nu\rho} - \frac{2}{d-2}\left(g^\alpha_{[\nu}R_{\rho]\mu} - g_{\mu[\nu}R^\alpha_{\rho]}\right) + \frac{2}{(d-1)(d-2)} g^\alpha_{[\nu} g_{\rho]\mu} R, \qquad (9.19)$$

where brackets on lower indices mean anti-symmetrized. Since the affine connection involves one derivative acting on the metric, it follows from Eq. (9.16) that the Riemann-Christoffel tensor (and all of its traces) always involves *two* derivatives acting on the metric.

The equations of motion of general relativity, the Einstein equations, can be obtained from the Einstein-Hilbert action,

$$S_{\text{EH}} = \frac{2}{\kappa^2} \int d^4x \sqrt{-g}\, R, \qquad (9.20)$$

where $\kappa = 1/M_{\text{Pl}}$ is the Planck length (in natural units, up to factors of 2 and π) and additional matter fields can be included in the "matter action"

$$S_{\text{matter}} = \int d^4x \sqrt{-g}\, \mathcal{L}_{\text{matter}} \qquad (9.21)$$

The Planck mass, which depends on the universal constant of Newton's Law of Gravity, defines the scale of quantum gravity. The total action of Eqs. (9.20-9.21) gives Einstein's equations,

$$R_{\mu\nu} - \frac{1}{2} g_{\mu\nu} R = 8\pi G_N\, T_{\mu\nu}, \qquad (9.22)$$

where $T_{\mu\nu}$ is the energy momentum tensor of matter, whose form we shall determine in the next section. These equations immediately tell us that it is the energy of the gravitating body, not its mass, that defines the curvature of the space-time around it.

It will be very convenient for us to use a background field method for dealing with quantum fluctuations of the gravitational field. Let us introduce a background metric, $\bar{g}_{\mu\nu}$. This metric is usually taken as a solution of classical Einstein equations for a given situation. The simplest solution, of course, is flat Minkowski space ($\bar{g}_{\mu\nu} = \eta_{\mu\nu}$), which we shall employ in this chapter; however, this is not the only possibility. Other options include the Schwarzschild or Kerr metrics that describe black holes, or the FRW metric of cosmology.

Whatever the background metric is, we can expand $g_{\mu\nu}$ about it,

$$g_{\mu\nu} \equiv \bar{g}_{\mu\nu} + \kappa h_{\mu\nu} \tag{9.23}$$

We shall raise and lower indices with the background metric, so up to $\mathcal{O}(\kappa^2)$ the contravariant metric tensor can be written as

$$g^{\mu\nu} = \bar{g}^{\mu\alpha}\bar{g}^{\nu\beta}g_{\alpha\beta} = \bar{g}^{\mu\nu} - \kappa h^{\mu\nu} + \kappa^2 h^\mu_\alpha h^{\nu\alpha} + \ldots. \tag{9.24}$$

Let us choose Minkowski space as our background, which is appropriate for describing weak gravitational fields. In this case the infinitesimal change of the coordinates in Eq. (9.12) takes the form of a "gauge transformation"

$$h_{\mu\nu}(x) \to h'_{\mu\nu}(x) = h_{\mu\nu}(x) - \partial_\mu \epsilon_\nu(x) - \partial_\nu \epsilon_\mu(x), \tag{9.25}$$

and expanding the integration measure for the action gives

$$\sqrt{-g} = 1 + \frac{1}{2}\kappa h - \frac{\kappa^2}{4} h_{\alpha\beta} P^{\alpha\beta,\mu\nu} h_{\mu\nu}, \tag{9.26}$$

where $h = h^\mu_\mu$ and

$$2P^{\alpha\beta,\mu\nu} = \eta^{\alpha\mu}\eta^{\beta\nu} + \eta^{\alpha\nu}\eta^{\beta\mu} - \eta^{\alpha\beta}\eta^{\mu\nu}, \tag{9.27}$$

We can further obtain the corresponding expressions for the weak-field limit of the affine connection,

$$\Gamma^\lambda_{\mu\nu} = \frac{1}{2}\left[\partial_\mu h^\lambda_\nu + \partial_\nu h^\lambda_\mu - \partial^\lambda h_{\mu\nu}\right], \tag{9.28}$$

and the Ricci scalar,

$$R = \Box h - \partial_\lambda \partial_\sigma h^{\lambda\sigma}. \tag{9.29}$$

Finally, in the weak-field limit the Euler-Lagrange equations of motion obtained from the Lagrangian of Eqs. (9.20-9.21) reduce to

$$\Box h_{\mu\nu} = -16\pi G\left(T_{\mu\nu} - \frac{1}{2}\eta_{\mu\nu}T^\alpha_\alpha\right). \tag{9.30}$$

These equations can be solved to get various solutions which depend on the form of the right-hand side. In regions of space where there is no matter, the equations simplify to $\Box h_{\mu\nu} = 0$ which is simply the wave equation in flat space. The solution is

$$h_{\mu\nu} = A\epsilon_{\mu\nu} e^{ik\cdot x} + \text{h.c.} \tag{9.31}$$

with $\eta_{\mu\nu}k^\mu k^\nu = 0$; this is consistent with a massless spin-two wave.

9.2.3 Gravity sources: energy-momentum tensor

Einstein's equations of Eq. (9.22) contain the source of gravity on the right-hand side, the energy-momentum tensor. This term results from minimizing the matter action with respect to $g_{\mu\nu}$:

$$\delta S = \int d^4x \sqrt{-g}\, \delta g_{\mu\nu} \times \frac{1}{\sqrt{-g}} \frac{\delta(\sqrt{-g}\, \mathcal{L}_{\text{matter}})}{\delta g^{\mu\nu}}$$
$$= \int d^4x \sqrt{-g} \left[-\frac{1}{2} T^{\mu\nu}\right] \delta g_{\mu\nu}. \qquad (9.32)$$

Using the identity

$$\delta\sqrt{-g} = \frac{1}{2}\sqrt{-g}\, g^{\mu\nu} \delta g_{\mu\nu}, \qquad (9.33)$$

we can show that the energy-momentum tensor of matter can be written as

$$T_{\mu\nu} = -\frac{2}{\sqrt{-g}} \frac{\partial}{\partial g_{\mu\nu}} \mathcal{L}_{\text{matter}}. \qquad (9.34)$$

Let us now use this general expression to study gravity-matter interactions, i.e. a consistent way to couple gravitational fields to external sources. The transformation properties of the matter fields can be written as

$$\phi'(x') = \phi(x) \qquad \text{scalar;}$$
$$A'^{\mu}(x') = \Lambda^{\mu}_{\nu}(x) A^{\nu}(x) \qquad \text{vector.} \qquad (9.35)$$

In this book we will be only concerned with the simplest applications of EFT for gravity (see [Donoghue and Holstein (2015)] for a summary of a growing list of applications). With this in mind, we will use flat space $\eta_{\mu\nu}$ as our background metric. This will simplify the resulting equations: for example, gravitational covariant derivatives will reduce to the ordinary derivatives.

In the *simplest* case, the real scalar field ϕ can be coupled to gravity via

$$S_{\text{matter, 0}} = \int d^4x \sqrt{-g} \left(\frac{g^{\mu\nu}}{2} \partial_\mu \phi \partial_\nu \phi - \frac{m^2}{2}\phi^2, \right). \qquad (9.36)$$

where, as noted above, in a general case of background metric that is not flat we would have to promote the ordinary derivatives to the covariant ones. Using Eqs.(9.23), (9.24) for the flat background metric, as well as Eq. (9.26), we can expand Eq. (9.36) in powers of κ [Donoghue (1994)],

$$S_{\text{matter, 0}} = \int d^4x \sqrt{-g} \left(\mathcal{L}^{(0)}_{\text{m, 0}} + \mathcal{L}^{(1)}_{\text{m, 0}} + \mathcal{L}^{(2)}_{\text{m, 0}} + ...\right), \qquad (9.37)$$

where we find that the leading term

$$\sqrt{-g}\, \mathcal{L}_{m,\,0}^{(0)} = \frac{1}{2}\partial_\mu\phi\partial^\mu\phi - \frac{m^2}{2}\phi^2, \qquad (9.38)$$

represents a familiar expression for the real scalar Lagrangian in flat spacetime. This Lagrangian can be used to calculate the energy-momentum tensor for scalar fields using Eq. (9.34). It also takes a familiar form,

$$T_{\mu\nu} = \partial_\mu\phi\partial_\nu\phi - \frac{1}{2}\eta_{\mu\nu}\left(\partial_\alpha\phi\partial^\alpha\phi - m^2\phi^2\right). \qquad (9.39)$$

At higher orders in κ the Lagrangian would look like

$$\sqrt{-g}\, \mathcal{L}_{m,\,0}^{(1)} = -\frac{\kappa}{2}\, h_{\mu\nu}T^{\mu\nu},$$

$$\sqrt{-g}\, \mathcal{L}_{m,\,0}^{(2)} = -\frac{\kappa^2}{2}\left[\frac{1}{8}\left(h^2 - 2h_{\mu\nu}h^{\mu\nu}\right)\left(\partial_\alpha\phi\partial^\alpha\phi - m^2\phi^2\right)\right. \qquad (9.40)$$

$$\left. + \left(h^{\mu\alpha}h_\alpha^\nu - \frac{1}{2}hh^{\mu\nu}\right)\partial_\mu\phi\partial_\nu\phi\right].$$

We finally note that gravity can also be a source for gravity. The corresponding expression for the energy-momentum tensor can be found in [Weinberg (1972); Donoghue and Holstein (2015)],

$$T_{\mu\nu}^{\mathrm{grav}} = -2h^{\alpha\beta}\left(\partial_\mu\partial_\nu h_{\alpha\beta} + \partial_\alpha\partial_\beta h_{\mu\nu} - \partial_\beta\partial_\mu h_{\nu\alpha} + \partial_\beta\partial_\nu h_{\mu\alpha}\right)$$

$$- 2\partial_\alpha h_{\beta\nu}\partial^\alpha h_\mu^\beta + \partial_\alpha h_{\beta\nu}\partial^\beta h_\mu^\alpha - \partial_\nu h_{\alpha\beta}\partial_\mu h^{\alpha\beta} - h_{\mu\nu}\Box h \qquad (9.41)$$

$$- h_{\mu\nu}\left(\partial_\alpha h_{\beta\gamma}\partial^\beta h^{\alpha\gamma} - \frac{3}{2}\partial_\alpha h_{\beta\gamma}\partial^\alpha h^{\beta\gamma} - h^{\alpha\beta}\Box h_{\alpha\beta}\right).$$

We shall use those Lagrangians, after introducing the appropriate quantization procedure, to obtain matter-graviton and graviton self-interaction vertices.

9.3 Building an effective field theory

Taking the EFT point of view, there is absolutely no reason for the effective Lagrangian of gravity to retain the Einstein-Hilbert structure of Eq. (9.20). In fact, all independent operators of increasing dimension allowed by the symmetries of the theory must be retained. For pure gravity that includes (to next order in κ) three more operators in addition to the ones included in Eq. (9.20),

$$S_{\mathrm{eff\ GR}} = \int d^4x\, \sqrt{-g}\, \left[\Lambda + \frac{2}{\kappa^2}R + c_1 R^2 + c_2 R^{\mu\nu}R_{\mu\nu} + \cdots\right]. \qquad (9.42)$$

The first term, Λ, represents the cosmological constant, $\lambda = -8\pi G_N \Lambda$ [Weinberg (1972); Carroll (2004)]. Its effects are very important for the cosmological evolution of the Universe, but due to its smallness, it is irrelevant for the gravitational experiments. The terms proportional to the (dimensionless) c_i represent the so-called $R + R^2$ theories. The effects of those higher terms were studied [Stelle (1978)] with the conclusion that for any reasonable value of c_i (quoted values are $c_{1,2} < 10^{74}$ [Stelle (1978)]) these terms will have negligible effect on any gravitational experiments. Thus, General Relativity safely deals only with the Einstein-Hilbert action as described in Eq. (9.20).

Yet, if we are to follow the principles of EFT, there is no reason to drop any of the terms in the Lagrangian of Eq. (9.42). In fact, we see that EFT logic works quite well in the case of low energy gravitational interactions. Since, as we noted before, curvature always contains two derivatives acting on the metric, in the low energy regime the terms in the action that involve "one power" of curvature would scale as p^2, where p is some momentum scale, $p \ll M_{\text{Pl}}$. Higher order terms in Eq. (9.42) involve more powers of curvature and, therefore, necessary involve more powers of p. At sufficiently small energies terms of order p^2 will always dominate those of order p^4 that come from the $c_{1,2}$ terms in Eq. (9.42). That is why [Stelle (1978)] found tiny effects associated with the "R^2" terms making the bounds on $c_{1,2}$ coefficients extremely poor.

The expansion of Eq. (9.42) looks like a conventional EFT expansion in inverse powers of the Planck mass, with the terms including zero, one and two curvature tensors having, respectively, zero, two and four derivatives acting on the metric. The gravity Lagrangian organized this way certainly bears some resemblance to the Lagrangian of chiral perturbation theory described in Chapter 4!

9.3.1 Quantization. Feynman rules

The quantization of gravity can be conveniently done with path integrals. Just like with any gauge theory, one cannot define a Green function without first "spoiling" the gauge invariance, i.e.: specifying a particular way of removing equivalent configurations of gauge fields from the path integral. This implies that gauge-fixing $\mathcal{L}_{\text{gf}}$ and ghost terms $\mathcal{L}_{\text{ghost}}$ need to be added to the Lagrangian,

$$\mathcal{L} = \mathcal{L}_{EH} + \mathcal{L}_{\text{matter}} + \mathcal{L}_{\text{gf}} + \mathcal{L}_{\text{ghost}}. \tag{9.43}$$

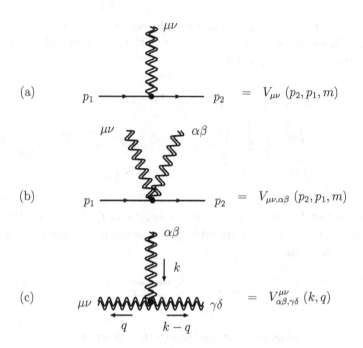

Fig. 9.1 (a) Matter-graviton, (b) matter-two gravitons, and (c) three graviton vertices in the EFT of Gravity. Gravitons are represented by doubly wavy lines. See text for the relevant expressions.

A popular choice of gauge in gravitational calculations is the so-called de Donder or harmonic gauge: $g^{\mu\nu}\Gamma^{\alpha}_{\mu\nu} = 0$. For weak fields, this becomes

$$G_\mu \equiv \partial^\lambda h_{\mu\lambda} - \frac{1}{2}\partial_\mu h^\lambda_\lambda = 0. \tag{9.44}$$

As we shall see later, the form of the graviton propagator in this gauge is especially simple. The gauge-fixing condition of Eq. (9.44) can be written as

$$\mathcal{L}_{\text{gf}} = -G_\mu G^\mu. \tag{9.45}$$

Notice that, unlike the usual gauge theories, the gauge condition of Eq. (9.44) contains a Lorentz index. That is, it encompasses four conditions at once. Because of this, the ghost field will carry a Lorentz index as well. In other words, while in the usual gauge theories ghosts are scalar particles that obey Fermi-Dirac statistics, gravitational ghosts are fermonic vector states. The ghost term is

$$\mathcal{L}_{\text{ghost}} = -\bar{c}^\mu \left[\frac{\delta G_\mu}{\delta \epsilon^\nu}\right] c^\nu. \tag{9.46}$$

After adding the the gauge fixing term, the action can be written in such a way that inversion of the kinetic term – and thus extraction of the graviton propagator – becomes possible,

$$S = \int d^4x \sqrt{-g}\mathcal{L} = S_{\text{ghost}} - \int d^4x G_\mu G^\mu$$
$$+ \int d^4x \left[-\frac{3}{2}\partial_\mu h_{\alpha\nu}\partial^\mu h^{\alpha\nu} + \partial_\alpha h_{\mu\nu}\partial^\nu h^{\mu\alpha} - h^{\mu\nu}\Box h_{\mu\nu} \right] \quad (9.47)$$
$$= \frac{1}{2}\int d^4x \, h_{\alpha\beta}(x) P^{\alpha\beta,\mu\nu} \Box h_{\mu\nu}(x) + S_{\text{ghost}},$$

where the last line is obtained after integrations by parts and $P^{\alpha\beta,\mu\nu}$ is defined in Eq. (9.27). The graviton propagator (in harmonic gauge) will then take the form,

$$D^{\alpha\beta,\mu\nu} = \frac{iP^{\alpha\beta,\mu\nu}}{p^2 + i\epsilon}, \quad (9.48)$$

where p is the momentum flowing through the graviton.

Using Eq. (9.40), Feynman rules for graviton-matter interaction vertices can be derived [Donoghue (1994)]. The one-graviton-matter interaction vertex (see Fig. 9.1(a))

$$V_{\mu\nu}(p_2, p_1, m) = -i\frac{\kappa}{2}\left[p_{1\mu}p_{2\nu} + p_{1\nu}p_{2\mu} - \eta_{\mu\nu}\left(p_1 \cdot p_2 - m^2\right)\right]. \quad (9.49)$$

The two-graviton-matter interaction vertex Fig. 9.1(b) would take the form

$$V_{\mu\nu,\alpha\beta}(p_2, p_1, m) = i\frac{\kappa^2}{2}\Big[2I_{\mu\nu,\gamma\delta}I^\delta_{\lambda,\alpha\beta}\left(p_1^\gamma p_2^\lambda + p_1^\lambda p_2^\gamma\right)$$
$$- \left(\eta_{\mu\nu}I_{\delta\lambda,\alpha\beta} + \eta_{\alpha\beta}I_{\delta\lambda,\mu\nu}\right)p_1^\delta p_2^\lambda \quad (9.50)$$
$$- P_{\mu\nu,\alpha\beta}(p_1 \cdot p_2 - m^2)\Big],$$

where the tensor $I_{\alpha\beta,\gamma\delta}$ is defined as

$$I_{\alpha\beta,\gamma\delta} = \frac{1}{2}\left(\eta_{\alpha\gamma}\eta_{\beta\delta} + \eta_{\alpha\delta}\eta_{\beta\gamma}\right), \quad (9.51)$$

Finally, we will also list an expression for the triple-graviton vertex

Fig. 9.1(c). Its derivation can be found in [Donoghue (1994)]

$$\begin{aligned}V^{\mu\nu}_{\alpha\beta,\gamma\delta}(k,q) = &\frac{i\kappa}{2}\Big[P_{\alpha\beta,\gamma\delta}\Big[k^\mu k^\nu + (k-q)^\mu(k-q)^\nu + q^\mu q^\nu - \frac{3}{2}\eta^{\mu\nu}q^2\Big]\\&+ 2q_\lambda q_\sigma Q^{\lambda\sigma,\mu\nu}_{\alpha\beta,\gamma\delta} + [q_\lambda q^\mu(\eta_{\alpha\beta}I^{\lambda\nu,}{}_{\gamma\delta} + \eta_{\gamma\delta}I^{\lambda\nu,}{}_{\alpha\beta})\\&+ q_\lambda q^\nu(\eta_{\alpha\beta}I^{\lambda\mu,}{}_{\gamma\delta} + \eta_{\gamma\delta}I^{\lambda\mu,}{}_{\alpha\beta}) - q^2(\eta_{\alpha\beta}I^{\mu\nu,}{}_{\gamma\delta} + \eta_{\gamma\delta}I^{\mu\nu,}{}_{\alpha\beta})\\&- \eta^{\mu\nu}q^\lambda q^\sigma(\eta_{\alpha\beta}I_{\gamma\delta,\lambda\sigma} + \eta_{\gamma\delta}I_{\alpha\beta,\lambda\sigma})]\\&+ [2q^\lambda(I^{\sigma\nu,}{}_{\alpha\beta}I_{\gamma\delta,\lambda\sigma}(k-q)^\mu + I^{\sigma\mu,}{}_{\alpha\beta}I_{\gamma\delta,\lambda\sigma}(k-q)^\nu\\&- I^{\sigma\nu,}{}_{\alpha\beta}I_{\gamma\delta,\lambda\sigma}k^\mu - I^{\sigma\mu,}{}_{\alpha\beta}I_{\gamma\delta,\lambda\sigma}k^\nu)\\&+ q^2(I^{\sigma\mu,}{}_{\alpha\beta}I_{\gamma\delta,\sigma}{}^\nu + I_{\alpha\beta,\sigma}{}^\nu I^{\sigma\mu,}{}_{\gamma\delta}) \quad (9.52)\\&+ \eta^{\mu\nu}q^\lambda q_\sigma(I_{\alpha\beta,\lambda\rho}I^{\rho\sigma,}{}_{\gamma\delta} + I_{\gamma\delta,\lambda\rho}I^{\rho\sigma,}{}_{\gamma\delta})]\\&+ [(k^2 + (k-q)^2)\Big[I^{\sigma\mu,}{}_{\alpha\beta}I_{\gamma\delta,\sigma}{}^\nu + I^{\sigma\nu,}{}_{\alpha\beta}I_{\gamma\delta,\sigma}{}^\mu - \frac{1}{2}\eta^{\mu\nu}P_{\alpha\beta,\gamma\delta}\Big]\\&- (k^2\eta_{\gamma\delta}I^{\mu\nu,}{}_{\alpha\beta} + (k-q)^2\eta_{\alpha\beta}I^{\mu\nu,}{}_{\gamma\delta})]\Big],\end{aligned}$$

where we also defined the short-cut notations,

$$Q^{\lambda\sigma,\mu\nu}_{\alpha\beta,\gamma\delta} = I^{\lambda\sigma,}{}_{\alpha\beta}I^{\mu\nu,}{}_{\gamma\delta} + I^{\lambda\sigma,}{}_{\gamma\delta}I^{\mu\nu,}{}_{\alpha\beta} - I^{\lambda\mu,}{}_{\alpha\beta}I^{\sigma\nu,}{}_{\gamma\delta} - I^{\sigma\nu,}{}_{\alpha\beta}I^{\lambda\mu,}{}_{\gamma\delta}. \quad (9.53)$$

We shall use some of those Feynman rules in the applications discussed below.

9.3.2 Quantum EFT for gravity

The above Feynman rules allow us to calculate classical and quantum corrections to many physical quantities. Although we already know that UV divergences are an expected occurrence in EFT, we would like to show how divergences known to plague quantum GR can be handled. We will illustrate how it works in a simple example [Donoghue (1994)].

Consider a four-graviton interaction vertex. A graviton-graviton scattering amplitude obtained from the Einstein-Hilbert action Eq. (9.20) would behave as

$$A_{\text{EH}} \sim \kappa^2 p^2, \quad (9.54)$$

where p is some (small) external graviton momentum. Using the more general Lagrangian of Eq. (9.42) gives as the next-order correction to the scattering amplitude that comes from the R^2 term and thus is of higher order in p,

$$A_{\text{EFT}} \sim c_1 \kappa^4 p^4. \quad (9.55)$$

How do loop amplitudes behave with respect to p? Using the lowest-order vertex (from Eq. (9.54)) gives for the one-loop correction,

$$A_{\text{loop}} \sim \int d^4 l (\kappa^2 l^2) \frac{1}{l^2} \frac{1}{(l-p)^2} (\kappa^2 l^2) \sim \kappa^4 I(p) \sim \kappa^4 p^4, \qquad (9.56)$$

where we are only interested in powers of external momentum, so we dropped all Lorentz indices, etc. $I(p)$ is the loop integral computed in dimensional regularization, so it only depends on the external momenta (see Appendix C). As we can see, the loop integral scales just like the next-to-leading term in Eq. (9.55). This implies that the renormalization program here is exactly the same as the one developed for the chiral perturbation theory in Chapter 4: divergences of a $\mathcal{O}(p^{2n})$ Lagrangian are cancelled by counter terms from the $\mathcal{O}(p^{2n+2})$ Lagrangian. This makes the quantum gravity calculations self-consistent.

Some pieces of this program have been known for a long time. For instance, one-loop corrections due to gravitons renormalize $c_{1,2}$ coefficients ['t Hooft and Veltman (1974)] to $c_{1,2}^{(r)}$

$$\begin{aligned} c_1^{(r)} &= c_1 + \frac{1}{960\pi^2} \frac{1}{\epsilon}, \\ c_2^{(r)} &= c_2 + \frac{1}{160\pi^2} \frac{1}{\epsilon}, \end{aligned} \qquad (9.57)$$

where minimal subtraction scheme was used. Other examples are also known [Donoghue and Holstein (2015)].

The analogy with χPT can be pushed even further with some rather interesting applications for quantum gravity. Both chiral perturbation theory and gravity deal with massless degrees of freedom, which generate non-analytic terms in the expressions for scattering amplitudes. For example, the loop expansion of a generic transition amplitude is

$$A(q^2) = Aq^2 \left[1 + \alpha \kappa^2 q^2 + \beta \kappa^2 q^2 \log(-q^2) + \cdots \right]. \qquad (9.58)$$

Contrary to the local analytic terms that contribute to the UV renormalization of the transition amplitude, logarithmic non-analytic terms $\log(-q^2)$ come from the infrared parts of the respective loops. These terms, along with the related ones, $m/\sqrt{-q^2}$ pick up imaginary parts for $q^2 > 0$ and are required by the unitarity of the S-matrix. Just like in chiral perturbation theory, for very small q^2 they could dominate over the analytic pieces of the transition amplitude and thus produce a somewhat large effect.

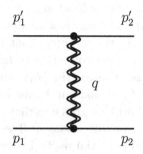

Fig. 9.2 Leading order graph for calculation of the Newton potential.

9.3.3 Newtonian potential

As an example of the described formalism, let us consider a simple application: a derivation of the Newton potential. We shall use the simplest approach to compute the potential in coordinate space. In this approach one studies the matrix element for the scattering of two test masses. In the non-relativistic limit this matrix element, after suitable change of normalization, corresponds to the quantum-mechanical transition amplitude. The Born approximation then tells us that the coordinate space potential can be obtained by a simple Fourier transform

$$V(r) = \int \frac{d^3q}{(2\pi)^3} e^{i\vec{q}\cdot\vec{r}} A(|\vec{q}|). \tag{9.59}$$

Let us implement this program at the leading nontrivial order in κ. We choose the normalization for the matter states as

$$\langle p'|p\rangle = 2E_p (2\pi)^3 \delta^3(p - p'). \tag{9.60}$$

The tree-level scattering amplitude can be obtained from Fig. 9.2 using the leading-order vertex diagram of Eq. (9.49) and the propagator of Eq. (9.48)

$$A(q) = V_{\mu\nu}(p_2, p_1, m) D^{\mu\nu,\alpha\beta}(p_2 - p_1) V_{\alpha\beta}(p'_2, p'_1, m). \tag{9.61}$$

We now have to take the static limit of Eq. (9.61). With $q = p_2 - p_1 \to (0, \vec{q})$ and $p_i \to (m_i, \vec{0})$ and changing to the non-relativistic normalization by dividing the expression of Eq. (9.61) by a factor of $2m_1 2m_2$, we obtain the momentum-space amplitude

$$A_{\mathrm{nr}}(q^2) = -\frac{\kappa^2}{8} \frac{m_1 m_2}{\vec{q}^2}, \tag{9.62}$$

which gives, from Eq. (9.59),

$$V(r) = -\frac{\kappa^2}{8} \int \frac{d^3q}{(2\pi)^3} e^{i\vec{q}\cdot\vec{r}} \frac{m_1 m_2}{\vec{q}^2} = -G_N \frac{m_1 m_2}{r}, \tag{9.63}$$

which is the familiar leading-order Newton potential.

This result is fairly simple and hardly warrants the development of heavy EFT machinery described in this chapter. But our EFT can be readily used to calculate a real quantum correction to the Newton potential. At one loop one would only need to use the Feynman rules presented in this chapter. The calculation of the quantum correction to Eq. (9.63) is rather straightforward, but the effort that goes into that calculation is truly heroic due to the proliferation of Lorentz indices. The complete result for the quantum-corrected Newton potential reads [Donoghue (1994)],

$$V(r) = -G_N \frac{m_1 m_2}{r} \left[1 - G_N \frac{m_1 + m_2}{rc^2} - \frac{127}{30\pi^2} \frac{G_N \hbar}{r^2 c^3} \right], \quad (9.64)$$

where the factors of $\hbar$ and c were reinstated to identify the true quantum effect (that is proportional to the $\hbar$). The fractional correction coming from the quantum correction in Eq. (9.64) computes to $\Delta V/V = [2.65 \times 10^{-35}/r(\text{meters})]^2$, which is unobservably small, even on subatomic scales.

9.3.4 *Postscript*

The search for quantum gravity has been a holy-grail quest among theoretical physicists since the days of Einstein. But while this goal is highly motivated by theoretical arguments, the fact is that quantum gravity will likely not be measured directly in a lab without a serious shift in our understanding. As we saw in the last section, quantum corrections to gravity are so incredibly small that they are simply not accessible to any experiment. In the EFT language, the issue is one of scale: the scale of quantum gravity is the Planck scale, which is 15 orders of magnitude higher than the electroweak scale. The reason for this huge discrepancy is called the "Hierarchy Problem" among high energy physicists, but whatever the resolution might be, chances are that quantum gravity will remain nothing more than a theoretical curiosity for the foreseeable future.

When scientists talk about quantum mechanics and gravity, they are often not referring to a quantum gravity theory, but rather to an ordinary quantum theory of matter that lives on a curved space-time. This basically corresponds to ordinary quantum field theory where $\bar{g}_{\mu\nu} \neq \eta_{\mu\nu}$ but something more complicated. This is how you would study issues such as Hawking radiation near a black hole, or quantum fluctuations in the Cosmic Microwave Background. But we often avoid the issues of the UV-incomplete GR by simply working in a semi-classical picture where the gravitational fields are given by Einstein's equations, while the matter fields are allowed to be treated in the full quantum field theory; see [Birrell and Davies (1984)] for examples of this approach.

9.4 Classical EFT: NRGR

We would like to talk about another approach that uses EFT techniques to solve a *classical* problem. We do not normally think about using EFT in classical physics, but it turns out that GR provides a very nice application of using many of the EFT tools while never using quantum mechanics! The EFT we want to discuss was originally proposed as a way to compute phase shifts in gravity waves coming from the inspiral phase of a binary black hole (or neutron star) system [Goldberger and Rothstein (2006a)], although there has been talk about applying similar ideas to other gravity-radiating systems as well.

In the particular example we want to consider, we need to compute phase shifts for these non-relativistic motions. In classical physics, the kinematic variables obey a simple relation that follows from the virial theorem:

$$v^2 \sim \frac{G_N m}{r}, \qquad (9.65)$$

so when dealing with nonrelativistic speeds, the quantity on the right hand side is small (in natural units). Historically, this kind of expansion was known as the *Post-Newtonian Expansion* (PN), and it is used in many places; for example, in the study of the precession of Mercury. However, the PN formalism runs into trouble. For example, at $\mathcal{O}(v^6)$ ("3PN") the terms in the expansion become infinite. This has been identified as coming from the breakdown of the point-particle approximation and the need to know the internal structure of the system.

Sound familiar?! In the EFT approach, these divergences are simply UV divergences that can be renormalized into effective couplings, pure and simple. Since the EFT you get is a nonrelativistic approximation to the classical theory of gravity, it has been given the rather oxymoronic name of *Non-Relativistic General Relativity* (or NRGR), in analogy with NRQCD.

What is even more amazing is that all of the EFT techniques we learned, such as power counting and RG analysis, can be used to handle these coefficients, just as though this were another quantum EFT. We compute (classical) gravity interactions with (classical) matter using Feynman diagrams, only *there are no loops,* which is very strange: how can you can get UV divergences in a theory with no loops?!

There are two reasons why you might want to study NRGR. The first reason is that experiments like LIGO will be sensitive to phase shifts in gravity waves,

$$\Delta\phi = \int \omega \, dt = \frac{2}{G_N m} \int dv \, v^3 \, \frac{dE/dv}{P(v)} \,, \qquad (9.66)$$

where E is the mechanical energy of a binary system radiating gravity waves (gravitons), and P is the power lost to radiation. Both of these quantities can be computed to high precision in NRGR, without any model dependence. This result can be derived using the definition of $P(t) = -dE/dt$ and $\omega = 2v^3/G_N m$, which follows from the virial theorem and the helicity-2 nature of the gravity wave.

It should be mentioned that, unlike the previous discussion where M_{Pl} was a huge number, it should now be thought of as being quite small. In fact, $M_{\text{Pl}} \sim 2.2$ μg, while the objects in question are black holes with $m \sim M_\odot \sim 10^{33}$ g. Thus $m/M_{\text{Pl}} \sim 10^{+39}$, which is by no means small! So this is a fundamentally different theory: the small expansion parameter is velocity, not coupling.

The second reason why NRGR is such a good theory to study: it is a beautiful example of just how powerful EFT techniques really are! So let us get started.

9.4.1 Setting up the problem

The goal of NRGR is to compute an effective action that describes a classical (massive) system such as a black hole binary as it emits gravitational radiation. We will start by matching the individual massive objects that make up the system onto a point-particle effective action, which encodes any internal structure as a collection of coefficients. This theory describes physics at the size-scale of the objects, which is taken to be the same order as the Schwarzschild radius $r_s \sim 2G_N m$. This theory is then matched onto a new effective theory consisting of a single object representing the entire system. This matching occurs at the size-scale of the orbit, r. If we wish to include RG effects, we can resum logarithms of r_s/r, which corresponds to resumming $\log(v^2)$ contributions. We will not discuss this resumming here, but focus on the matching. Once we have an effective theory of a single body, computing E and P in Eq. (9.66) can be done using standard field theory techniques.

To set up the problem correctly, we must first identify the relevant degrees of freedom. These are given by the following:

(1) A classical source is described by a worldline $x^\mu(\lambda)$, which is a 4-vector function of an affine parameter λ. We will often simply use $\lambda = x^0$, the object's time.
(2) Since the object can have spin, we must parametrize the object's *orientation* in space. This can be done by specifying a local frame $e^\mu_a(\lambda)$ defined at each point on the worldline. In the usual language of rigid body dynamics, this represents the "body frame" of the object. The dynamics of spin were worked out in [Porto (2006)], and we refer the reader to this paper. To keep things simple, we will ignore spin degrees of freedom in this book.
(3) We need a gravitational field parametrized by a metric. We will continue to use Minkowski space as our background, so

$$g_{\mu\nu} = \eta_{\mu\nu} + \frac{1}{M_{\rm Pl}} h_{\mu\nu}. \tag{9.67}$$

The tensor field $h_{\mu\nu}$ is responsible for graviton emission, and therefore represents the gravitational radiation. Recall that the Planck mass in the denominator does not necessarily imply that the $h_{\mu\nu}$ term is small: it is simply there to make the field canonically normalized with mass dimension 1.

In addition, we need to enforce all the symmetries of GR, which are

(1) Diffeomorphism invariance: this is the "gauge" symmetry given by Eqs. (9.12) and (9.25).
(2) Reparametrization invariance: worldlines should be invariant under changes in affine parameter: $\lambda \to \lambda'$.

The most general action of these parameters consistent with these symmetries is simply the usual "point-particle" action of GR, $S_{\rm EH} + S_{pp}$, where the Einstein-Hilbert action is given in Eq. (9.20) and

$$S_{pp} = \sum_a \left\{ m_a \int d\tau_a + c_R^{(a)} \int d\tau_a \, R(x_a) + c_V^{(a)} \int d\tau_a \, R_{\mu\nu}(x_a) \dot{x}_a^\mu \dot{x}_a^\nu \right.$$
$$\left. + c_E^{(a)} \int d\tau_a \, E_{\mu\nu}(x_a) E^{\mu\nu}(x_a) + c_B^{(a)} \int d\tau_a \, B_{\mu\nu}(x_a) B^{\mu\nu}(x_a) + \cdots \right\}.$$
$$\tag{9.68}$$

In this expression we have introduced the gravito-electric and gravito-magnetic fields

$$E_{\mu\nu} = C_{\mu\alpha\nu\beta}\dot{x}_a^\alpha \dot{x}_a^\beta \qquad (9.69)$$

$$B_{\mu\nu} = \frac{1}{2}\varepsilon_{\mu\alpha\beta\rho}C^{\alpha\beta}{}_{\nu\sigma}\dot{x}_a^\rho \dot{x}_a^\sigma, \qquad (9.70)$$

and $d\tau = ds$ is the proper time for our particle[2] from Eq. (9.3). The first term ("mass insertion term") is the usual action for a point particle. The "c-terms" encode finite-size effects that would otherwise be lost in the point-particle approximation. These coefficients will be used to absorb the divergences. They are not all independent, and in fact c_V and c_R can be removed by a field redefinition since we are working on a flat space-time background. However, since it is also possible to apply NRGR to cases where we are not on a flat background, we will leave them in.

The c_E and c_B terms are present and important. Ricci curvature represents deviations of the geodesics from straight lines, while Weyl curvature represents tidal forces. It should not be surprising, therefore, that c_E and c_B play an important role when discussing dissipative effects. However, they only start contributing at a rather high order in the velocity expansion $\mathcal{O}(v^{10})$, so we will not worry about them here. Dissipative effects in NRGR were discussed in [Goldberger and Rothstein (2006b)], and the reader is encouraged to investigate the results.

9.4.2 Gravition modes

Only "soft gravitons" with momenta that scale like v/r can interact with our system. However, when trying to compute the dynamics holding the system together, it is convenient to include another auxiliary mode responsible for the binding force, called "potential gravitons". This force can be thought of as acting over distances of order the size of the system (r), and therefore gravitons mediating these forces can carry momenta that scale like $1/r$. However, these modes would continue to transfer energies of order v/r. Therefore they cannot be on-shell gravitons. When we integrate them out of the theory, we end up with a set of contact interactions between the two particles that make up the binary system.

We start with the action in Eq. (9.68) which describes two worldlines and gravitons. Set the metric to the form:

$$g_{\mu\nu} = \eta_{\mu\nu} + \frac{1}{M_{\text{Pl}}}\left(H_{\mu\nu} + \bar{h}_{\mu\nu}\right), \qquad (9.71)$$

---

[2]Remember that "particles" in this EFT are black holes or neutron stars!

where $H_{\mu\nu}$ are the potential gravitons, and $\overline{h}_{\mu\nu}$ are the (soft) radiation gravitons. Potential modes describe gravitons with energy-momenta of the form $(k^0, \vec{k}) \sim (v/r, 1/r)$ and are necessarily off-shell. Radiation modes, on the other hand, describe gravitons with energy-momenta of the form $(k^0, \vec{k}) \sim (v/r, v/r)$; these are the modes that can propagate to infinity and represent the gravitational waves reaching detectors such as LIGO.

Since potential gravitons carry large spacial momenta, we can repeat the techniques of Type-II and -III EFT and pull out an eikonal factor (this basically corresponds to performing a spacial-Fourier transform)

$$H_{\mu\nu}(t, \vec{x}) = \int \frac{d^{d-1}k}{(2\pi)^{d-1}} e^{i\vec{k}\cdot\vec{x}} H_{\vec{k}\mu\nu}(t), \qquad (9.72)$$

where we write the integral in d dimensions in order to use dimensional regularization to tame any divergent integrals. Now that we have made the large momenta explicit, when derivatives act on any field, they will always scale like v/r, and any large momenta will be included explicitly. If one expands the quadratic terms for the metric keeping only the leading terms in v, you can show that

$$\langle H_{\vec{k}\mu\nu}(x^0) H_{\vec{q}\alpha\beta}(y^0)\rangle = (2\pi)^3 \delta^{(3)}(\vec{k}+\vec{q})\delta(x^0-y^0)\frac{-iP_{\mu\nu,\alpha\beta}}{\vec{k}^2}, \qquad (9.73)$$

$$\langle \overline{h}_{\mu\nu}(x)\overline{h}_{\alpha\beta}(y)\rangle = D_F(x-y)P_{\mu\nu,\alpha\beta}, \qquad (9.74)$$

up to choice of gauge, where D_F is the usual Feynman (scalar) propagator used throughout the book, and $P_{\mu\nu,\alpha\beta}$ is given in Eq. (9.27). Notice that the potential graviton "propagator" represents an instantaneous interaction, as you would expect in the nonrelativistic limit. From these correlation functions and the rules for momentum/energy scaling, we can show that

$$H_{\vec{k}\mu\nu} \sim r^2\sqrt{v}, \qquad (9.75)$$

$$\overline{h}_{\mu\nu} \sim v/r. \qquad (9.76)$$

We also need to know how to power count mass insertions. In particular, how should m/M_{Pl} scale? To answer that we return to the virial theorem in Eq. (9.65) and identify $G_N \simeq M_{\text{Pl}}^{-2}$. If we recall ordinary mechanics, the angular momentum of the system is $L \sim mvr$, so that with a small amount of algebra we find

$$m/M_{\text{Pl}} \sim \sqrt{Lv} \qquad (9.77)$$

So now all of our quantities have definite scalings with the small velocity, except that we also have dependence on a new quantity, angular momentum. Actually, this is not a surprise when you remember that we are dealing with classical physics, and there are hidden factors of $\hbar$ in our formalism. It turns out that, just like $\hbar$ in ordinary QFT, L (or more precisely $\hbar/L$) is our loop-counting parameter. It can be shown that all tree level diagrams scale like L, so that the one-loop corrections (quantum gravity) are suppressed by $\hbar/L$, which is ludicrously small: remember, these are black-hole sources!

This, then, is the meaning of the statement that we need only do classical (read: "tree-level") calculations. Divergences then come from the integrals on the potential graviton propagators mediating instantaneous nonlocal interactions, rather than from loop integrals.

Notice that by "loops" we only mean loops of gravitons. *Classical worldlines cannot form loops.* Therefore, in diagrams where gravitons connect to a worldline in multiple places, that is *not* a loop diagram, since the worldline does not have a propagator. Rather, you should think of each place where a graviton touches a worldline as a vertex whose Feynman rule involves the worldline insertion x^μ. We will see how this works in a sample calculation.

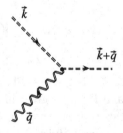

Fig. 9.3 Vertex with two potential gravitons (dashed lines) and one radiation graviton (wavy line).

We have one more complication that must be dealt with. Consider a process where a potential graviton (with momentum $\vec{k}$) absorbs a radiation graviton (with momentum $\vec{q}$). The potential graviton will come out of the interaction carrying momentum $\vec{k} + \vec{q}$, as seen in Fig. 9.3; but this does not have a proper scaling with velocity, since $\vec{k} \sim 1/r$ while $\vec{q} \sim v/r$. This means that we should drop the $\vec{q}$ from the outgoing potential graviton momentum. But while that is okay at leading order in the power counting, it violates conservation of momentum at higher orders.

We can understand the solution to this problem if we look closer at the propagator of the outgoing potential graviton, keep the small $\vec{q}$, and do a

Taylor expansion:

$$\frac{1}{(\vec{k}+\vec{q})^2} = \frac{1}{\vec{k}^2}\left(1 + \frac{2\vec{q}\cdot\vec{k}}{\vec{k}^2} + \cdots\right). \quad (9.78)$$

This is nothing more than the multipole expansion from electrodynamics! We can reproduce this expansion at the operator level if we perform the multipole expansion on the radiation graviton field itself:

$$\overline{h}_{\mu\nu}(t,\vec{x}) = \overline{h}_{\mu\nu}(t,\vec{x}_0) + (\vec{x}(t) - \vec{x}_0)^i \nabla_i \overline{h}_{\mu\nu}(t,\vec{x}_0) + \cdots, \quad (9.79)$$

where $\vec{x}_0$ is a convenient origin for the system, such as the center of mass of the binary. Now if we plug in this expansion everywhere we see a radiation graviton, and apply the rule that radiation gravitons cannot impart momentum to potential gravitons, we have successfully solved the problem of mixing contributions from different orders in power counting.

9.4.3 Feynman rules

We are now in a position to expand the action in Eq. (9.68) up to v^2 terms in the power counting. Choosing to work in the frame of the moving object (so $d\tau = dt$) we find:

$$S_1 = m\int dt \left\{\frac{1}{2}\vec{v}^2 + \frac{1}{8}\vec{v}^4 - \frac{1}{2M_{\text{Pl}}}h_{00} - \frac{1}{2M_{\text{Pl}}}h_{0i}\vec{v}_i \right.$$
$$\left. - \frac{1}{4M_{\text{Pl}}}h_{00}\vec{v}^2 - \frac{1}{2M_{\text{Pl}}}h_{ij}\vec{v}_i\vec{v}_j\right\}, \quad (9.80)$$

where $h_{\mu\nu} = \overline{h}_{\mu\nu} + \int \frac{d^3\vec{k}}{(2\pi)^3} e^{i\vec{k}\cdot\vec{x}} H_{\vec{k}\mu\nu}$. Notice that the first two terms (free motion) scale like L, while the other terms, describing couplings to gravitons, scales like $\sqrt{L}$ (this action contains radiation terms up to $\sqrt{L v^5}$). When we integrate out the potential gravitons, these terms will generate particle-particle interactions that scale with a full factor of L; radiation gravitons, however, are power suppressed (compared to free motion), which is expected. Since radiation comes with square-roots of velocity, corresponding to odd-powers of velocity in the final result, PN analyses often refer to radiation occurring at "2.5PN" (i.e.: $\mathcal{O}(v^5)$). We will see this explicitly below.

Now potential gravitons, since they are off-shell, can only appear as internal lines of a graph. The worldline variables $\vec{x}$ (and $\vec{v}$) should be thought of as part of the Feynman vertices, *not* as propagating degrees of freedom. This is the big difference between ordinary QFT and this EFT:

when determining the topology of the Feynman graph, erase the worldlines and only look at the graviton lines. The only purpose for the worldline in the graph is to tell you that each vertex on that worldline corresponds to using the same $\vec{x}$, evaluated at different times.

For example, when calculating the one-potential graviton exchange between two masses $(m_1, \vec{x}_1)$, $(m_2, \vec{x}_2)$, the leading Feynman vertex comes from the third term in Eq. (9.80). The diagram is the same as Fig. 9.2, except the exchanged particle is a potential graviton; we find

$$i\Delta S_{\text{eff}} = \int \frac{d^3\vec{k}}{(2\pi)^3} \int \frac{d^3\vec{q}}{(2\pi)^3} \left[\frac{-im_1}{2M_{\text{Pl}}} \int dt_1 \; e^{i\vec{k}\cdot\vec{x}_1}\right]$$
$$\times \left[(2\pi)^3 \delta^{(3)}(\vec{k}+\vec{q})\delta(t_1-t_2)\frac{-i}{\vec{k}^2}P_{00;00}\right] \times \left[\frac{-im_2}{2M_{\text{Pl}}} \int dt_2 \; e^{i\vec{q}\cdot\vec{x}_2}\right]$$
$$= \frac{m_1 m_2}{4M_{\text{Pl}}^2} \int dt \int \frac{d^3\vec{k}}{(2\pi)^3} e^{i\vec{k}\cdot(\vec{x}_1(t)-\vec{x}_2(t))} \frac{(+i)}{2\vec{k}^2}$$
$$= \frac{im_1 m_2}{8M_{\text{Pl}}^2} \int dt \frac{1}{(4\pi)|\vec{x}_1(t)-\vec{x}_2(t)|} = i\int dt \Big(-V_N(|\vec{x}_{12}|)\Big).$$
(9.81)

You can now see how "potential gravitons" got their name! Note that this potential scales as Lv^0, giving us Newton's potential at leading order in the power counting, as it should be. Inserting higher-order vertices in the same graph, as well as including leading-order vertices in diagrams with more interactions, gives us the leading correction, known as the *Einstein-Infeld-Hoffmann* potential. We leave this as an exercise: it is a good way to practice the power counting and computing these kinds of diagrams; see Fig. 9.4.

9.4.4 Gravitational radiation

Now we would like to know how gravitons can be emitted to provide radiation. This means we need to compute terms in the EFT that contain radiation gravitons as external states. When working on binary systems, we expect the dominant contributions to come not from individual objects, but from the *entire system* radiating; this means that we should match our 2-body system onto an effective 1-body system at a scale $\mu \sim 1/r$, where r is the size of the binary system. In such cases, it makes sense to choose the origin of our calculation to be the Center of Mass, and let $M = m_1 + m_2$ be the total mass of the system. In Feynman Diagrams, we will draw the worldlines of these effective-one-body objects with a double line. Then the

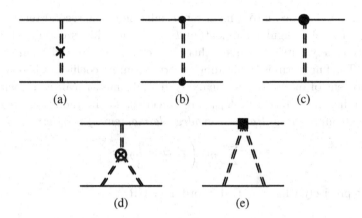

Fig. 9.4 Diagrams responsible for the $\mathcal{O}(v^2)$ EIH Potential. Vertices with no symbol are given by the Feynman rule used in Eq. (9.81). (a) Potential graviton kinetic term insertion. (b) Two $\mathcal{O}(v)$ vertices from Eq. (9.80). (c) One $\mathcal{O}(v^2)$ vertex from Eq. (9.80). (d) Three potential graviton self-coupling. (e) Two potential gravitons coupling to a worldline. Not shown are the permutation diagrams for (c)-(e). See text and Exercise 3.

effective action for the theory below this orbital scale is given by three terms:

$$S_0 = -\frac{M}{2M_{\text{Pl}}} \int dt\, \bar{h}_{00}, \tag{9.82}$$

$$S_1 = -\frac{1}{2M_{\text{Pl}}} \int dt \left[M\vec{X}_{\text{CM}} \cdot \nabla \bar{h}_{00} + 2\vec{P}^i_{\text{CM}} \bar{h}_{0i} \right], \tag{9.83}$$

$$S_2 = \frac{1}{2M_{\text{Pl}}} \int dt \left\{ -\bar{h}_{00} \left[\frac{1}{2}M\vec{v}^2 - \frac{Gm_1 m_2}{|\vec{x}_{12}|} \right] + \varepsilon_{ijk} \vec{L}_i \partial_j \bar{h}_{0k} \right.$$
$$+ \sum_a m_a \vec{x}_a^i \vec{x}_a^j \bar{R}_{0i0j}. \tag{9.84}$$

In terms of power counting, $S_0 \sim \sqrt{Lv}$, $S_1 \sim v\sqrt{Lv}$, and $S_2 \sim v^2\sqrt{Lv}$. The action S_0 would give us "scalar graviton" radiation, which should not be allowed. However, this term couples to mass, which is conserved in NRGR; thus this term cannot give radiation. This is basically the usual story from gauge theories. The next term S_1 is related to Center of Mass kinematics. If we choose to work in the CM frame, $\vec{X}_{\text{CM}} = \vec{P}_{\text{CM}} = 0$ and therefore S_1 cannot contribute to gauge-invariant processes. Indeed, this is also expected, since we know helicity-2 waves cannot be dipole-radiated.

This means that S_2, especially the term involving the linearized Riemann tensor for the radiation graviton ($\bar{R}_{0i0j}$), is the leading source for radiation. The first term represents energy T^{00} coupling to the graviton, and can be absorbed into a mass renormalization; the second term is proportional to the angular momentum $\vec{L}$, but since this is conserved in the absence of any outside torques, this term cannot create radiation, similar to S_0. The final term is the leading contribution; its coefficient is related to the moment of inertia of the binary system. In fact, it can be shown that the leading order terms in $\overline{R}_{0i0j}$ are traceless, so the radiation gravitons actually couple to the *traceless quadrupole moment of inertia*:

$$Q_{ij} = \sum_a m_a \left(\vec{x}_a^i \vec{x}_a^j - \frac{1}{3} \delta_{ij} \vec{x}_a^2 \right). \tag{9.85}$$

This is precisely what you find in ordinary GR.

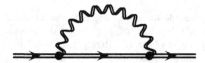

Fig. 9.5 Diagram whose imaginary part gives the power emitted by gravitational radiation. This diagram is computed in the one-body effective theory below the scale $1/r$, hence the double-line on the worldline. Note that this diagram is *not* a loop diagram, since the worldline is not a propagator.

We can most easily compute the rate of gravitational radiation by computing the imaginary part of the effective action, since by the optical theorem:

$$d\Gamma = \lim_{T \to \infty} \frac{2}{T} \text{Im } S_{\text{eff}}, \tag{9.86}$$

where $d\Gamma$ is the differential rate graviton emission rate; see Fig. 9.5. The classical power spectrum is then given by $dP = E \, d\Gamma$, as explained in [Goldberger and Rothstein (2006a)]. Unlike ordinary QFT, however, the imaginary part of diagrams like Fig. 9.5 are relatively easy to compute, because, as we said before, the worldline is not a propagator in a loop! The imaginary part therefore comes strictly form Cauchy's formula for the propagator:

$$\text{Im } S_{\text{eff}} = \frac{1}{2} \left[\frac{i}{2M_{\text{Pl}}} \int dt \, Q^{ij} \right] \left[\frac{i}{2M_{\text{Pl}}} \int dt' \, Q^{kl}(t') \right]$$
$$\times \text{Im} \langle \overline{R}_{0i0j}(t) \overline{R}_{0i0j}(t') \rangle, \tag{9.87}$$

where the averaging over the linearized Riemann tensors involves the propagator of the radiation gluons from Eq. (9.74), which has an imaginary part from the $+i\varepsilon$ convention. Doing the integral over the resulting delta-function gives us

$$d\Gamma = \frac{1}{T}\frac{1}{160\pi M_{\rm Pl}^2} E^5 |Q^{ij}(E)|^2\, dE, \qquad (9.88)$$

where $Q^{ij}(E)$ is the Fourier transform of the quadrupole moment. To get the power radiated, we use the formula $dP = E\, d\Gamma$, and after returning to time-space from energy-space we get the final answer:

$$P = \frac{G_N}{5\pi}\langle \dddot{Q}^{ij}\dddot{Q}_{ij}\rangle, \qquad (9.89)$$

where the brackets represent time-averaging. Since it involves two insertions of S_2 and no loops, it scales like Lv^5 (or 2.5PN, in the Post-Newtonian order-counting). This result is well known to students of GR.

9.4.5 Renormalization

Before ending our discussion of NRGR, let us see how divergences can appear in our theory, and how renormalization methods can help us make sense of these divergences. We already pointed out that there are no loops in any calculation (they would be suppressed by the tiny loop factor $\hbar/L$), which is simply a statement that this is a *classical* theory. Nonetheless, there are divergent integrals that appear, coming either from the integrals over potential graviton momenta, or else coming from the integrals of the radiation graviton propagators. Many of these divergences are "trivial," in the sense that they are power-law divergences and therefore can be reabsorbed into things like the mass; better still, when we use DimReg to tame these divergences (putting aside the challenge of maintaining general coordinate invariance while changing the spacetime dimension), power law divergences explicitly vanish as usual; see Appendix C for details. However, it is also true that there are log-divergences, and these could manifest as something physical, such as the running of couplings. We would like to see how this plays out.

We mentioned earlier that at 3PN (v^6 in the power counting) calculations of gravity waves from inspiraling black holes gave divergent results; these divergences have been identified as coming from a breakdown in the point-particle approximation. It is instructive to see how this same result occurs in NRGR, and therefore what must be done to solve it.

The divergence in question comes from diagrams like those in Fig. 9.6, where one radiation graviton is emitted but there are multiple internal

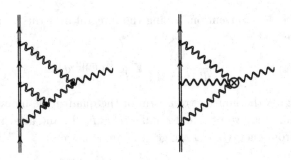

Fig. 9.6 Logarithmically divergent contributions at $\mathcal{O}(v^6)$ to radiation. There are also two permutations of the three graviton vertex in the first diagram.

graviton lines. If you sum up all the contributions, you would find that these diagrams generate the effective action:

$$\Delta S = \frac{1}{M_{\rm Pl}} \int \frac{d^4k}{(2\pi)^4} \bar{h}_{\mu\nu}(-k)(2\pi)\delta(k^0) \Big[A(k^2\eta^{\mu\nu} - k^\mu k^\nu) - Bk^2 v^\mu v^\nu \Big], \tag{9.90}$$

where

$$A = \frac{1}{48 M_{\rm Pl}} \left(-\frac{iM}{2M_{\rm Pl}} \right)^3 I_0(\vec{k})(1 - 2\epsilon) + c_R, \tag{9.91}$$

$$B = \frac{1}{24 M_{\rm Pl}} \left(-\frac{iM}{2M_{\rm Pl}} \right)^3 I_0(\vec{k}) \left(1 - \frac{21}{2}\epsilon \right) + \frac{1}{2} c_V, \tag{9.92}$$

and

$$I_0(\vec{k}) = \frac{1}{32\pi^2} \left[\frac{1}{2\epsilon} - \gamma + 3 - \log\left(\frac{\vec{k}^2}{4\pi\mu^2} \right) \right] + \mathcal{O}(\epsilon). \tag{9.93}$$

The $1/\epsilon$ divergences can then be absorbed into c_R and c_V, at the expense of introducing a dependence on scale. By insisting the action be renormalization-scale independent we find:

$$\mu \frac{dc_R}{d\mu} = -\frac{1}{6} G_N^2 M^3, \tag{9.94}$$

$$\mu \frac{dc_V}{d\mu} = +\frac{2}{3} G_N^2 M^3. \tag{9.95}$$

This, along with matching conditions at $1/r$, the scale where we match onto an effective one-body theory, give us everything we need to solve the problem. While we might need a model to determine these matching conditions, we can understand much better where the divergences are coming from, and how to interpret them: they're simply a running-coupling effect!

Now in this case, the constants $c_{R,V}$ are unphysical: they can be removed by a field redefinition and a change of coordinates. However, similar renormalizations occur with other constants such as $c_{E,B}$ which *are* physical, as well as in more general problems than the flat-space background. The idea that EFT tools such as regularization and renormalization can play a role in a classical field theory is very exciting!

9.5 Notes for further reading

Quantum gravity is a huge topic, so we would not even attempt to describe it in detail. We do invite readers to venture into the depths of various approaches to describe the problems of quantum description of gravity.

Good introduction to General Relativity can be found in a number of excellent texts. We would recommend [Carroll (2004)] and [Weinberg (1972)] for concise review.

A good introductions to string theory is given by B. Zwiebach [Zwiebach (2004)]. A more detailed approach, which requires solid mathematical base, can be found in Polchinski's books [Polchinski (1998a,b)]. Other approaches to quantum gravity can be found in a number of reviews: loop quantum gravity is discussed in [Rovelli (2011)], induced gravity in [Visser (2002)], asymptotically safe gravity in [Niedermaier and Reuter (2006); Weinberg (1980)], while casual dynamical triangulation in [Ambjorn *et al.* (2006)]. These texts contain extensive reference lists, including original papers on the subject.

Further information about effective field theory for quantum gravity can be found in a number of excellent papers [Donoghue and Holstein (2015); Donoghue (1995)], as well as in [Burgess (2004)]. Applications of EFT techniques to other problems in General Relativity can be found in [Rothstein (2014)].

NRGR was first worked out in a long but very interesting paper [Goldberger and Rothstein (2006a)]. One year after this paper came out, [Porto (2006)] showed how to incorporate spin and angular momenta of the bodies into the picture. Around the same time, [Goldberger and Rothstein (2006b)] showed how to use the $c_{E,B}$ operators to include absorption and dissipation, which are important for black holes. Since then, the ideas of NRGR have proven useful to many researchers working in inflation, problems of gravitational radiation-reaction, and computing higher and higher terms in the PN expansion, to name a few. For an excellent review of

Problems for Chapter 9

(1) **More on Christoffel symbols**
    Derive the explicit form of the Christoffel symbol by requiring the vanishing of the covariant derivative acting on the metric tensor, i.e. by considering
    $$D_\lambda g_{\mu\nu} = \frac{\partial g_{\mu\nu}}{\partial x_\lambda} - g_{\alpha\nu}\Gamma^\alpha_{\mu\lambda} - g_{\mu\alpha}\Gamma^\alpha_{\nu\lambda} = 0, \qquad (9.96)$$
    and symmetrizing with respect to lower indices.

(2) **Effective Lagrangian for gravitational interactions**
    Explain why the term $c_3 R_{\mu\nu\alpha\beta} R^{\mu\nu\alpha\beta}$ is missing in Eq. (9.42) in our four dimensional space-time, even though it contains the same number of derivatives as $c_1 R^2$ and $c_2 R_{\mu\nu} R^{\mu\nu}$. What about other terms, such as $c_4 \Box R$?

(3) **Einstein-Infeld-Hoffmann potential**
    The leading (Lv^2) corrections to the Newtonian potential were initially computed in [Einstein et al. (1938)][3]
    $$V_{\rm EIH} = -\frac{G_N m_1 m_2}{2|\vec{x}_1 - \vec{x}_2|} \left[3(\vec{v}_1^2 + \vec{v}_2^2) - 7(\vec{v}_1 \cdot \vec{v}_2) - \frac{(\vec{v}_1 \cdot \vec{x}_{12})(\vec{v}_2 \cdot \vec{x}_{12})}{|\vec{x}_1 - \vec{x}_2|^2} \right]$$
    $$+ \frac{G_N^2 m_1 m_2 (m_1 + m_2)}{2|\vec{x}_1 - \vec{x}_2|^2}. \qquad (9.97)$$
    Derive this expression in NRGR. You will need to compute the diagrams in Fig. 9.4. You will also need the (power suppressed) "kinetic" terms for the potential graviton:
    $$\Delta S_{H{\rm kin}} = \frac{1}{2} \int dt \int \frac{d^3 k}{(2\pi)^3} \left[\partial_0 H_{\vec{k}\mu\nu} \partial_0 H^{\mu\nu}_{-\vec{k}} - \frac{1}{2} \partial_0 H^\mu_{\vec{k}\mu} \partial_0 H^\mu_{-\vec{k}\mu} \right]. \qquad (9.98)$$
    The three-potential graviton vertex, as well as the two potential gravitons coupling to a worldline, can be read off of the vertices in Fig. (9.1) suitably recast for potential gravitons.

---

[3] The observant reader might have noticed a factor of 2 difference between this expression and the relativistic correction (second term) in Eq. (9.64). This is a very subtle and interesting ambiguity in our choice of coordinate system ("gauge"). The expression here is the full expression, while Eq. (9.64) is in the static limit ($\vec{v}_i = 0$), where Harmonic gauge is not enough to fix all of the coordinates. See [Bjerrum-Bohr et al. (2003)] for a nice explanation of this problem.

Chapter 10

Outlook

In this book, we have delved deeply into the practical applications of effective field theory, studying problems varying from pion physics and superconductors to gravitational radiation. And yet, when we think of all the ways EFT can be used to solve real-world problems, it seems like we have hardly scratched the surface. The methods we have discussed in these pages are so far reaching that they can find application in almost any context. The true beauty of EFT is its ability to approach a problem in a *model-independent way*, properly identifying the relevant operators, and cordoning off any additional assumptions into coefficients that can be measured, calculated or in some way quantified. Intuitively, it is the most sensible way to approach a problem in physics; logically, it is the most robust.

Time and space give us a very limited list of topics that we can discuss in detail in the book, but to give you a sense for some other applications for these ideas, we wanted to use these last pages to share some quick words on how the EFT approach has been used to great success in other active research areas. This list is far from complete, and we cannot do these topics justice. But we hope to inspire you to keep studying.

10.1 Supersymmetry

One of the most popular ideas for new physics beyond the electroweak symmetry breaking scale is supersymmetry. Originally motivated by string theory, this additional space-time symmetry says that for every integer-spin (boson) degree of freedom, there exists a half-integer-spin (fermion) degree of freedom, and vice versa. These pairs of particles are to have the same masses and charges, so since we do not see bosonic electrons in nature, the symmetry must be broken.

This idea found a home with particle physicists looking for a way to explain why the Higgs boson was so light. Being a neutral scalar field, quantum corrections to its mass sample energies all the way up to the highest known scales, and therefore it should be very massive; this is known as the "Fine-tuning Problem" among particle physicists. Pairing it with a fermion, however, allows a chiral symmetry to stabilize the Higgs mass at a low scale without any need to fine-tune cancellations between various quantum effects.

But putting the phenomenology questions aside, the extra constraints supersymmetry imposes on theoretical models gives the subject a new layer of interest. Supersymmetry puts very strong constraints on the kinds of operators that one can write down, such as "holomorphy" (meaning you cannot put Φ and $\Phi^\dagger$ in the same operator); it also limits the renormalization behavior of the theory as well. We have already seen how the Higgs scalar mass is protected from quantum corrections; in fact, many coupling constants do not get renormalized at all in supersymmetric theories, and the beta function for a supersymmetric gauge coupling has been completely worked out in general.

The big game in the supersymmetry model-building world is to try and find the best way to break supersymmetry while maintaining as many of these theoretical benefits as possible. This is precisely where "supersymmetric EFT" comes in. Most models set the physics of supersymmetry breaking at a very high scale, such as the Planck scale (or at least several orders of magnitude above the weak scale) and then the effects trickle down through renormalization effects like running couplings and masses. This is an ideally suited application of EFT techniques, with the added tools of how to write supersymmetric operators.

10.2 Extra dimensions

Another idea that has been popular for many years is the notion that we live in more than the 3+1 dimensions of space-time. The idea goes all the way back to T. Kaluza who, in 1921, showed that a fifth dimension could explain the phenomenon of electromagnetism; the idea was extended to a quantum world by O. Klein five years later, the resulting theory known as "Kaluza-Klein extra dimensions". Nearly eighty years later, models like the Randall-Sundrum model, originally meant to be a way to separate scales in supersymmetry breaking, suddenly took on a life of their own when it was realized that they solve the hierarchy problem without supersymmetry at all! The idea is that gravity, whose degrees of freedom work at energy scales near the Planck scale, acts in all dimensions of space-time, while gauge interactions like the electroweak theory are limited in how far they could penetrate the extra dimensions. This led to the notion of an effective theory where "scale" was replaced by "location in the extra dimension," and the problem became one of geometry. Gravity was weak simply because of Gauss's Law: gravitational flux can fill more area since it can move in more directions, and therefore it weakens more quickly.

The situation became even more exciting after J. Maldecena published his conjecture suggesting that a theory of gravity in extra dimensions (string theory) could be dual to a four-dimensional *conformal* field theory, that is, a theory without any scale. Conformal field theories are the opposite of effective field theories, since they predict the same physics at all scales. However, our world is not conformally invariant, so that means there must be something that breaks the conformal symmetry, such as a "brane" somewhere in the extra dimension. At this point, EFT can take over!

10.3 Technicolor and compositeness

Effective field theory in its current form has its origins in pion physics of the 1960s. Chiral symmetry breaking led us to identify the pions as pseudo-NG bosons that mediated the strong nuclear force. Later, with the proposal and discovery of quarks and QCD, we were able to UV-complete this picture. But the beauty of the chiral Lagrangian was that it did not need a UV completion to be useful.

It is not surprising, therefore, that many have tried to "repeat the joke", this time at the electroweak scale. Another chiral symmetry breaking occurs among another class of strongly interacting particles, given the fanciful name of "techniquarks". The NG bosons that follow are identified as the "techni-pions", which would provide the degrees of freedom that are eaten by the W and Z bosons. There is no Higgs boson, but just as in QCD, there would be higher states of particles that must exist. The search was on to try and find the "techni-rho" particles, but with no success as of the time of this books publication.

Because the theory is strongly coupled, we are limited in what we can calculate, just like with ordinary pions. But we have an advantage: the details of how chiral symmetry breaking occurs are irrelevant – the chiral

Lagrangian is a universal description of our system. Therefore, to make predictions in the world of techni-particles, all one needs to do is to "scale-up" known quantities from pion physics. Non-universal corrections might change details, but we should be able to make faithful order-of-magnitude predictions of matrix elements, masses and decay widths.

Sadly for this theory, it has not had much luck on the experimental side, but people are still looking. There have been many attempts to improve on the program, and many still like it since it does not have fine-tuning problems prevalent with the Higgs scalar boson. Nonetheless, with a Higgs boson discovered at the LHC at 125 GeV, looking pretty much like a standard model CP-even scalar field, technicolor theories are in trouble.

10.4 High-T_c superconductivity

We do not want to close without mentioning one of the biggest questions on the minds of many condensed matter physicists: the mechanism of high-temperature superconductivity. Cuprate superconductors transition at temperatures above Nitrogen's boiling point, and more superconductors are discovered in stranger places like iron. We are pushed to our limits to explain what mechanisms could be responsible for these transitions. Fortunately, we do not need to know the details to understand the phenomenology. With the help of EFT, we can isolate the non-universal effects from the universal ones, and put constraints on the resulting coefficients that are sensitive to the underlying (UV) physics. This gives us a very useful handle for theorists to try and match to their models, and avoids the confusing reports where a model predicts something that is model-independent!

With that, we leave you to explore the wonders that Effective Field Theory can reveal to you!

Appendix A

Review of Group Theory

When trying to construct an effective field theory, the first step is to identify the relevant degrees of freedom and any symmetries that relate them. Once you have done that, you will be in a position to think about how these degrees of freedom are allowed to interact with each other. In this appendix we will not go deeply into the rich and fascinating structure of the theory of groups, but provide enough information about the subject that is needed to understand basic constructions used in this book. After a brief review of the basics, we will discuss the nature of representation theory of Lie groups, which is the chief tool to codify symmetries in field theory.

A.1 General definitions

We will begin by displaying a bunch of useful definitions and theorems about group theory in general.

Definition A.1. A **group** G is a set and an operation (usually called "multiplication") such that four axioms hold:

(1) **CLOSURE:** If $a, b \in G$, then $ab \in G$.
(2) **ASSOCIATIVE:** $\forall a, b, c \in G$, $(ab)c = a(bc)$.
(3) **IDENTITY:** There exists a unique element $e \in G$ such that $\forall a \in G$, $ae = ea = a$.
(4) **INVERSES:** $\forall a \in G$, there exists a unique element, denoted a^{-1}, such that $aa^{-1} = a^{-1}a = e$.

If, in addition, the multiplication is commutative ($\forall a, b \in G$: $ab = ba$), then the group is called an **abelian** group; otherwise it is called **non-abelian**.

Definition A.2. $H \subset G$ is a **subgroup** of G if it is a subset that satisfies the above axioms of a group.

Definition A.3. If G, G' are groups and $\phi : G \to G'$ is a map that preserves multiplication: $\phi(ab) = \phi(a)\phi(b)$ $\forall a, b \in G$, then ϕ is called a **homomorphism**:

- If ϕ is one to one, it is a **monomorphism**.
- If ϕ is onto, it is an **epimorphism**.
- If ϕ is bijective, it is an **isomorphism**.
- If ϕ is an isomorphism of G into itself, it is an **automorphism**.

Notice that isomorphisms are equivalence relations. We will denote isomorphic groups as $G \cong H$.

Definition A.4. If $\phi : G \to G'$ is a group homomorphism, then the set

$$\text{Ker } \phi \equiv \{g \in G | \phi(g) = e'\} \,, \tag{A.1}$$

is called the **kernal** of ϕ. It is easy to prove that it is a subgroup of G for *any* group homomorphism.

Theorem A.1. *ϕ is an isomorphism if and only if* Ker ϕ *is trivial. That is: it is the group of a single element (the identity).*

Definition A.5. Let $H \subset G$ be a subgroup. The *set*

$$G/H \equiv \{gh | g \in G, h \in H\} \,, \tag{A.2}$$

called "G mod H" is the set of **left cosets of H in G**. An element of this set is itself a set

$$gH = [g] = \{gh | h \in H\} \,. \tag{A.3}$$

One can also define **right cosets** similarly (denoted Hg).

Definition A.6. $H \subset G$ subset is called a **normal** or **invariant subgroup** if and only if $gH = Hg$ $\forall g \in G$. Equivalently:

$$\forall g \in G, h \in H : \; ghg^{-1} \in H \,. \tag{A.4}$$

Normal subgroups are denoted $H \triangleleft G$.

Notice that ghg^{-1} need not equal h – as long as the product remains in H, that is enough. In particular, the definition of normal subgroups does *not* imply commutativity! Although when a group happens to be abelian, it certainly satisfies this condition, therefore we have

Theorem A.2. *All subgroups of abelian groups are normal subgroups.*

Normal subgroups are very important in group theory and in physics. Here are three vitally important theorems about them that explain why this is so.

Theorem A.3. *If $H \triangleleft G$ then G/H is a subgroup of G, called a* **factor group**. *The converse is also true.*

Theorem A.4. *If $H \triangleleft G$, then there exists a homomorphism ϕ such that $H \equiv \text{Ker } \phi$. Thus, one can think of "G/H" as "collapsing" the subgroup H to the identity.*

Definition A.7. $G_1 \times G_2 = \{g_1 g_2 | g_i \in G_i\}$ is an **internal direct product** of G_1 and G_2. $G_1 \otimes G_2 = \{(g_1, g_2) | g_i \in G_i\}$ is the **external direct product** of G_1 and G_2.

Theorem A.5. *If $H \subset G$, then $G \cong (G/H) \times H$. Similarly, if $G \cong G_1 \otimes G_2$, then $G_1, G_2 \triangleleft G$. If $H \triangleleft G$, then so is $G/H \triangleleft G$.*

This last theorem is the central idea behind the CCWZ construction in Section 2.7. It says that all elements of a group can be represented by a product of a (left or right)-coset (an element of G/H) times an element of the group H.

Definition A.8. If G has no normal subgroups, it is called a **simple** group; and if it does not have any *abelian* normal subgroups, it is called a **semisimple** group. Notice that this does *not* mean that it has no general subgroups.

A.2 Continuous groups

Discrete group theory is a beautiful subject in its own right, and is dramatically useful in so many areas of physics such as the physics of crystals. There is also the theory of "continuous groups." These are, roughly speaking, groups with elements that differ continuously (in the calculus sense) from one another. That is: all group elements can be written as $g(\vec{\theta})$ for some set of quantities $\vec{\theta}$, where g is a *continuous* function in these parameters.

Definition A.9. Let $G = \{g(\vec{\theta})\}$ be a continuous group.

(1) If the vector of parameters $\vec{\theta}$ is a finite n dimensional vector, then G is called a **finite group of dimension** n. We will sometimes denote n as $\dim(G)$.
(2) If each θ_i only assumes values over a compact space (*e.g.*: a closed and bounded interval) then G is called a **compact group**.

From the above definition of a continuous group, it follows that if $\vec{\theta}_1, \vec{\theta}_2$ are continuous parameters that describe the group elements $g(\vec{\theta}_1), g(\vec{\theta}_2)$, then by the closure axiom:

$$g(\vec{\theta}_1) \cdot g(\vec{\theta}_2) = g[F(\vec{\theta}_1, \vec{\theta}_2)],$$

where F is itself continuous.

Definition A.10. If the multiplication function $F(\vec{\theta}_1, \vec{\theta}_2)$ described above is *analytic*[1] as well as continuous, then the group is called a **Lie Group**.

Because of this additional condition, we can consider "small" parameters and rely on the first few terms of the expansion:

$$\boxed{g(\vec{\theta}) = 1 + it^a \theta^a + \mathcal{O}(\theta^2)}$$

where we have normalized $g(0) = e$ (the identity, represented by the number 1) and it^a are the linear coefficients of the expansion, called the **group generators** for reasons that will become apparent shortly.

What we have shown is that a Lie group is a group that is also a *differentiable manifold*. It would take us too far afield to discuss all the rich and fascinating implications of this statement, but one very useful consequence of this is that there exists a very useful and special choice of coordinates on the Lie group manifold called **normal coordinates**, defined by an object called the **exponential map**. Consequently, we can parametrize all of the group elements as

$$g(\vec{\theta}) = e^{i\theta^a t^a}.$$

In this notation, it is also clear that $g(\vec{\theta})^{-1} = e^{-i\theta^a t^a}$. The proof of the existence of this parametrization is a standard exercise in the theory of differential manifolds; we strongly encourage you to look it up on your own.

---

[1] Recall that in real analysis: a function is analytic at x if it has a converging Taylor expansion at x.

It turns out that the structure of the group can be completely determined by working out the structure of the linear coefficients t^a, up to some topological questions that are not important at this stage. To see how, consider the group commutator, defined as:

$$[g(\vec{\theta}_1), g(\vec{\theta}_2)] \equiv g(\vec{\theta}_1) \cdot g(\vec{\theta}_2) \cdot g(\vec{\theta}_1)^{-1} \cdot g(\vec{\theta}_2)^{-1} = g(\vec{\theta}_3) , \qquad (A.5)$$

where the last equality follows from closure of the group. Expanding the group elements for small $\vec{\theta}_i$, we get:

$$g(\vec{\theta}_3) = 1 + it^c \theta_3^c + \mathcal{O}(\theta_3^2) = 1 + \theta_1^a \theta_2^b [t^a, t^b] + \cdots \qquad (A.6)$$

where we have kept the lowest nontrivial term on each side and used the standard notation: $[t^a, t^b] \equiv t^a t^b - t^b t^a$. This means that we need $\theta_3 \sim \theta_1 \theta_2$ to be consistent, and therefore $\theta_3^2 \sim \theta_{1,2}^4$, so it is safe to truncate the series as we did. Comparing both sides we find:

$$\boxed{[t^a, t^b] = i f^{abc} t^c} \qquad (A.7)$$

for some $\mathcal{O}(1)$ coefficients f^{abc}. These are called the **structure constants** and they uniquely determine the group, at least locally.

To summarize: a Lie group structure is determined by a **Lie Algebra**:

Definition A.11. An **algebra** is a vector space with a product. A **Lie algebra** has the added condition that this product is differentiable.

For the Lie Algebra, this product is given by the commutator $[t^a, t^b]$, where the t^a form a basis for the algebra.

You all know examples of commutative algebras: $\mathbb{R}$ and $\mathbb{C}$. An example of a well-known noncommutative algebra is $\mathbb{R}^3$ with the product of two vectors being the ordinary cross product. This is an example of a Lie algbra, in fact, with $f^{abc} = \epsilon^{abc}$; this algebra is called A_1 for the experts.

The generators t^a can always be represented by matrices, and we can define an inner product on the Lie algebra by using the Trace operation. There is a certain arbitrariness to this. Let us define:

$$\text{Tr}\{t^a t^b\} = T \delta^{ab}. \qquad (A.8)$$

T can be any number, in principle. However, once we chose it for one particular representation, its value is fixed for all representations, as we will see. This number is called the **Dynkin index**, and we will definite it more carefully in Section A.5.

In general the Lie algebra is not commutative, otherwise it would be a pretty boring subject! However, there are some structure constants that do vanish, and this leads us to the following important definition:

Definition A.12. Let $C_G \equiv \{t^a | [t^a, t^b] = 0\}$ – that is, this is the set of generators that mutually commute with each other (although not generally with the whole group). Borrowing from our experience from quantum mechanics, these are the generators that are simultaneously diagonalizable. These are called the **Cartan generators** of the algebra. It can be shown that they generate a subalgebra, called the Cartan subalgebra. The dimension of this subalgebra is called the **rank** of G.

There are only a finite number of "families" of Lie algebras of a particular type. In particular: for any given rank n, there are only four (or sometimes five) simple, compact Lie algebras. They fall into four families: A_n, B_n, C_n, D_n, as well as five "exceptional" algebras:

- A_n generate $SU(n+1)$.
- B_n generate $SO(2n+1)$.
- C_n generate $Sp(2n)$ [sometimes just called $Sp(n)$ – be careful to know which convention is being used.]
- D_n generate $SO(2n)$ for $n \geq 2$.
- G_2, F_4, E_6, E_7 and E_8 are the "exceptional" Lie algebras.

In addition, there are isomorphisms between some of the algebras and the groups they generate:

$$A_1 = B_1 = C_1 \longrightarrow SU(2) \cong SO(3) \cong Sp(2)$$
$$D_2 = A_1 \times A_1 \longrightarrow SO(4) \cong SU(2) \times SU(2)$$
$$B_2 = C_2 \longrightarrow SO(5) \cong Sp(4)$$
$$A_3 = D_3 \longrightarrow SU(4) \cong SO(6)$$

The group isomorphisms above are true up to global issues. All other simple, compact Lie algebras and groups are then unique, up to isomorphisms.

A.3 Representation theory of Lie groups

Group theory is all well and good, a cute exercise in abstract mathematics. The million dollar question is: how do we use this information to organize a physical calculation?

Here we generally denote $\Gamma_N \equiv \{D(g) | g \in G\}$; $D(g)$ is an $N \times N$ matrix.

Definition A.13. A homomorphic map of a group G to the group of $N \times N$ matrices with matrix multiplication Γ_N is called a **matrix representation** of G. N is called the **dimension of the representation**.

- If this representation map is monomorphic, Γ_N is called a **faithful representation**.
- If $D(g)$ is a unitary matrix $\forall g \in G$, Γ_N is called a **unitary representation**.

Examples:

(1) $D(g) = 1 \quad \forall g \in G$ is the **trivial representation**. It is not faithful, but is still very important.
(2) $[t^a]^{bc} \equiv -if^{abc}$ is called the **adjoint** (or **regular**) **representation**. It is a faithful representation. It is also true that $\dim(\text{Adj}) = \dim(G)$.
(3) Consider the Lie group (algebra) $SU(N)$ and construct the N dimensional representation. It is a theorem that this is the smallest faithful representation you can construct. It is called the **fundamental** (or **defining**) **representation**.

The defining representation is not unique. In fact:

Theorem A.6. *For a group of rank n, there are n defining representations.*

For example, $SU(2)$, generated by A_1, has only one fundamental representation (the 2×2 unitary matrices with unit determinant), while $SU(3)$, generated by A_2, has *two* defining representations, each 3-dimensional.

Definition A.14. Let Γ_N be the N dimensional representation of G. Then the set

$$\overline{\Gamma}_N \equiv \{[D(g)]^* | g \in G, \ D(g) \in \Gamma_N\},$$

is called the **conjugate representation of** Γ_N. If $\Gamma_N = \overline{\Gamma}_N$, then Γ_N is called a **real representation**.

There is an ambiguity about which of $\Gamma_N, \overline{\Gamma}_N$ should be called the "conjugate" representation, since clearly $\overline{\overline{\Gamma}}_N = \Gamma_N$. We will fix this ambiguity shortly.

Now we are going to make a vital definition:

Definition A.15. A **group action** on a set S is a map $G \times S \to S$, generally written $\hat{g}s$ with $g \in G$, $s \in S$ such that the following axioms hold:

(1) **IDENTITY:** $\hat{e}s = s \quad \forall s \in S$;
(2) **ASSOCIATIVE:** $(\hat{g}_1\hat{g}_2)s = \hat{g}_1(\hat{g}_2 s)$.

The point of group theory in physics is to define a Lie group action on a vector space whose dimension is the dimension of the representation. The goal of this group action is to implement a rule for performing symmetry transformations on the vector space. So if Γ_N is the N dimensional representation of G, then an N vector x^a $(a = 1, \ldots n)$ can transform as

$$x^a \to [D(g)]^a_{.b} x^b.$$

where Einstein notation is assumed. Note the placement of indices. Similarly, we have under the conjugate representation $\overline{\Gamma}_N$

$$y_a \to [D(g)^*]_a^{.b} y_b.$$

Keeping track of covariant and contravariant indices helps keep track of conjugate representations and prevents you from making mistakes, just like in General Relativity.

Definition A.16. A representation is **reducible** if every matrix can be written as a block-diagonal matrix[2]. We can write $\Gamma_N = \Gamma_{N_1} \oplus \Gamma_{N_1} \oplus \cdots \oplus \Gamma_{N_p}$, where $\sum_{k=1}^p N_k = N$. If a representation cannot be so decomposed, it is called **irreducible**.

It is clear that if you know of all of the irreducible representations of a group, then you are in good shape, since any reducible representation can always be decomposed into a Kroneker sum of irreducible representations.

We can also consider the product of representations:

$$\Gamma_{N_1} \otimes \Gamma_{N_2} \equiv \{[D_1(g)]_{ab}[D_2(g)]_{cd} | \forall g \in G\}.$$

Such objects are generally reducible, and we would like to understand how to decompose products like this into irreducible representations:

$$\Gamma_{N_1} \otimes \Gamma_{N_2} = \Gamma_{N_3} \oplus \Gamma_{N_4} \oplus \cdots \oplus \Gamma_{N_p} \qquad \text{where} \quad \sum_{k=3}^p N_k = N_1 \cdot N_2.$$

Theorem A.7. *For a vector in an (irreducible) representation, the elements of the vector are described by the n eigenvalues of the Cartan generators.*

This is just a generalization of the spin algebra from quantum mechanics. $SU(2)$ $(= A_1)$ is a rank 1 algebra, and inside a given representation, the elements are indexed by the eigenvalues of the one Cartan generator (J_3).

---

[2]Technically, this is a stronger condition called **completely reducible**; but for everything we will be doing, all reducible representations are completely reducible.

These eigenvalues are called **weights**. This suggests a very powerful tool for constructing representations: Let $\{\vec{e}_1, \cdots, \vec{e}_n\}$ be a set of orthonormal vectors (n =rank(G)) that transform in the different defining representations:

- $\vec{e}_1$ is the fundamental representation (Γ_1).
- $\vec{e}_{n-i+1}$ transforms in the conjugate representation of $\vec{e}_i$.

This set of vectors provide a basis for any quantity that transforms under any representation of the Lie group, generally called a **tensor operator**:

$$\hat{O} = O_{a_1 \cdots a_i; b_1 \cdots b_j; \cdots}(e_1^{a_1} \cdots e_1^{a_i})(e_2^{b_1} \cdots e_2^{b_j}) \cdots \quad (A.9)$$

where all indices of a given letter are symmetrized and all conjugate indices are trace free. All tensor operators are products of copies of the defining representations:

$$g : O_{a_1 \cdots; b_1 \cdots; \cdots} \to [D_1(g)]_{a_1 \bar{a}_1} \cdots [D_2(g)]_{b_1 \bar{b}_1} \cdots O_{\bar{a}_1 \cdots; \bar{b}_1 \cdots; \cdots} \quad (A.10)$$

In particular, an operator that transforms under an irreducible representation of G is described completely as an operator with i indices of type 1, j indices of type 2, etc. In summary: *Irreducible representations of a Lie group of rank n are described by n positive integers.*

Let us see some examples:

(1) $SU(2)$ – This is a rank 1 group, so irreducible representations are defined by a single integer:
   - (1) : ξ^a
   - (4) : ξ^{abcd}

(2) $SU(3)$ – Rank 2, so it is described by two integers:
   - (0,0) : Trivial rep
   - (1,0) : Fundamental (ξ^a)
   - (0,1) : Antifundamental ($\bar{\xi}_a$)
   - (1,1) : Adjoint (ξ^a_b)

(3) $SU(4)$ – Rank 3, needing three integers:
   - (0,0,0) : Trivial
   - (1,0,0) : Fundamental (ξ^a)
   - (0,0,1) : Antifundamental ($\bar{\xi}_a$)
   - (1,0,1) : Adjoint (ξ^a_b)
   - (0,1,0) : A new defining representation with one index[3] ($\xi^{\hat{a}}$).

---

[3] There is no single accepted notation for these kinds of indices that we are aware of, so we put a hat over it. As we will see below, it is a real representation so it does not matter if you put it upper or lower, but in general it would.

From these examples, we can see a few patterns:

(1) $(0,\cdots,0)$ is always the trivial representation.
(2) $(1,0,\cdots,0)$ is the fundamental representation.
(3) $(1,0,\cdots,0,1)$ is the adjoint representation.
(4) $\overline{(a,b,\cdots,c)} = (c,\cdots,b,a)$.

From this last point, we see that the adjoints are always real representations. So is $(0,1,0)$ in $SU(4)$.

A.4 Young Tableaux for $SU(N)$

There is a very powerful and useful notation for keeping track of representations, and also for decomposing products of representations. This is the technology of **Young Tableaux**, named after an accountant who developed them as a means of keeping track of permutations. We, however, will exclusively use them for describing representations of the $SU(N)$ groups. They can also be used for any other group with only slight modification of the rules.

Consider $SU(N)$, which has rank $N-1$. As we have seen above, an irreducible representation of this group is described by $N-1$ integers: $(a_1, a_2, \ldots, a_{N-1})$. To construct a Young tableau for this representation, draw a_1 boxes in a row, going right to left; then draw a_2 columns of two boxes, and keep going until you have a_{N-1} columns of $N-1$ boxes on the left side of the figure. Here are some examples:

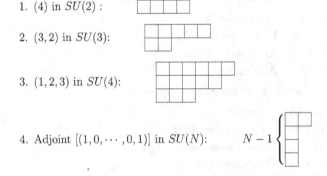

1. (4) in $SU(2)$:

2. (3, 2) in $SU(3)$:

3. (1, 2, 3) in $SU(4)$:

4. Adjoint $[(1,0,\cdots,0,1)]$ in $SU(N)$: $N-1\Big\{$

Young tableaux are wonderous devices – not only are they a powerful mnemonic for keeping track of representations, but they allow you to compute the dimensions of the representations almost without effort! Consider an arbitrary Young tableau describing a representation of $SU(N)$. To compute the dimension, there are two steps:

(1) In the top left box, put an N. Going down the column, label the boxes counting down. Going across the rows left to right, count up. Take the product of all of these numbers. Call it F.
(2) Starting on the right side of the tableau, draw all the "hooks" that you can – these are lines that go right to left, and make a single, hard turn south and continue down. Count the number of boxes each hook passes through, and take the product of all of these numbers. Call it H.

Then we have the simple result:

$$\boxed{\dim(\text{Y.T.}) = \frac{F}{H}} \qquad (A.11)$$

Here are the dimensions of the Young Tableaux we saw above:

1. (4) in $SU(2)$: $\boxed{2\,3\,4\,5}$ $\dim = \frac{2\cdot 3\cdot 4\cdot 5}{1\cdot 2\cdot 3\cdot 4} = 5$.

2. (3,2) in $SU(3)$: $\begin{array}{l}\boxed{3\,4\,5\,6\,7}\\\boxed{2\,3}\end{array}$ $\dim = \frac{3\cdot 4\cdot 5\cdot 6\cdot 7\cdot 2\cdot 3}{1\cdot 2\cdot 3\cdot 5\cdot 6\cdot 1\cdot 2} = 42$.

3. (0,1,0) in $SU(4)$: $\begin{array}{l}\boxed{4}\\\boxed{3}\end{array}$ $\dim = \frac{4\cdot 3}{2\cdot 1} = 6$.

4. Adjoint $[(1,0,\cdots,0,1)]$ in $SU(N)$: $N-1\left\{\begin{array}{l}\boxed{N\;\;}\\\boxed{\vdots}\\\boxed{2}\end{array}\right.$ $\dim = \frac{(N+1)!}{N\cdot(N-2)!} = N^2-1$.

A word on notation: often a representation is described by its dimension. For example, in $SU(3)$, the $(1,0)$ is sometimes called the **3** while the $(0,1)$ is called the **3̄**; in $SU(4)$, the $(0,1,0)$, which is a real representation, is called the **6**($=\bar{\mathbf{6}}$). This is a very common notation, but unfortunately it can be a little confusing, since there could be more than one representation with the same dimension. So you must be careful to know what convention the author is using when you see such a labeling of dimensions. The tradition is to use the conventions of [Slansky (1981)].

To turn a representation to the conjugate representation, all you need to do is replace each column of x boxes with a column of $N-x$ boxes, since that corresponds to reflecting the integers in the weight vector; then you must left-justify the resulting tableau. To decide whether a rep should be barred or not, we use the rule: *If the largest integer is to the left (least* number of long columns in the Young tableau), the representation is not barred. So for example, in $SU(3)$:

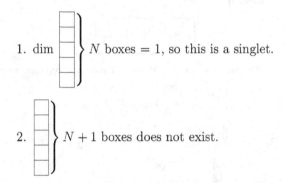

Now we are in a position to understand how to decompose products of representations in terms of a direct sum of irreducible representations. Again, Young tableaux make the whole process deliciously simple! As a first step, take a moment to notice a few facts about columns in a typical Young tableau in $SU(N)$:

1. dim } N boxes = 1, so this is a singlet.

2. } $N+1$ boxes does not exist.

This means that when we construct Young tableaux and end up with N boxes in a column, we can drop the whole column (since it is just 1) and we *never* can construct a Young tableau wiht more than N boxes in a column.

It is sufficient to consider the product of two irreducible representations here, since more general products can be built up by associativity. Also note that this product is commutative, so we can arrange the Young tableaux in any order that we wish. So, to compute $\Gamma_1 \otimes \Gamma_2$:

(1) Chose the Young tableau with the least number of boxes (let it be Γ_2) and place an "a" in every box in the first row, a "b" in every box in the second row, etc.
(2) Start adjoining "a" boxes onto Γ_1's tableau to the **right and below**, being careful to maintain the justification rules, and with **no more than one "a" in each column**. This last rule is there to avoid both symmetrizing and antisymmetrizing in the same index (which would give zero).
(3) Now do the same thing with the "b" boxes with one additional rule:

the number of as above and to the right of the first b must be greater than or equal to the total number of bs. This rule is there to avoid double counting.

(4) Repeat with $c, d, \ldots$ with Step 3 being generalized accordingly.

(5) Now go through the list of tableaux, killing off any vanishing diagrams or singlet columns as mentioned above. What is left is your answer!

Let us try our hand at this with two examples from $SU(3)$:

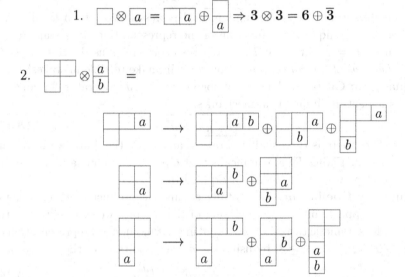

The first diagram in each row violates the rule about as before bs, as does the second diagram in the third row. The last diagram in this chain has four boxes in a column, and since this is $SU(3)$ this diagram vanishes. Finally, the last diagram in the first two rows have singlet columns that we can cross out. So our final result is:

$$(1,1) \otimes (0,1) = (1,2) \oplus (2,0) \oplus (0,1)$$
$$8 \otimes \bar{3} = \overline{15} \oplus 6 \oplus \bar{3}$$

where we write the results in three ways. Notice how we identified conjugate representations consistent with our rules given earlier.

A.5 Group theory coefficients

There are some special products of generators of Lie groups that often appear when performing calculations. Through the orthonormality condition in Eq. (A.8), we can write very general rules for what these products are. In particular, there are three quantities of practical interest:

Dynkin Index: This is the result from Eq. (A.8):

$$\boxed{\text{Tr}[t_R^a t_R^b] = T_R \delta^{ab}} \qquad (A.12)$$

If we specify T_F in the fundamental representation, then the Dynkin index is unique and a function of the representation. It is standard to use $T_F = \frac{1}{2}$, although $T_F = 1$ is also sometimes used. *Be careful to know what the convention is,* or you will make terrible mistakes!

Quadratic Casimir: This is analogous to the J^2 operator for angular momentum in quantum mechanics:

$$\boxed{t_R^a t_R^a = C_R \mathbf{1}_R} \qquad (A.13)$$

where there is the usual sum over a and $\mathbf{1}_R$ is the $\dim R \times \dim R$ unit matrix. Fixing T_F above means that C_R is also unique and a function of R, as we will see below in Eq (A.16).

Anomaly Coefficient: The totally symmetric product of three generators appears often. From closure of the algebra, we can define the totally symmetric three-index symbol in the fundamental representation of $SU(N)$, with $N \geq 3$, by using the *anti*-commutator of the generators:

$$[t_F^b, t_F^c]_+ = \frac{1}{N}\delta^{ab} + d^{abc} t_F^c \;. \qquad (A.14)$$

The numbers d^{abc} are not fundamental: they can be computed in terms of the structure constants – but they are a useful shortcut. Using this tool we can compute for the totally symmetric product of three generators in any representation:

$$\boxed{\text{Tr}\{t_R^a [t_R^b, t_R^c]_+\} = A_R T_F d^{abc}} \qquad (A.15)$$

The A_R are called anomaly coefficients and can be computed or looked up in a table. Notice that they are normalized so that $A_F = 1$. Also notice that the existence of a nontrivial d^{abc} is only true for $SU(N)$ groups for $N \geq 3$, so anomaly coefficients are nontrivial only for these groups. In particular, $Sp(N)$ and $SO(N)$ (except $SO(6)$, which is isomorphic to $SU(4)$) have vanishing anomaly coefficients, as does $SU(2)$. This means, among other things, that gauge theories based on these groups are automatically anomaly-free; see Section 2.9.

The Dynkin index, quadratic Casimir and anomaly coefficient can all be looked up in tables for any irreducible representation of $SU(N)$, and for the other groups as well. We also have a very useful formula relating the Casimir and Dynkin index:

$$C_R = \left(\frac{\dim(G)}{\dim(R)}\right) T_R .\qquad(\text{A.16})$$

This formula immediately gives us for $SU(N)$

$$C_F = \frac{N^2-1}{2N}, \qquad C_A = N.\qquad(\text{A.17})$$

It is also clear from the rules of the Trace operation that:

(1) $T_{R_1 \oplus R_2} = T_{R_1} + T_{R_2}$,
(2) $T_{R_1 \otimes R_2} = (\dim R_1) T_{R_2} + (\dim R_2) T_{R_1}$.

The anomaly coefficients have similar rules, with the extra useful rule that:

$$A_{\overline{R}} = -A_R .\qquad(\text{A.18})$$

In particular, this means that an object that transforms in a real representation (either an irreducible real representation like the Adjoint representation or a vector-like representation $R \oplus \overline{R}$) automatically has a vanishing anomaly coefficient. This is another way to see that $SU(2)$, $SO(N)$ and $Sp(N)$ algebras must have vanishing anomaly coefficients, since they only admit real (or pseudo-real) representations.

A.6   Notes for further reading

We only touched briefly on some of the more important and relevant aspects of group theory. The subject is vast and timeless. An excellent introduction to the subject and its relevance to physics can be found in [Chesnut (1974)]. This book focuses on finite groups, which are relevant to crystal structures, for example. For more information on Lie Groups and Algebras, you cannot beat two classics, [Georgi (1999)] and [Cahn (1985)]. The classic review by [Slansky (1981)] is a necessary component to everyone's library, as it contains all of the Lie group and algebra representations and rules that you ever need to model-build, all in one place.

Appendix B

Short Review of QED and QCD

This Appendix contains a review of the main features of Quantum Electrodynamics (QED) and Quantum Chromodynamics (QCD). It is not intended as a place to first get acquainted with those theories – there are fantastic books that accomplish this [Weinberg (1996); Peskin and Schroeder (1995); Donoghue et al. (1992)]. While in principle any good course in basic quantum field theory contains at least some introduction to QED and QCD, we would like to illuminate some of the main features of those theories, emphasizing those that are useful for understanding the material presented in this book.

The free fermion (quark) Lagrangian, whose Euler-Lagrange equation of motion produces a Dirac equation for a free field,

$$\mathcal{L} = \bar{\psi}(x)(i\slashed{\partial} - m)\psi(x) \tag{B.1}$$

can be used to build QED and QCD depending on a requirement of what local gauge group Eq. (B.1) is invariant.

B.1 Quantum electrodynamics

Quantum Electrodynamics is a theory of electric charge, carried by quarks and leptons, which is based on an abelian group $U(1)$. The Lagrangian of QED is built by requiring that the free fermion Lagrangian of Eq. (B.1) is invariant under a local (gauge) transformation $\psi(x) \to \psi'(x) = V(x)\psi(x)$, which is simply a phase rotation, $V(x) = \exp(i\alpha(x))$. This results in an introduction of a covariant derivative,

$$D_\mu = \partial_\mu - ieQA_\mu, \tag{B.2}$$

which transforms exactly like the fermion field. Here we introduced Q being the charge of the fermion's field relative to electric charge e (i.e. electron charge is $Q_e = -1$, up-quark charge is $Q_u = +2/3$, etc.), with $e = |e|$ being positive, i.e. charge of the positron. Substituting the ordinary derivative with a covariant derivative in Eq. (B.1) leads to the term describing photon-fermion interactions.

In order to build a kinetic term for the photon field we need to build an electromagnetc field strength tensor. This can be done by calculating a commutator of covariant derivatives [Peskin and Schroeder (1995)], which leads to

$$F_{\mu\nu} = \partial_\mu A_\nu - \partial_\nu A_\mu. \tag{B.3}$$

The electromagnetic field strength tensor transforms as $F_{\mu\nu} \to V F_{\mu\nu} V^{-1}$ under QED gauge transformations. Maxwell's equations would follow from the QED Lagrangian if a kinetic term is written as

$$\mathcal{L}_{kin} = -\frac{1}{4} F_{\mu\nu} F^{\mu\nu}. \tag{B.4}$$

Gathering all the terms together, the Lagrangian of QED takes the form,

$$\mathcal{L} = \bar{\psi}(x)(i\slashed{D} - m)\psi(x) - \frac{1}{4} F_{\mu\nu} F^{\mu\nu}. \tag{B.5}$$

The Lagrangian of Eq. (B.5) implies the set of Feynman rules given in Fig. B.1.

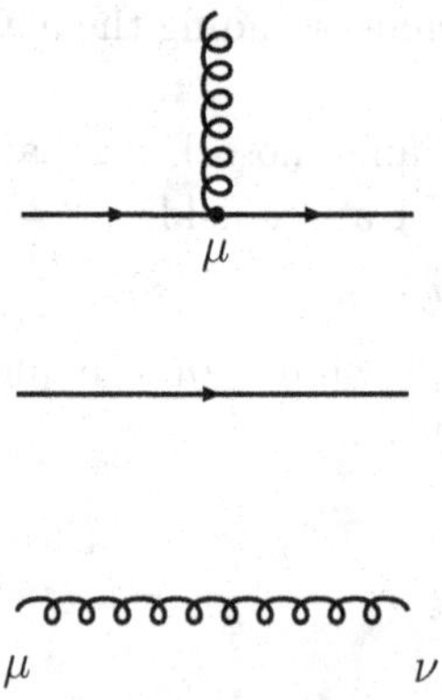

(a) $\quad = ieQ\gamma^\mu$

(b) $\quad = \dfrac{i(\slashed{p} + m)}{p^2 - m^2 + i\epsilon}$

(c) $\quad = \dfrac{(-i)}{k^2 + i\epsilon}\left[g_{\mu\nu} - (1-\xi)\dfrac{k_\mu k_\nu}{k^2}\right]$

Fig. B.1 Feynman rules for QED: (a) photon-fermion vertex, (b) fermion propagator, and (c) photon propagator.

B.2 Quantum chromodynamics

Quantum Chromodynamics is a non-abelian theory of *color* charge, which is carried by quarks and gluons, based on a non-abelian $SU(3)_c$ group.

B.2.1 QCD Lagrangian and Feynman rules

In QCD, fermions (quarks) of the type (or *flavor*) f transform according to the fundamental representation of $SU(3)_c$, i.e. they carry a single color index, $\psi_j^{(f)}$. Since color interactions do not change a quark's flavor, the flavor index is often suppressed. In fact, in this book (except for this Appendix), we shall simply denote the flavor of the quark by calling it by the letter representing its flavor. For example, fermion field $s_j(x)$ would denote a strange quark of color j.

The Lagrangian of QCD is built by requiring that the free quark Lagrangian of Eq. (B.1) is invariant under a local (gauge) transformation $\psi_i(x) \to \psi_i'(x) = V_{ij}(x)\psi_j(x)$. Note that $\bar\psi(x) \to \bar\psi'(x) = \bar\psi(x)V(x)^{-1}$. We know that this requirement results in the introduction of a (matrix) covariant derivative,

$$D_\mu = \partial_\mu - igA_\mu^a t^a, \tag{B.6}$$

where, as usual, explicit reference to color indices (as well as the unit matrix multiplying the partial derivative) is not shown. For infinitesimal transformations,

$$V(x) = 1 + i\alpha^a(x)t^a + \mathcal{O}(\alpha^2). \tag{B.7}$$

The matrices t^a are expressed in terms of Gell-Mann matrices λ^a as $t^a = \lambda^a/2$. There are eight matrices t^a, so there are eight gauge fields A_μ^a. The matrix of the gluon fields $A_\mu \equiv A_\mu^a t^a$ transforms according to the adjoint representation of $SU(3)_c$ plus an inhomogeneous term,

$$A_\mu \to A_\mu' = V(x)A_\mu(x)V^{-1}(x) - \frac{i}{g}(\partial_\mu V)V^{-1} \text{ or}$$

$$A_\mu'^a = A_\mu^a + \frac{1}{g}D_\mu^{ab}\alpha^b(x), \tag{B.8}$$

where D_μ^{ab} is a covariant derivative acting on an object in the adjoint representation of the color group as

$$D_\mu^{ab} = \delta^{ab}\partial_\mu - ig(t^c)^{ab}A_\mu^c, \tag{B.9}$$

with $(t^c)^{ab} = -if^{abc}$. The covariant derivative acting on ψ_j transforms, by construction, the same way as ψ_j,

$$D_\mu'\psi = VD_\mu\psi \text{ or } (\partial_\mu - igA_\mu')V\psi = V(\partial_\mu - igA_\mu)\psi. \tag{B.10}$$

Just like in QED, we can build a *gluon field strength* tensor,[4]

$$G_{\mu\nu} \equiv G_{\mu\nu}^a t^a = \partial_\mu A_\nu - \partial_\nu A_\mu - ig[A_\mu, A_\nu], \tag{B.11}$$

---
[4] It is the commutator in Eq. (B.11) that makes a world of difference between QED and QCD – contrary to the photons, gluons are self-interacting, which changes the way the theory behaves in the ultraviolet.

which transforms as $G_{\mu\nu} \to VG_{\mu\nu}V^{-1}$. In terms of components,
$$G^a_{\mu\nu} = \partial_\mu A^a_\nu - \partial_\nu A^a_\mu + gf^{abc}A^b_\mu A^c_\nu. \tag{B.12}$$
The component field strength tensor transforms as $G^a_{\mu\nu} \to V^{ab}G^b_{\mu\nu}$. The component equations above are easily obtained from Eq. (B.12) using formulas from Appendix A.

The transformation properties of the $G_{\mu\nu}$ imply that there are actually *two* renormalizable, gauge-invariant combinations of the gluon field strength tensors,
$$\text{Tr } G_{\mu\nu}G^{\mu\nu} \quad \text{and} \quad \text{Tr } G_{\mu\nu}\widetilde{G}^{\mu\nu} = \frac{1}{2}\epsilon^{\mu\nu\alpha\beta}\text{Tr } G_{\mu\nu}G_{\alpha\beta}. \tag{B.13}$$
Therefore, both of them can be used to build a QCD Lagrangian. However, only the first combination is used,
$$\mathcal{L}_{QCD} = \sum_\psi \bar{\psi}\left[i\slashed{D} - m_\psi\right]\psi - \frac{1}{4}G^a_{\mu\nu}G^{a\mu\nu} + \mathcal{L}_{GF}. \tag{B.14}$$
Experimentally, QCD interactions conserve parity, so we dropped parity-violating combination $\text{Tr } G_{\mu\nu}\widetilde{G}^{\mu\nu}$, but invite the reader to consult [Weinberg (1996); Peskin and Schroeder (1995); Donoghue *et al.* (1992)] for a fascinating discussion of this term.

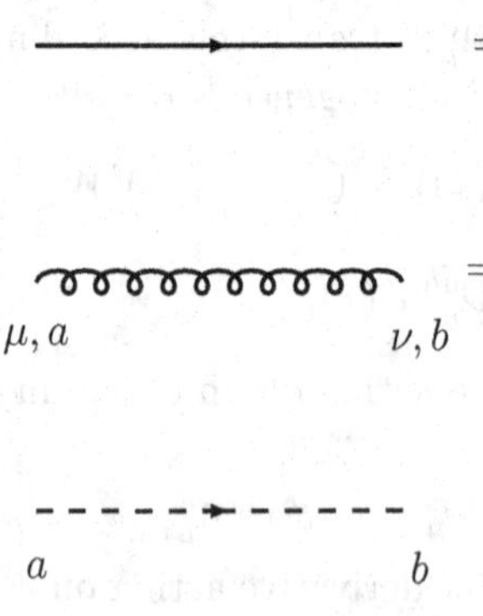

Quark: $\qquad\qquad = \dfrac{i\left(\slashed{p}+m\right)}{p^2 - m^2 + i\epsilon}$

Gluon: $\qquad\qquad = \dfrac{-i\delta^{ab}}{k^2 + i\epsilon}\left[g_{\mu\nu} - (1-\xi)\dfrac{k_\mu k_\nu}{k^2}\right]$
$\mu, a \qquad\qquad \nu, b$

Ghost: $\qquad\qquad = \dfrac{i\delta^{ab}}{k^2 + i\epsilon}$
$a \qquad\qquad b$

Fig. B.2 Feynman rules for QCD: propagators.

We also introduced a *gauge-fixing/ghost* term $\mathcal{L}_{GF}$ in Eq. (B.14). Its origin lies in the fact that a massless gluon has only two degrees of freedom, while A_μ contains four. Moreover, gauge symmetry tells us that there are classes of physically equivalent gluon vector potentials. This poses a certain problem in writing the generating functional of QCD, which is needed for quantization, as summing over infinite gauge-equivalent configurations of A_μ leads to infinities. This is solved by introducing a gauge-fixing term in the QCD Lagrangian, while canceling the contributions of equivalent gauge configurations from internal lines of Feynman diagrams by using the Faddeev-Popov trick – i.e. by including "fake" scalar anti commuting particles called ghosts c^a. Referring the reader to [Weinberg (1996); Peskin and Schroeder (1995); Donoghue et al. (1992)] for more details, we simply write $\mathcal{L}_{GF}$ out,

$$\mathcal{L}_{GF} = -\frac{1}{2\xi}\left(\partial^\mu A_\mu^a\right)^2 + \left(\partial^\mu \bar{c}^a\right) D_\mu^{ab} c^b. \quad \text{(B.15)}$$

The new terms of Eq. (B.15) introduce new Feynman rules involving ghosts, quarks and gluons. Only gauges where ghost fields couple to gluons (like Feynman gauge with $\xi = 1$) do you need to explicitly consider Feynman diagrams that involve ghosts when calculating physical processes. Physical gauges like Coulomb gauge or fixed-point gauge $x_\mu A^\mu = 0$ do not require special considerations of the ghost fields.

The Lagrangian Eq. (B.14) leads to a set of Feynman rules that can be used for calculations of transition amplitudes. We shall not derive them here, instead referring the reader to [Weinberg (1996); Peskin and Schroeder (1995); Donoghue et al. (1992)]. But we shall list them, as they will be useful for the future discussion. Inversion of the kinetic terms for quarks, gluons and ghosts in Eq. (B.14) leads to the propagators given in Fig. B.2. Vertices for various interactions are given in Fig. B.3, where we defined

$$V_{g3}^{\mu\nu\alpha}(p_1, p_2, p_3) = (p_3 - p_2)^\mu g^{\nu\alpha} + (p_1 - p_3)^\nu g^{\alpha\mu} + (p_2 - p_1)^\alpha g^{\mu\nu}. \quad \text{(B.16)}$$

It is understood that all momenta are *incoming* into the vertex. While it is conventional to follow "clockwise rotation" about the vertex while writing the Feynman diagram, changing the direction of "rotation" will change both color band tensor indices, leaving the value of the vertex unchanged. The four-gluon vertex is

$$\begin{aligned} V_{g4}^{\mu\nu\alpha\beta}(a,b,c,d) = -ig^2[&f^{abe}f^{cde}\left(g^{\mu\alpha}g^{\nu\beta} - g^{\mu\beta}g^{\nu\alpha}\right) \\ +&f^{ace}f^{bde}\left(g^{\mu\nu}g^{\alpha\beta} - g^{\mu\beta}g^{\alpha\nu}\right) \\ +&f^{ade}f^{bce}\left(g^{\mu\nu}g^{\beta\alpha} - g^{\mu\alpha}g^{\beta\nu}\right)] \end{aligned} \quad \text{(B.17)}$$

Notice that in practical calculations it is convenient to have separate color and tensor structure of Feynman rules, as it helps to automate the calculations of Feynman diagrams. While all other vertices can in fact be factorized into the product of tensors acting in color space and tensors acting in Minkowski space-time, as it can be seen from Eq. (B.17), it is not possible to do so for the four-gluon vertex.

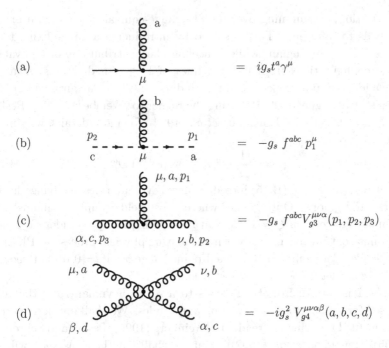

Fig. B.3 Feynman rules for QCD: vertices for (a) quark-gluon, (b) ghost-gluon, (c) three-gluon, and (d) four gluon interactions. See text for the definitions of $V_{g3}^{\mu\nu\alpha}$ and $V_{g4}^{\mu\nu\alpha\beta}$.

Yet, not all is lost! As shown in [Grozin (2005)], the following trick can be employed. A fictitious non-propagating particle can be introduced that can only interact with gluons in the manner shown in Fig. B.4, where it is denoted by a double dashed line. Since the fictitious particle does

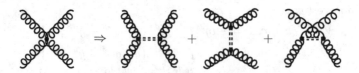

Fig. B.4 Substitution rules for a four-gluon vertex.

not propagate, its propagator is proportional to a delta-function in coordinate space. The vertex function is given by a factorizable combination of color and Minkowski tensors. The "Feynman rules" for those are given in Fig. B.5. Clearly, each four-gluon vertex now produces three separate

Feynman graphs, which makes the calculation longer! However, the convenience of a separation of color and Minkowski degrees of freedom pays off, especially since it is a computer that does the calculations of all diagrams.

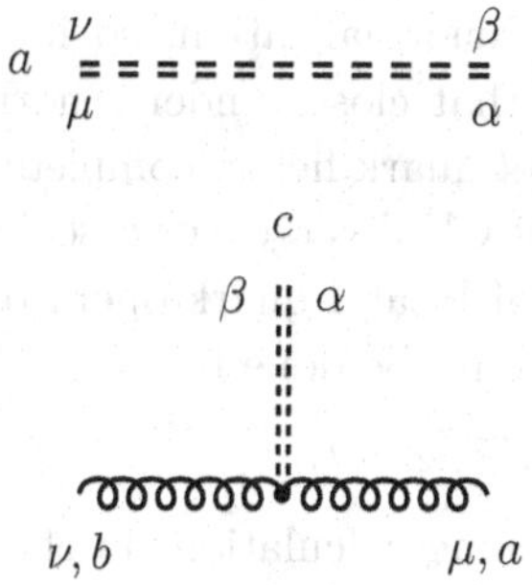

(a) $$a \overset{\nu}{\underset{\mu}{=\!=\!=\!=\!=\!=\!=\!=\!=\!=}} \overset{\beta}{\underset{\alpha}{b}} \;=\; \frac{i}{2}\delta^{ab}\left(g^{\mu\alpha}g^{\nu\beta} - g^{\mu\beta}g^{\nu\alpha}\right)$$

(b) $$\text{(diagram with legs } c, \beta\alpha; \nu,b; \mu,a\text{)} \;=\; i\sqrt{2}g_s \, f^{abc} g^{\mu\alpha} g^{\nu\beta}$$

Fig. B.5 "Feynman rules" for the substitution of the four-gluon vertex.

These Feynman rules can be used to calculate all needed perturbative QCD processes. In particular, loop corrections to various quantities can be evaluated. As those graphs are divergent, various counter terms need to be calculated. The renormalized fields and constants in QCD are related by

$$\psi_0 = Z_\psi^{1/2}\psi, \; A_0^\mu = Z_A^{1/2}A^\mu, \; m_0 = Z_m m, \; g_{s0} = Z_g^{1/2}g_s. \tag{B.18}$$

where the bare fields are denoted by the index 0. A similar relation is also available for the gauge parameter $\xi_0 = Z_A \xi$. QCD counter-terms, as well as many other interesting quantities, can be calculated using the renormalization constants in Eq. (B.18). All needed renormalization constants can be calculated given the information in this Appendix (or obtained from [Weinberg (1996); Peskin and Schroeder (1995); Donoghue et al. (1992)]). For example, at one loop, in dimensional regularization with MS-like subtraction, the quark field renormalization constant takes the form

$$Z_\psi = 1 - C_F \frac{\alpha_s}{4\pi\epsilon}, \tag{B.19}$$

while mass renormalization constant is

$$Z_m = 1 - 3C_F \frac{\alpha_s}{4\pi\epsilon}. \tag{B.20}$$

Note that, as usual, $\alpha_s = g^2/4\pi$, $C_F = 4/3$, and $d = 4 - 2\epsilon$. Notice that since renormalization constants are flavor-blind and only depend on poles in ϵ, in practical calculations γ-algebra can be done directly in four dimensions.

Composite operators require additional renormalization factors. For example, bare operators $O_i^{(0)}$ are related to renormalized operators O_i via

$$O_i^{(0)} = Z_{ij} O_j. \tag{B.21}$$

Here Z_{ij} is a *matrix* of renormalization constants. This equation reminds us that operators can mix under renormalization, so it is important to find a complete basis of operators that closes under renormalization. For the composite operators built out of quark fields, complete removal of UV divergences also requires removal of UV divergences associated with quark fields. For example, if Q_i is a set of local 4-quark operators of dimension six, its complete renormalization would be done by

$$Q_i^{(0)} = Z_\psi^2 Z_{ij} Q_j. \tag{B.22}$$

This relation is important for matching calculations in effective field theories.

In effective theories with QCD one can also view the relation of Eq. (B.22) from a different point of view. An effective Hamiltonian describing a certain physical process would often come in a factorized form, where all effective operators Q_i would be multiplied by corresponding Wilson coefficients C_i, so

$$\mathcal{H}_{eff} \sim \sum_i C_i Q_i, \tag{B.23}$$

where we suppressed all possible constants that multiply the sum. For a practical example of Q_i built out of four quark fields we can view C_i's as coupling constants, which can be renormalized by

$$C_{i0} = Z_{ij}^C C_j, \tag{B.24}$$

where the C_{i0} represent "bare" Wilson coefficients. In this case a term $C_i Q_i$ gets renormalized as

$$Z_\psi^2 Z_{ij}^C C_j Q_i^{(0)} \tag{B.25}$$

where by $Q_i^{(0)}$ we denoted operators built out of "bare" quark fields ψ_0. The same finite result can be obtained by renormalizing the operators,

$$Z_\psi^2 C_j Z_{ji}^{-1} Q_i^{(0)} \tag{B.26}$$

Comparing the last two equations we conclude that

$$Z_{ij}^C = Z_{ji}^{-1}. \tag{B.27}$$

We shall use this simple result in chapter 4 to resum large logarithms that appear in matching calculations of QCD-corrected Fermi theory.

B.2.2 Symmetries of the QCD Lagrangian

Under various assumptions, the QCD Lagrangian of Eq. (B.14) possesses a wide variety of symmetries. Below we shall only discuss those symmetries that will be useful for us in this book. Some are rather trivial, like a global $U(1)$-rotation of all quark fields, which corresponds to baryon charge.

Flavor symmetries. Isospin. As it turns out, the three lightest quarks (u, d, and s) have masses whose values are numerically close to each other. This is especially true for the up and down quarks with $m_u \approx 2$ MeV and $m_d \approx 5$ MeV. In the *theoretical* limit where all three quark masses are the same, QCD has a vectorial global $SU(3)$ flavor symmetry,[5]

$$\psi^{(f)}(x) \equiv \begin{pmatrix} u \\ d \\ s \end{pmatrix} \to \psi^{(f)\prime}(x) = \exp\left(-i\theta^a t^a\right) \psi^{(f)}(x), \tag{B.28}$$

where $t^a = \lambda^a/2$ acts on *flavor* indices. It is also often useful to consider $SU(2)$ subgroups of flavor $SU(3)$: isospin (or I-spin), which works in the limit $m_u = m_d \neq m_s$, U-spin, for which $m_d = m_s \neq m_u$, and V-spin, where $m_u = m_s \neq m_d$. Phenomenologically, isospin has been the most successful of the three, as it is only broken by the mass difference of the up and down quarks, which is rather small compared to m_s, as well as equally small electromagnetic effects due to the difference in the charges of the quarks.

Chiral symmetry. If some quark masses are zero (which might be a good approximation for u, d, and, less so, s-quarks), QCD Lagrangian becomes invariant under *chiral* symmetry,

$$\psi \to \exp\left(i\frac{\alpha}{2}\gamma_5\right)\psi. \tag{B.29}$$

It would then be equivalent to rewrite the Lagrangian of these masses quarks in terms of the *chiral* fields,

$$\psi_{L,R}(x) = \frac{1 \mp \gamma_5}{2}\psi(x), \tag{B.30}$$

which, for the the quark part of Eq. (B.14), would mean that

$$\mathcal{L}_\psi = \bar{\psi}_L i \slashed{D} \psi_L + \bar{\psi}_R i \slashed{D} \psi_R. \tag{B.31}$$

Chiral symmetry plays an important role in QCD. It is broken explicitly by the mass terms,

$$m^{(f)}\bar{\psi}^{(f)}\psi^{(f)} = m\left(\bar{\psi}_L^{(f)}\psi_R^{(f)} + \bar{\psi}_R^{(f)}\psi_L^{(f)}\right), \tag{B.32}$$

---
[5]Not to be confused with the $SU(3)_c$ gauge symmetry.

as they connect left-handed and right-handed quark fields. Thus, in the *theoretical limit* where all three lightest quarks (u, d, and s) are treated as massless, QCD has a larger symmetry group, $SU(3)_L \times SU(3)_R$. Mass terms for the three quarks break this symmetry down to a vectorial flavor $SU(3)_V$ discussed earlier, $SU(3)_L \times SU(3)_R \to SU(3)_V$. This pattern of chiral symmetry breaking is important for building chiral perturbation theory in Chapter 4.

Conformal symmetry. It is interesting to note that if *all* quark masses are set to zero, QCD acquires yet another symmetry. In this situation QCD does not have a single dimensional parameters, so it is invariant under scale transformations with scale parameter λ,

$$x^\mu \to \lambda x^\mu, \quad \psi \to \lambda^{-3/2}\psi, \quad A_\mu \to \lambda^{-1}A_\mu. \tag{B.33}$$

The reader will undoubtedly notice that the exponents of λ above correspond to negative of engineering dimensions of the fields.

The classical massless QCD is thus invariant under the *conformal group* of transformations, which include above scale transformations, Lorentz transformations, translations, and a *special conformal transformation*,

$$x^\mu \to \frac{x^\mu + a^\mu x^2}{1 + 2a \cdot x + a^2 x^2}, \tag{B.34}$$

which is obtained by performing inversion $x^\mu \to x^\mu/x^2$, translation by a and then another inversion. This symmetry is *anomalous*, due to the running of the gauge coupling constant, which introduces a scale Λ_{QCD} via dimensional transmutation.

In addition to the continuous symmetries discussed above, QCD is also found to be separately invariant under parity P, charge C and time-reversal T, as well as any possible combinations of those symmetries. Finally, remnants of the gauge symmetry can be found in a non-linear relation between transformations of the gauge and ghost fields. This is the BRST symmetry.

QCD, as a theory of strong interactions, works fantastically. It has *asymptotic freedom* observed in high energy collisions, and is renormalizable. It has passed (so far) all tests at high energy accelerators. It might have one big problem, though: it is written in terms of the *wrong* degrees of freedom! All effects of strong interactions are observed in *hadrons*, i.e. bound states of quarks and gluons. This is the essence of the non-perturbative phenomenon of *confinement* and is a subject of intensive theoretical investigations.

Appendix C

Useful Features of Dimensional Regularization

C.1 Overview of dimensional regularization

Dimensional regularization (DR) is a wonderful regularization scheme that achieves control over divergent integrals by computing them in $d = 4 - 2\epsilon$ dimensions. Here ϵ can be thought as an arbitrary parameter that is often taken as small. The integrals are analytically continued to d-dimensions, which requires introduction of an arbitrary mass scale μ,

$$\int d^4 k \to \mu^{2\epsilon} \int d^{4-2\epsilon} k = \mu^{2\epsilon} \int d^d k. \tag{C.1}$$

An important feature of dimensional regularization is that this mass scale does not appear as a cut-off that can spoil momentum-dependent power counting of effective theory. Both ultraviolet (UV) and infrared (IR) infinities appear as poles in ϵ and can be subtracted (redefined into relevant physical parameters) using a scheme of choice. One such convenient scheme is $\overline{MS}$, where one subtracts the poles in ϵ and an additional constant $\log(4\pi) - \gamma$ appearing from the expansion of Γ-functions. This scheme is the one that is most often used by effective field theorists (except for the cases when it is not!), but it is by no means unique.

Here we will not attempt to introduce the techniques of dimensional regularization, which is usually covered in any course in quantum field theory. But we are happy to recommend a review paper by G. Leibbrandt [Leibbrandt (1975)], where this technique is described rather concisely. Instead, we will provide some reference formulas needed to perform calculations described in this book and elaborate on some points of this technique that is relevant for calculations in effective field theories.

While there is no *unique* regularization scheme for dealing with divergent integrals appearing in calculations of quantum corrections in field theory

(Nature does not care which method we use to calculate the integrals), dimensional regularization is certainly one of the most convenient schemes we know. It helps to regularize both UV and IR divergencies without introducing a hard scale or breaking important symmetries of the theory (such as gauge or chiral). In that sense, DR is the preferred scheme for EFT calculations; it is often implied that DR is included in the definition of EFT.

C.2 Useful formulas

Dimensional regularization is often used in conjunction with Feynman parameterization of integrals for combining denominators.

$$\frac{1}{A_1 A_2 ... A_n} = \int_0^1 dx_1 dx_2 ... dx_n \frac{(n-1)!\, \delta(1 - x_1 - ... - x_n)}{(x_1 A_1 + x_2 A_2 + ... + x_n A_n)^n}, \quad (C.2)$$

where $\delta(1 - x_1 - ... - x_n)$ is a Dirac delta function. Some particular cases are used most often, so we provide those particular examples here. Two denominators can be combined as

$$\frac{1}{A_1 A_2} = \int_0^1 \frac{dx}{(A_2 - x(A_2 - A_1))^2}. \quad (C.3)$$

Taking a derivative with respect to A_1 or A_2 several times, we obtain

$$\frac{1}{A_1^n A_2^m} = \frac{\Gamma(n+m)}{\Gamma(n)\Gamma(m)} \int_0^1 dx \frac{x^{n-1}(1-x)^{m-1}}{(A_2 - x(A_2 - A_1))^{n+m}}, \quad (C.4)$$

which is also often used. Three denominators can be combined as

$$\frac{1}{A_1 A_2 A_3} = \int_0^1 dx_1 \int_0^{1-x_1} \frac{2\, dx_2}{(A_3 - x_1(A_3 - A_1) - x_2(A_3 - A_2))^3}. \quad (C.5)$$

Similar formulas can be obtained for different powers of A_i. We note that the parameterization above is not unique and sometimes not the most convenient one. For example, in HQET another formula is often used,

$$\frac{1}{A_1^n A_2^m} = \frac{2^m \Gamma(n+m)}{\Gamma(n)\Gamma(m)} \int_0^\infty dx \frac{x^{m-1}}{(A_1 + 2x A_2)^{n+m}}. \quad (C.6)$$

This formula is useful when the denominators have different dimensions, such as a heavy-quark propagator and a gluon propagator. In that case, the dummy integration variable caries dimensions and is integrated to infinity rather than over a finite range.

Let us also present several useful formulas needed for isolation of $1/\epsilon$ poles in calculations of one-loop integrals. The following tensor integral often appears in calculations of radiative corrections in effective theories,

$$\int \frac{d^d s}{(2\pi)^d} \left(\frac{\omega}{v \cdot s + \omega}\right)^\beta \frac{s^{\mu_1}...s^{\mu_n}}{(-s^2)^\alpha} = i\,(4\pi)^{d/2}\,I_n(\alpha,\beta)$$
$$\times (-2\omega)^{d-2\alpha+n} K^{\mu_1...\mu_n}(v;\alpha), \quad \text{(C.7)}$$

where we followed notations of [Amoros et al. (1997)] and denoted

$$I_n(\alpha,\beta) = \frac{\Gamma(d/2 - \alpha + n)\Gamma(2\alpha + \beta - d - n)}{\Gamma(\alpha)\Gamma(\beta)}. \quad \text{(C.8)}$$

The tensor structure can be written as follows,

$$K^{\mu_1...\mu_n}(v;\alpha) = \sum_{j=0}^{[n/2]} (-1)^{n-j} C(n,j;\alpha) \quad \text{(C.9)}$$
$$\times \sum_{\nu_i = \sigma(\mu_i)}{}' g^{\nu_1\nu_2}...g^{\nu_{2j-1}\nu_{2j}} v^{\nu_{2j+1}}...v^{\nu_n},$$

where $[n/2]$ denotes the largest integer that is less than or equal to $n/2$, and the primed sums denote a summation over all permutations that lead to a different assignment of indices. Finally,

$$C(n,j;\alpha) = \prod_{k=1}^{j} \frac{1}{d + 2(n - k - \alpha)}. \quad \text{(C.10)}$$

Standard d-dimensional integrals also often appear in EFT calculations,

$$\int \frac{d^d k}{(2\pi)^d} \frac{1}{[k^2 - \Lambda^2]^n} = \frac{i(-1)^n}{(4\pi)^{d/2}} \frac{\Gamma(n - d/2)}{\Gamma(n)} \left[\frac{1}{\Lambda^2}\right]^{n-d/2}, \quad \text{(C.11)}$$

$$\int \frac{d^d k}{(2\pi)^d} \frac{k^\mu k^\nu}{[k^2 - \Lambda^2]^n} = \frac{i(-1)^{n-1}}{(4\pi)^{d/2}} \frac{g^{\mu\nu}}{2} \frac{\Gamma(n - 1 - d/2)}{\Gamma(n)} \left[\frac{1}{\Lambda^2}\right]^{n-1-d/2},$$

where $\Lambda = \Lambda(x_i, m_j^2, p_k \cdot p_n, ...)$ is a function of Feynman parameters x_i, masses m_j^2, and/or external momenta p_k. More identities can be obtained by tracing the indices with $g_{\mu\nu}$, one just has to remember that in d-dimensions $g^{\mu\nu} g_{\mu\nu} = d$. Also, rotational symmetry implies that one can always substitute $k^\mu k^\nu \to g^{\mu\nu} k^2/d$ under the integral. Integrals with more momenta in the numerator can be found in appendices of standard quantum field theory textbooks, such as [Peskin and Schroeder (1995)] or [Donoghue et al. (1992)].

Ultraviolet poles in ϵ appear from expansion of Gamma functions of the type appearing in Eq. (C.11) around their poles at negative integers $n \in N$,

$$\Gamma(-n+\epsilon) = \frac{(-1)^n}{n!}\left[\frac{1}{\epsilon} + \sum_{k=1}^{n}\frac{1}{k} - \gamma\right] + \mathcal{O}(\epsilon), \qquad \text{(C.12)}$$

where $\gamma \approx 0.5772$ is the Euler-Mascheroni constant. More identities can be obtained from the ones above by using one of the properties of the Gamma function, $\Gamma(z+1) = z\Gamma(z)$. It is also useful to remembers that

$$\begin{aligned}\Gamma(n) &= (n-1)! \\ \Gamma(1) &= 1 \\ \Gamma(2) &= 1 \\ \Gamma(1/2) &= \sqrt{\pi}.\end{aligned} \qquad \text{(C.13)}$$

Finally, most expansions of Feynman integrals around poles in ϵ are obtained using the following trick,

$$a^b = \exp\log a^b = \exp(b\log(a)) = 1 + b\log(a) + \dots \qquad \text{(C.14)}$$

For example, in the expansion of the Feynman integral

$$\mu^{4-d}\left(\frac{1}{\Lambda^2}\right)^{2-d/2} \approx 1 - \epsilon\log\frac{\Lambda^2}{\mu^2} \qquad \text{(C.15)}$$

for $4 - d = 2\epsilon$, where μ is some scale.

C.3 Dimensional regularization vs other schemes

There is nothing really mysterious on how divergent integrals are regulated in dimensional regularization. In order to show this for the UV divergencies, let us consider a generic[6] one loop integral:

$$I = \int \frac{d^4k}{(2\pi)^4} \frac{1}{(P_1^2 - m_1^2)\cdots(P_\alpha^2 - m_\alpha^2)}, \qquad \text{(C.16)}$$

where $P_i = \sum_{j=1}^{i-1}p_j - k$. After combining the denominators with Feynman parameters and shifting the loop momenum in the usual way, we have

$$I = \int [dx] \int \frac{d^4k}{(2\pi)^4} \frac{1}{(k^2 + A^2)^\alpha}, \qquad \text{(C.17)}$$

where A^2 is a function of all the invariants $p_i \cdot p_j$, as well as various masses and Feynman parameters, and $\int[dx]$ is the Feynman parameter measure.

---
[6] We shall ignore the possibility that there is momentum in the numerator.

This integral may or may not be divergent, depending on the value of α. When applying dimensional regularization, we let $4 \to d \equiv 4 - 2\epsilon = 4 + \delta$, where we introduce $\delta = -2\epsilon$ to simplify notation below. We also must introduce a factor of $\mu^{2\epsilon}$ to maintain dimensions in the integral. The dimensionally continued integral is then

$$I(\epsilon) = \frac{C(\epsilon)}{\mu^{-2\epsilon}} \int [dx] \int \frac{d^{4-2\epsilon}k}{(2\pi)^{4-2\epsilon}} \frac{1}{(k_4^2 + k_\epsilon^2 + A^2)^\alpha}, \qquad (C.18)$$

where we split k^2 into a four dimensional part and an 2ϵ dimensional part, and $C(0) = 1$.

Normally we would do the integral in d dimensions, expand in ϵ and isolate divergences; now we are going to integrate over the ϵ dimensions explicitly. Note that in general,

$$\int \frac{d^d k}{(2\pi)^d} \frac{(k^2)^\beta}{(k^2 - A^2)^\alpha} \qquad (C.19)$$

$$= \frac{i(-1)^{\alpha+\beta}(A^2)^{\beta-\alpha+d/2}}{(4\pi)^{d/2}} \frac{\Gamma(\beta+d/2)\Gamma(\alpha-\beta-d/2)}{\Gamma(d/2)\Gamma(\alpha)}.$$

Thus we have

$$I(\epsilon) = C(\epsilon) \int [dx] \int \frac{d^\delta \kappa}{(2\pi)^\delta} \int \frac{d^4 k}{(2\pi)^4} \frac{1}{(\kappa^2 + k^2 + A^2)^\alpha}$$

$$= r(\delta) \int [dx] \int \frac{d^4 k}{(2\pi)^4} \frac{1}{(k^2 + A^2)^\alpha} \left(\frac{k^2 + A^2}{4\pi\mu^2}\right)^{\delta/2}, \qquad (C.20)$$

where $r(\delta) = C(-\delta/2)\Gamma(\alpha - \delta/2)/\Gamma(\alpha)$. Notice that $r(0) = 1$, and that $\delta = -2\epsilon$ is always negative, so the factor in parentheses serves to regulate the UV divergence. This factor is analogous to the regulating factor you would get if you used a Pauli-Villars regulator in four dimensions.

C.4 Advanced features: scaleless integrals

One convenient feature of dimensional regularization with a mass-independent subtraction scheme like $\overline{MS}$ is the fact that all scaleless one-loop integrals vanish. This is a very convenient feature, which is quite useful when performing matching between full and effective field theories. Scaleless integrals often appear when Green's functions in effective theories are expanded in the external momenta.

While this fact can be argued on general grounds, let us illustrate it with a simple example. Consider the following scaleless integral,

$$I = \int \frac{d^d k}{(2\pi)^d} \frac{1}{k^4}. \qquad (C.21)$$

Naively, this integral is divergent in four dimensions. However, introducing a scale m and breaking it up into two parts, one of which is UV divergent, and the other one is IR divergent, we get

$$I = I_1 - I_2 = \int \frac{d^d k}{(2\pi)^d} \frac{1}{k^2(k^2 - m^2)} - \int \frac{d^d k}{(2\pi)^d} \frac{m^2}{k^4(k^2 - m^2)}. \quad \text{(C.22)}$$

The first integral's divergence comes from the integration over k, so we denote it by the index ϵ_{UV},

$$\begin{aligned} I_1 &= \frac{(i)\Gamma(2 - d/2)}{(4\pi)^{d/2}} \int_0^1 \frac{dx}{(xm^2)^{2-d/2}} \\ &= \frac{i}{(4\pi)^2} \left(\frac{1}{\epsilon_{UV}} - \gamma + \log 4\pi + 1 - \log m^2 \right) + \mathcal{O}(\epsilon_{UV}) \quad \text{(C.23)} \end{aligned}$$

The second integral's divergence comes from the integration over a Feynman parameter x (see below) and is of the IR nature, so we shall denote it by the index in ϵ_{IR}

$$\begin{aligned} I_2 &= \frac{(-i)\Gamma(3 - d/2)}{(4\pi)^{d/2}} m^2 \int_0^1 \frac{dx(1-x)}{(xm^2)^{3-d/2}} \\ &= \frac{i}{(4\pi)^2} \left(\frac{1}{\epsilon_{IR}} - \gamma + \log 4\pi + 1 - \log m^2 \right) + \mathcal{O}(\epsilon_{IR}) \quad \text{(C.24)} \end{aligned}$$

As a result,

$$I = \int \frac{d^d k}{(2\pi)^d} \frac{1}{k^4} = \frac{i}{16\pi^2} \left[\frac{1}{\epsilon_{UV}} - \frac{1}{\epsilon_{IR}} \right] = 0, \quad \text{(C.25)}$$

because in dimensional regularization $\epsilon_{IR} = \epsilon_{UV} \equiv \epsilon$ (notice that the m does not appear in the final expression). The relation of Eq. (C.25) is sometimes referred to as Veltman's formula (actually, it is a particular case for $n = 2$).

This result is true for any scaleless integral calculated in dimensional regularization, so

$$\int \frac{d^d k}{(2\pi)^d} f(k^2, k \cdot v) = 0. \quad \text{(C.26)}$$

This is a very useful feature of dimensional regularization that is used quite often.

C.5 Advanced features: integration by parts

One of the most powerful methods associated with multi loop integration in dimensional regularization is the method of integration by parts (IBP) developed by Chetyrkin and Tkachov [Chetyrkin and Tkachov (1981)]. It rests on a simple observation that, for massless integrals, the integral of a total derivative is zero. That is to say,

$$\int d^d k \frac{\partial}{\partial k^\mu} f(p,...) = 0. \tag{C.27}$$

It can be viewed as a consequence of translational invariance of dimensionally regularized integrals in k-space. Recall that we are dealing with convergent integrals in d-dimensions! Those who prefer coordinate representations of Feynman integrals can check that the statement of Eq. (C.27), taken after Fourier transformation,

$$\frac{1}{q^{2n}} = \frac{\Gamma(d/2-n)}{(2\pi)^{d/2}\Gamma(n)} \int d^d y \frac{e^{iqy}}{(y^2)^{d/2-n}}, \tag{C.28}$$

simply corresponds to to the fact that $(y_1 - y_n)^\mu + (y_n - y_{n-1})^\mu + ... + (y_2 - y_1)^\mu = 0$ in the integrand of the coordinate representation of the same Feynman diagram. Here y_i is the position of ith vertex in the coordinate representation of a Feynman diagram.

If we integrate by parts any integrals that appear in out computations with the constraint of Eq. (C.27), we are able to write the original integrands in terms of sums of other integrals. Now, the same procedure is applicable to each integral in the sum as well. So, at the end of the day, one is able to write a *recurrence relation*, expressing a loop integral of arbitrary complexity in terms of a few *master integrals*, i.e. simpler integrals that can be computed by other means – and, more importantly, tabulated for future use! This means that, once all master integrals are available, computation of classes of multi loop integrals becomes an algebraic procedure that can be implemented in a computer algorithm. This makes it a very powerful tool in the evaluation of multi loop integrals in EFTs.

Let us consider an example of how this is done [Chetyrkin and Tkachov (1981)]. Consider the diagram in Fig. C.1. This is one of the diagrams contributing to the two loop calculation of self energy in ϕ^3 theory for a field with external momentum q. This diagram is expressed by the following integral in d-dimensions,

$$I(q;1,1,1,1,1) = \int \frac{d^d k\, d^d p}{(p+k+q)^2(p+q)^2 p^2 (p+k)^2 k^2}, \tag{C.29}$$

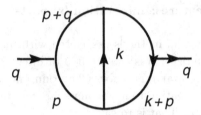

Fig. C.1 Sample two loop diagram for illustration of IBP method.

to which we apply IBP method. While this notation is somewhat strange, it really means that the integral in Eq. (C.29) is a particular case of a more general integral,

$$I(q;\alpha_1,\alpha_2,\alpha_3,\alpha_4,\alpha_5) = \int \frac{d^d k\, d^d p}{(p+k+q)^{2\alpha_1}(p+q)^{2\alpha_2}p^{2\alpha_3}(p+k)^{2\alpha_4}k^{2\alpha_5}}. \quad (C.30)$$

Now according to Eq. (C.27),

$$\int d^d k\, d^d p\, \frac{\partial}{\partial k^\mu} \frac{k^\mu}{(p+k+q)^2(p+q)^2 p^2 (p+k)^2 k^2} = 0. \quad (C.31)$$

Taking this derivative we notice that

$$0 = \int d^d k\, d^d p\, \frac{1}{(p+q)^2 p^2} \left[d - \frac{2k\cdot k}{k^4(p+k+q)^2(p+k)^2} \right.$$
$$\left. - \frac{2k\cdot(p+k+q)}{k^2(p+k+q)^4(p+k)^2} - \frac{2k\cdot(p+k)}{k^2(p+k+q)^2(p+k)^4} \right], \quad (C.32)$$

where we took derivatives of the numerator and three terms in the denominator containing k, respectively. Noting that

$$2k\cdot(p+k) = (k+p)^2 - p^2 + k^2,$$
$$2k\cdot(p+k+q) = (p+k+q)^2 - (p+q)^2 + k^2, \quad (C.33)$$

we can rewrite this relation in a much simplified form,

$$(d-4)\, I(q;1,1,1,1,1) + I(q;1,1,0,2,1) - I(q;1,1,1,2,0) \quad (C.34)$$
$$+ I(q;2,0,1,1,1) - I(q;2,1,1,1,0) = 0.$$

Defining "raising" and "lowering" operators $\mathbf{1}^\pm I(q;\alpha_1,\alpha_2,\alpha_3,\alpha_4,\alpha_5) = I(q;\alpha_1\pm 1,\alpha_2,\alpha_3,\alpha_4,\alpha_5)$, $\mathbf{2}^\pm I(q;\alpha_1,\alpha_2,\alpha_3,\alpha_4,\alpha_5) = I(q;\alpha_1,\alpha_2\pm 1,\alpha_3,\alpha_4,\alpha_5)$, and similarly for other indices α_i, we arrive at the following algebraic recurrence relation,

$$\left[(d-4) + (\mathbf{3}^- - \mathbf{5}^-)\mathbf{4}^+ + (\mathbf{2}^- - \mathbf{5}^-)\mathbf{1}^+\right] I(q;1,1,1,1,1) = 0. \quad (C.35)$$

This relation is the final goal of our example. The integrals to which the diagram is reduced can be computed and tabulated [Chetyrkin and Tkachov (1981)]. Since "removing" the propagator in Eq. (C.35) amounts to a delta-function in coordinate space, i.e. contracting a propagator to the point, while "squaring" the propagator amounts to, at least in ϕ^3 theory, insertion of ϕ^2 operator (which we can denote by a cross on a propagator), the relation Eq. (C.35) has a ready graphical interpretation given in Fig. C.2.

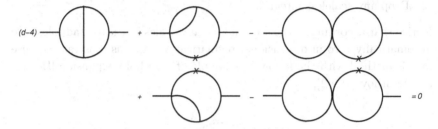

Fig. C.2   Diagram reduced by application of IBP method.

A computer realization of the IBP algorithm is now easy: we simply apply recurrence relation above to the original integral and all subsequent obtained integrals. The procedure terminates once all integrals are reduced to the tabulated master integrals or new master integrals are discovered. In the latter case those are computed and tabulated. And then the procedure repeated. This is it! One has to notice though that in modern multi loop calculations it is common to see thousands of integrals, so the use of a rather powerful computer is essential.

The last comment on the IBP method that we provide here is a reminder that raising the power of denominators can bring new IR divergencies. Those need to be removed, for example, with help of R^* method [Chetyrkin and Tkachov (1982); Chetyrkin and Smirnov (1984)]. Examples on how this is done in matching calculations can be found, e.g., in [Amoros et al. (1997); Becher et al. (2001)].

C.6   Advanced features: method of regions

Another useful method of evaluation of integrals in effective theories involves expanding the integrands in regions of integration where such an expansion is justified. This method goes under the name of *method (or strategy) of regions* [Beneke and Smirnov (1998)]. This method is very useful for calculation of matching coefficients between full and effective theories.

The strategy employed by the method of regions is very simple. Any asymptotic expansion of a Feynman integral can be done in there steps,

(1) Identify relevant and important regions of integration momentum,
(2) Integrate the expanded expression, separately for each expanded region, *but over the whole range of integration* of the original integral,
(3) Drop any scaleless integrals.

Notice that according to the section C.4, any scales integrals (tadpoles) are automatically zero in dimensional regularization. Let us illustrate the use of this method with the following example of threshold expansion [Beneke and Smirnov (1998)].

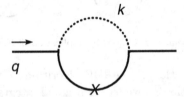

Fig. C.3   Sample one loop diagram for illustration of the method of regions.

Consider a Feynman integral that appears from the calculation of the Feynman graph presented in Fig. C.3. Let us suppose that dashed line in Fig. C.3 represents a massless scalar particle, while the solid line represents a particle with mass m and the cross on solid line represents mass insertion. External momentum q flows through this diagram. Then this diagram, up to some constants, would correspond to the following Feynman integral in d-dimensions,

$$I(q,m) = \int \frac{d^d k}{(2\pi)^d} \frac{1}{k^2((k-q)^2 - m^2)^2}. \tag{C.36}$$

Suppose that we need to evaluate this integral in the *threshold* region $q^2 \approx m^2$. In this case, it is more convenient to rewrite as

$$I(q,m) = \int \frac{d^d k}{(2\pi)^d} \frac{1}{k^2(k^2 - 2k \cdot q - y)^2}, \tag{C.37}$$

where $y = m^2 - q^2 \to 0$ in the threshold region that we are interested in. According to the method of regions, we are supposed to identify relevant regions, then perform relevant expansions and integrations.

In the integral of Eq. (C.37) there are three relevant regions,

(1) *hard*, where $|k| \sim |q|$,
(2) *soft*, where $|k| \sim \sqrt{y}$, and
(3) *ultrasoft*, where $|k| \to 0$

We shall consider all those limits in turn. At the end, according to the adopted strategy,

$$I(q,m) = I_h(q,m) + I_s(q,m) + I_{us}(q,m). \qquad (C.38)$$

Consider the *hard* region first. In this region the following relation is true for the denominator of the integrand, $k^2 \sim 2k \cdot q \ll y$. Using Eq. (C.4) for $n=2$ and $m=1$,

$$\begin{aligned} I_h(q,m) &= \int \frac{d^d k}{(2\pi)^d} \frac{1}{k^2 (k^2 - 2k \cdot q)^2} \\ &= 2 \int_0^1 x \, dx \int \frac{d^d \tilde{k}}{(2\pi)^d} \frac{1}{(\tilde{k}^2 - x^2 q^2)^3} \qquad (C.39) \\ &= \frac{i}{(4\pi)^{d/2}} \frac{\Gamma(1+\epsilon)}{2\epsilon} \frac{1}{(q^2)^{1+\epsilon}}. \end{aligned}$$

Now let us consider the *soft* region. In this region $|k| \sim \sqrt{y}$, so the integral of Eq. (C.37) is then

$$I_s(q,m) = \int \frac{d^d k}{(2\pi)^d} \frac{1}{k^2(-2k \cdot q)^2} = 0. \qquad (C.40)$$

This is so because we encountered *scaleless* integral, which is automatically zero in dimensional regularization. Finally, there is an *ultrasoft* region where $|k| \sim 0$. We again get a nontrivial integral, which can be written as

$$I_{us}(q,m) = \int \frac{d^d k}{(2\pi)^d} \frac{1}{k^2(-2k \cdot q - y)^2}. \qquad (C.41)$$

Once again, using Eq. (C.4) for $n=1$ and $m=2$,

$$\begin{aligned} I_{us}(q,m) &= 2 \int_0^1 dx \, \frac{1-x}{x^3} \int \frac{d^d \tilde{k}}{(2\pi)^d} \frac{1}{(\tilde{k}^2 - \Lambda^2)^3} \qquad (C.42) \\ &= \frac{-i}{(4\pi)^{d/2}} \frac{\Gamma(1-\epsilon)}{y^{2\epsilon}} \frac{\Gamma(2\epsilon)}{(q^2)^{1-\epsilon}}. \end{aligned}$$

Putting all these expressions together and expanding in ϵ gives us

$$I(q,m) = \frac{i}{(4\pi)^2} \frac{1}{q^2} \log \frac{y}{q^2} + ..., \qquad (C.43)$$

which is the same result as if we'd have obtained if we first calculated the diagram and *then* expanded it in the threshold region. The difference is that we did what effective field theories are designed to do: instead of solving one complicated problem we solved several simple ones!

Real-life applications of the strategy of regions might not be that straightforward due to the presence of several other regions, such as a *collinear* one (or several collinear ones), which often happens in SCET calculations. To introduce most often encountered regions, let us assume that, in some process, there are two relevant scales, incoming momentum p (as well as the outgoing momentum $p' \sim p$) and a transferred momentum $Q = \sqrt{-(p-p')^2}$, which is the only large scale in our problem. Let us define a dimensionless variable $\lambda^2 = p^2/Q^2$. Our integration momentum k can be conveniently written in terms of the light cone variables $k^\mu = (k^0 + k^3, k^0 - k^3, \vec{k}_\perp)$. Then, the following *regions* are often encountered,

(1) *hard*, where $k \sim (1, 1, 1)\, Q$,
(2) *left collinear*, where $k \sim (1, \lambda^2, \lambda)\, Q$,
(3) *right collinear*, where $k \sim (\lambda^2, 1, \lambda)\, Q$, and
(4) *soft* or *ultrasoft*, where $k \sim (\lambda^2, \lambda^2, \lambda^2)\, Q$

As one can readily check, denominations of these regions are consistent with the ones used in our example above. Note that the definition of λ could vary in different effective theories.

Note that other effective theories might have different relevant regions as well. In a sense, the search for relevant regions of an integral often involves some art in an otherwise rigorous science of effective field theories. Yet, with some experience – and using some common-sense tests – evaluation of EFT integrals can be greatly simplified. Some checks include looking for $1/\epsilon$ poles that should not be there (which is usually indicative of a missing region), doing numerical checks, or running other explicit tests where the results are either known or could be obtained easily.

Bibliography

AlFiky, M. T., Gabbiani, F. and Petrov, A. A. (2006). X(3872): Hadronic molecules in effective field theory, *Phys.Lett.* **B640**, pp. 238–245, doi: 10.1016/j.physletb.2006.07.069.

Altarelli, G. and Parisi, G. (1977). Asymptotic Freedom in Parton Language, *Nucl. Phys.* **B126**, p. 298, doi:10.1016/0550-3213(77)90384-4.

Ambjorn, J., Jurkiewicz, J. and Loll, R., "Quantum Gravity, or The Art of Building Spacetime," in Oriti, D. (ed.): *Approaches to Quantum Gravity*, 341–359, [hep-th/0604212].

Amoros, G., Beneke, M. and Neubert, M. (1997). Two loop anomalous dimension of the chromomagnetic moment of a heavy quark, *Phys.Lett.* **B401**, pp. 81–90, doi:10.1016/S0370-2693(97)00345-6.

Anastassov, A. *et al.* (2002). First measurement of $\Gamma(D^{*+})$ and precision measurement of $m_{D^{*+}} - m_{D^0}$, *Phys.Rev.* **D65**, p. 032003, doi:10.1103/PhysRevD.65.032003.

Anderson, P. W. (1963). Plasmons, Gauge Invariance, and Mass, *Phys.Rev.* **130**, pp. 439–442, doi:10.1103/PhysRev.130.439.

Appelquist, T. and Carazzone, J. (1975). Infrared Singularities and Massive Fields, *Phys.Rev.* **D11**, p. 2856, doi:10.1103/PhysRevD.11.2856.

Appelquist, T., Dine, M. and Muzinich, I. (1977). The Static Potential in Quantum Chromodynamics, *Phys.Lett.* **B69**, p. 231, doi:10.1016/0370-2693(77)90651-7.

Bauer, C. W., Fleming, S. and Luke, M. E. (2000). Summing Sudakov logarithms in $B \to X_s\gamma$ in effective field theory, *Phys. Rev.* **D63**, p. 014006, doi: 10.1103/PhysRevD.63.014006.

Bauer, C. W., Fleming, S., Pirjol, D., Rothstein, I. Z. and Stewart, I. W. (2002a). Hard scattering factorization from effective field theory, *Phys. Rev.* **D66**, p. 014017, doi:10.1103/PhysRevD.66.014017.

Bauer, C. W., Fleming, S., Pirjol, D. and Stewart, I. W. (2001a). An Effective field theory for collinear and soft gluons: Heavy to light decays, *Phys. Rev.* **D63**, p. 114020, doi:10.1103/PhysRevD.63.114020.

Bauer, C. W., Lee, C., Manohar, A. V. and Wise, M. B. (2004). Enhanced nonperturbative effects in Z decays to hadrons, *Phys. Rev.* **D70**, p. 034014, doi:10.1103/PhysRevD.70.034014.

Bauer, C. W., Pirjol, D. and Stewart, I. W. (2001b). A Proof of factorization for $B \to D\pi$, *Phys. Rev. Lett.* **87**, p. 201806, doi:10.1103/PhysRevLett.87.201806.

Bauer, C. W., Pirjol, D. and Stewart, I. W. (2002b). Soft collinear factorization in effective field theory, *Phys. Rev.* **D65**, p. 054022, doi:10.1103/PhysRevD.65.054022.

Bauer, C. W., Pirjol, D. and Stewart, I. W. (2003). Factorization and endpoint singularities in heavy to light decays, *Phys. Rev.* **D67**, p. 071502, doi:10.1103/PhysRevD.67.071502.

Bauer, C. W. and Schwartz, M. D. (2007). Event Generation from Effective Field Theory, *Phys. Rev.* **D76**, p. 074004, doi:10.1103/PhysRevD.76.074004.

Bauer, C. W. and Stewart, I. W. (2001). Invariant operators in collinear effective theory, *Phys. Lett.* **B516**, pp. 134–142, doi:10.1016/S0370-2693(01)00902-9.

Becher, T., Broggio, A. and Ferroglia, A. (2014). Introduction to Soft-Collinear Effective Theory, doi:10.1007/978-3-319-14848-9.

Becher, T., Neubert, M. and Pecjak, B. D. (2007). Factorization and Momentum-Space Resummation in Deep-Inelastic Scattering, *JHEP* **01**, p. 076, doi:10.1088/1126-6708/2007/01/076.

Becher, T., Neubert, M. and Petrov, A. A. (2001). Two loop renormalization of heavy light currents at order $1/m_Q$ in the heavy quark expansion, *Nucl.Phys.* **B611**, pp. 367–382, doi:10.1016/S0550-3213(01)00341-8.

Beneke, M., Buchalla, G., Greub, C., Lenz, A. and Nierste, U. (1999). Next-to-leading order QCD corrections to the lifetime difference of B_s mesons, *Phys.Lett.* **B459**, pp. 631–640, doi:10.1016/S0370-2693(99)00684-X.

Beneke, M., Chapovsky, A. P., Diehl, M. and Feldmann, T. (2002). Soft collinear effective theory and heavy to light currents beyond leading power, *Nucl. Phys.* **B643**, pp. 431–476, doi:10.1016/S0550-3213(02)00687-9.

Beneke, M. and Feldmann, T. (2003). Multipole expanded soft collinear effective theory with nonAbelian gauge symmetry, *Phys. Lett.* **B553**, pp. 267–276, doi:10.1016/S0370-2693(02)03204-5.

Beneke, M. and Smirnov, V. A. (1998). Asymptotic expansion of Feynman integrals near threshold, *Nucl.Phys.* **B522**, pp. 321–344, doi:10.1016/S0550-3213(98)00138-2.

Benson, D., Bigi, I. I., Mannel, T. and Uraltsev, N., "Imprecated, yet impeccable: On the theoretical evaluation of $\Gamma(B \to X_c \ell \nu)$," *Nucl. Phys. B* **665**, 367 (2003).

Benzke, M., Lee, S. J., Neubert, M. and Paz, G. (2010). Factorization at Subleading Power and Irreducible Uncertainties in $\bar{B} \to X_s \gamma$ Decay, *JHEP* **08**, p. 099, doi:10.1007/JHEP08(2010)099.

Bernard, V., Kaiser, N., Kambor, J. and Meissner, U. G. (1992). Chiral structure of the nucleon, *Nucl.Phys.* **B388**, pp. 315–345, doi:10.1016/0550-3213(92)90615-I.

Bertone, G., Hooper, D. and Silk, J. (2005). Particle dark matter: Evidence, candidates and constraints, *Phys. Rept.* **405**, pp. 279–390, doi:10.1016/j.physrep.2004.08.031.

Bigi, I. I., Shifman, M. A. and Uraltsev, N. (1997). Aspects of heavy quark theory, *Ann.Rev.Nucl.Part.Sci.* **47**, pp. 591–661, doi:10.1146/annurev.nucl. 47.1.591.

Bigi, I. I., Shifman, M. A., Uraltsev, N. and Vainshtein, A. I. (1993). QCD predictions for lepton spectra in inclusive heavy flavor decays, *Phys.Rev.Lett.* **71**, pp. 496–499, doi:10.1103/PhysRevLett.71.496.

Birrell, N. D. and Davies, P. C. W. (1984). *Quantum Fields in Curved Space*, Cambridge Monographs on Mathematical Physics (Cambridge Univ. Press, Cambridge, UK), ISBN 0521278589, 9780521278584, 9780521278584.

Bjerrum-Bohr, N. E. J., Donoghue, J. F. and Holstein, B. R. (2003). Quantum gravitational corrections to the nonrelativistic scattering potential of two masses, *Phys. Rev.* **D67**, p. 084033, doi:10.1103/PhysRevD.71.069903,10. 1103/PhysRevD.67.084033, [Erratum: Phys. Rev.D71,069903(2005)].

Bjorken, J. D. (1969). Asymptotic Sum Rules at Infinite Momentum, *Phys. Rev.* **179**, pp. 1547–1553, doi:10.1103/PhysRev.179.1547.

Bodwin, G. T., Braaten, E. and Lepage, G. P. (1992). Rigorous QCD predictions for decays of P wave quarkonia, *Phys.Rev.* **D46**, pp. 1914–1918, doi:10. 1103/PhysRevD.46.R1914.

Bodwin, G. T., Braaten, E. and Lepage, G. P. (1995). Rigorous QCD analysis of inclusive annihilation and production of heavy quarkonium, *Phys.Rev.* **D51**, pp. 1125–1171, doi:10.1103/PhysRevD.55.5853,10.1103/PhysRevD. 51.1125.

Braaten, E., Fleming, S. and Yuan, T. C. (1996). Production of heavy quarkonium in high-energy colliders, *Ann.Rev.Nucl.Part.Sci.* **46**, pp. 197–235, doi:10. 1146/annurev.nucl.46.1.197.

Braaten, E. and Hammer, H.-W. (2006). Universality in few-body systems with large scattering length, *Phys.Rept.* **428**, pp. 259–390, doi:10.1016/j.physrep. 2006.03.001.

Brambilla, N., Pineda, A., Soto, J. and Vairo, A. (2000). Potential NRQCD: An Effective theory for heavy quarkonium, *Nucl.Phys.* **B566**, p. 275, doi: 10.1016/S0550-3213(99)00693-8.

Brambilla, N., Pineda, A., Soto, J. and Vairo, A. (2005). Effective field theories for heavy quarkonium, *Rev. Mod. Phys.* **77**, p. 1423, doi:10.1103/RevModPhys. 77.1423.

Brambilla, N. e. a. (2011). Heavy quarkonium: progress, puzzles, and opportunities, *Eur.Phys.J.* **C71**, p. 1534, doi:10.1140/epjc/s10052-010-1534-9.

Buchalla, G., Buras, A. J. and Lautenbacher, M. E. (1996). Weak decays beyond leading logarithms, *Rev.Mod.Phys.* **68**, pp. 1125–1144, doi:10.1103/ RevModPhys.68.1125.

Buchmuller, W. and Wyler, D. (1986). Effective Lagrangian Analysis of New Interactions and Flavor Conservation, *Nucl. Phys.* **B268**, pp. 621–653, doi: 10.1016/0550-3213(86)90262-2.

Buras, A. J., "Weak Hamiltonian, CP violation and rare decays," hep-ph/9806471.

Burdman, G. and Donoghue, J. F. (1992). Union of chiral and heavy quark symmetries, *Phys.Lett.* **B280**, pp. 287–291, doi:10.1016/0370-2693(92)90068-F.

Burgess, C. P. (2004). Quantum gravity in everyday life: General relativity as an effective field theory, *Living Rev. Rel.* **7**, pp. 5–56, doi:10.12942/lrr-2004-5.

Cahn, R. (1985). *Semisimple Lie Algebras and their Representations* (Dover Publications), ISBN 10: 0486449998, 174 pages.

Callan, J., Curtis G., Coleman, S. R., Wess, J. and Zumino, B. (1969). Structure of phenomenological Lagrangians. 2. *Phys.Rev.* **177**, pp. 2247–2250, doi: 10.1103/PhysRev.177.2247.

Carroll, S. M. (2004). *Spacetime and geometry: An introduction to general relativity* (Addison-Wesley), ISBN 0805387323, 513 pages.

Caswell, W. and Lepage, G. (1986). Effective Lagrangians for Bound State Problems in QED, QCD, and Other Field Theories, *Phys.Lett.* **B167**, p. 437, doi:10.1016/0370-2693(86)91297-9.

Chaikin, P. M. and Lubensky, T. C. (2014). *Principles of Condensed Matter Physics* (Cambridge University Press), ISBN 1139648616, 9781139648615, 721 pages.

Chay, J., Georgi, H. and Grinstein, B. (1990). Lepton energy distributions in heavy meson decays from QCD, *Phys.Lett.* **B247**, pp. 399–405, doi:10.1016/0370-2693(90)90916-T.

Chesnut, D. (1974). *Finite Groups and Quantum Theory* (John Wiley and Sons), ISBN-10: 0471154458, 272 pages.

Chetyrkin, K. and Smirnov, V. A. (1984). R* operation corrected, *Phys.Lett.* **B144**, pp. 419–424, doi:10.1016/0370-2693(84)91291-7.

Chetyrkin, K. and Tkachov, F. (1981). Integration by Parts: The Algorithm to Calculate beta Functions in 4 Loops, *Nucl.Phys.* **B192**, pp. 159–204, doi: 10.1016/0550-3213(81)90199-1.

Chetyrkin, K. and Tkachov, F. (1982). Infrared R operation and ultraviolet counterterms in the MS scheme, *Phys.Lett.* **B114**, pp. 340–344, doi: 10.1016/0370-2693(82)90358-6.

Cheung, W. M.-Y., Luke, M. and Zuberi, S. (2009). Phase Space and Jet Definitions in SCET, *Phys. Rev.* **D80**, p. 114021, doi:10.1103/PhysRevD.80.114021.

Cirigliano, V. and Ramsey-Musolf, M. J. (2013). Low Energy Probes of Physics Beyond the Standard Model, *Prog. Part. Nucl. Phys.* **71**, pp. 2–20, doi: 10.1016/j.ppnp.2013.03.002.

Ciuchini, M., Franco, E., Lubicz, V. and Mescia, F. (2002). Next-to-leading order QCD corrections to spectator effects in lifetimes of beauty hadrons, *Nucl.Phys.* **B625**, pp. 211–238, doi:10.1016/S0550-3213(02)00006-8.

Coleman, S. (1985). *Aspects of Symmetry: Selected Erice Lectures* (Cambridge University Press), ISBN-10: 0521318270, 420 pages.

Coleman, S. R. and Weinberg, E. J. (1973). Radiative Corrections as the Origin of Spontaneous Symmetry Breaking, *Phys.Rev.* **D7**, pp. 1888–1910, doi: 10.1103/PhysRevD.7.1888.

Coleman, S. R., Wess, J. and Zumino, B. (1969). Structure of phenomenological Lagrangians. 1. *Phys.Rev.* **177**, pp. 2239–2247, doi:10.1103/PhysRev.177.2239.

Collins, J. C. and Soper, D. E. (1982). Parton Distribution and Decay Functions, *Nucl. Phys.* **B194**, p. 445, doi:10.1016/0550-3213(82)90021-9.

Dokshitzer, Y. L. (1977). Calculation of the Structure Functions for Deep Inelastic Scattering and e+ e- Annihilation by Perturbation Theory in Quantum Chromodynamics. *Sov. Phys. JETP* **46**, pp. 641–653, [Zh. Eksp. Teor. Fiz.73,1216(1977)].

Donoghue, J., Golowich, E. and Holstein, B. R. (1992). Dynamics of the Standard Model, *Camb.Monogr.Part.Phys.Nucl.Phys.Cosmol.* **2**, pp. 1–540.

Donoghue, J. F. (1994). General relativity as an effective field theory: The leading quantum corrections, *Phys.Rev.* **D50**, pp. 3874–3888, doi:10.1103/PhysRevD.50.3874.

Donoghue, J. F., "Introduction to the effective field theory description of gravity," gr-qc/9512024.

Donoghue J. F. and Holstein, B. R., "Low Energy Theorems of Quantum Gravity from Effective Field Theory," *J. Phys. G* **42**, No. 10, 103102 (2015).

Duehrssen-Debling, M., Mendes, A. T., Falkowski, A. and Isidori, G. (2015). Higgs Basis: Proposal for an EFT basis choice for LHC HXSWG, .

Eichten, E. and Feinberg, F. (1981). Spin Dependent Forces in QCD, *Phys.Rev.* **D23**, p. 2724, doi:10.1103/PhysRevD.23.2724.

Eichten, E. and Hill, B. R. (1990). An Effective Field Theory for the Calculation of Matrix Elements Involving Heavy Quarks, *Phys.Lett.* **B234**, p. 511, doi:10.1016/0370-2693(90)92049-O.

Einstein, A., Infeld, L. and Hoffmann, B. (1938). The Gravitational equations and the problem of motion, *Annals Math.* **39**, pp. 65–100, doi:10.2307/1968714.

Englert, F. and Brout, R. (1964). Broken Symmetry and the Mass of Gauge Vector Mesons, *Phys.Rev.Lett.* **13**, pp. 321–323, doi:10.1103/PhysRevLett.13.321.

Falk, A. F. and Grinstein, B. (1990). Power corrections to leading logs and their application to heavy quark decays, *Phys.Lett.* **B247**, pp. 406–411, doi:10.1016/0370-2693(90)90917-U.

Fischler, W. (1977). Quark - anti-Quark Potential in QCD, *Nucl.Phys.* **B129**, pp. 157–174, doi:10.1016/0550-3213(77)90026-8.

Fleming, S., Kusunoki, M., Mehen, T. and van Kolck, U. (2007). Pion interactions in the $X(3872)$, *Phys.Rev.* **D76**, p. 034006, doi:10.1103/PhysRevD.76.034006.

Fleming, S. and Maksymyk, I. (1996). Hadronic ψ production calculated in the NRQCD factorization formalism, *Phys.Rev.* **D54**, pp. 3608–3618, doi:10.1103/PhysRevD.54.3608.

Freedman, S. M. and Luke, M. (2012). SCET, QCD and Wilson Lines, *Phys. Rev.* **D85**, p. 014003, doi:10.1103/PhysRevD.85.014003.

Fujikawa, K. (1979). Path Integral Measure for Gauge Invariant Fermion Theories, *Phys.Rev.Lett.* **42**, p. 1195, doi:10.1103/PhysRevLett.42.1195.

Gabbiani, F., Onishchenko, A. I. and Petrov, A. A. (2004). Spectator effects and lifetimes of heavy hadrons, *Phys.Rev.* **D70**, p. 094031, doi:10.1103/PhysRevD.70.094031.

Gasser, J., Sainio, M. and Svarc, A. (1988). Nucleons with Chiral Loops, *Nucl.Phys.* **B307**, p. 779, doi:10.1016/0550-3213(88)90108-3.

Georgi, H. (1984). *Weak Interactions and Modern Particle Theory* (Benjamin/Cummings Pub. Co.), ISBN 0805331638, 9780805331639, 165 pages.

Georgi, H. (1990). An Effective Field Theory for Heavy Quarks at Low-energies, *Phys.Lett.* **B240**, pp. 447–450, doi:10.1016/0370-2693(90)91128-X.

Georgi, H. (1999). *Lie Algebras in Particle Physics* (Westview Press) ISBN-10: 0738202339, 344 pages.

Giudice, G. F., Grojean, C., Pomarol, A. and Rattazzi, R. (2007). The Strongly-Interacting Light Higgs, *JHEP* **06**, p. 045, doi:10.1088/1126-6708/2007/06/045.

Goldberger, W. D. (2007). Les Houches lectures on effective field theories and gravitational radiation, in *Les Houches Summer School - Session 86: Particle Physics and Cosmology: The Fabric of Spacetime Les Houches, France, July 31-August 25, 2006*.

Goldberger, W. D. and Rothstein, I. Z. (2006a). An Effective field theory of gravity for extended objects, *Phys. Rev.* **D73**, p. 104029, doi:10.1103/PhysRevD.73.104029.

Goldberger, W. D. and Rothstein, I. Z. (2006b). Dissipative effects in the worldline approach to black hole dynamics, *Phys. Rev.* **D73**, p. 104030, doi:10.1103/PhysRevD.73.104030.

Goodman, J., Ibe, M., Rajaraman, A., Shepherd, W., Tait, T. M. P. and Yu, H.-B. (2010). Constraints on Dark Matter from Colliders, *Phys. Rev.* **D82**, p. 116010, doi:10.1103/PhysRevD.82.116010.

Gribov, V. N. and Lipatov, L. N. (1972). Deep inelastic e p scattering in perturbation theory, *Sov. J. Nucl. Phys.* **15**, pp. 438–450, [Yad. Fiz.15,781(1972)].

Grinstein, B., "An Introduction to heavy mesons," hep-ph/9508227.

Grinstein, B., Jenkins, E. E., Manohar, A. V., Savage, M. J. and Wise, M. B. (1992). Chiral perturbation theory for f_{D_s}/f_D and B_{B_s}/B_B, *Nucl.Phys.* **B380**, pp. 369–376, doi:10.1016/0550-3213(92)90248-A.

Grozin, A., *Lectures on QED and QCD*, in "Grozin, Andrey: Lectures on QED and QCD", 1–156, [hep-ph/0508242].

Grzadkowski, B., Iskrzynski, M., Misiak, M. and Rosiek, J. (2010). Dimension-Six Terms in the Standard Model Lagrangian, *JHEP* **10**, p. 085, doi:10.1007/JHEP10(2010)085.

Gupta, S. N. (1954). Gravitation and Electromagnetism, *Phys.Rev.* **96**, pp. 1683–1685, doi:10.1103/PhysRev.96.1683.

Guralnik, G., Hagen, C. and Kibble, T. (1964). Global Conservation Laws and Massless Particles, *Phys.Rev.Lett.* **13**, pp. 585–587, doi:10.1103/PhysRevLett.13.585.

Heinonen, J., Hill, R. J. and Solon, M. P. (2012). Lorentz invariance in heavy particle effective theories, *Phys.Rev.* **D86**, p. 094020, doi:10.1103/PhysRevD.86.094020.

Higgs, P. W. (1964). Broken Symmetries and the Masses of Gauge Bosons, *Phys.Rev.Lett.* **13**, pp. 508–509, doi:10.1103/PhysRevLett.13.508.

Isgur, N. and Wise, M. B. (1989). Weak Decays of Heavy Mesons in the Static Quark Approximation, *Phys.Lett.* **B232**, pp. 113–117, doi:10.1016/0370-2693(89)90566-2.

Itzykson, C. and Zuber J. (2006). *Quantum Field Theory* (Dover Publications), ISBN-10: 0486445682, 752 pages.

Jenkins, E. E. and Manohar, A. V. (1991). Baryon chiral perturbation theory using a heavy fermion Lagrangian, *Phys.Lett.* **B255**, pp. 558–562, doi:10.1016/0370-2693(91)90266-S.

Jenkins, E. E., Manohar, A. V. and Trott, M. (2013). Renormalization Group Evolution of the Standard Model Dimension Six Operators I: Formalism and lambda Dependence, *JHEP* **1310**, p. 087, doi:10.1007/JHEP10(2013)087.

Ji, X.-D. and Musolf, M. (1991). Subleading logarithmic mass dependence in heavy meson form-factors, *Phys.Lett.* **B257**, pp. 409–413, doi:10.1016/0370-2693(91)91916-J.

Kaplan, D. B. (1995). Effective field theories, in *Beyond the standard model 5. Proceedings, 5th Conference, Balholm, Norway, April 29-May 4, 1997*.

Kaplan, D. B., Savage, M. J. and Wise, M. B. (1996). Nucleon - nucleon scattering from effective field theory, *Nucl.Phys.* **B478**, pp. 629–659, doi:10.1016/0550-3213(96)00357-4.

Kaplan, D. B., Savage, M. J. and Wise, M. B. (1998a). A New expansion for nucleon-nucleon interactions, *Phys.Lett.* **B424**, pp. 390–396, doi:10.1016/S0370-2693(98)00210-X.

Kaplan, D. B., Savage, M. J. and Wise, M. B. (1998b). Two nucleon systems from effective field theory, *Nucl.Phys.* **B534**, pp. 329–355, doi:10.1016/S0550-3213(98)00440-4.

Kronfeld, A. S. (2002). Uses of effective field theory in lattice QCD: Chapter 39 in At the Frontiers of Particle Physics, Handbook of QCD, hep-lat/0205021.

Lees, J. et al. (2013). Measurement of the $D^*(2010)^+$ meson width and the $D^*(2010)^+ - D^0$ mass difference, *Phys.Rev.Lett.* **111**, 11, p. 111801, doi:10.1103/PhysRevLett.111.111801,10.1103/PhysRevLett.111.169902.

Leibbrandt, G. (1975). Introduction to the Technique of Dimensional Regularization, *Rev.Mod.Phys.* **47**, p. 849, doi:10.1103/RevModPhys.47.849.

Lenz, A. and Nierste, U. (2007). Theoretical update of $B_s - \bar{B}_s$ mixing, *JHEP* **0706**, p. 072, doi:10.1088/1126-6708/2007/06/072.

Lepage, G. P., Magnea, L., Nakhleh, C., Magnea, U. and Hornbostel, K. (1992). Improved nonrelativistic QCD for heavy quark physics, *Phys.Rev.* **D46**, pp. 4052–4067, doi:10.1103/PhysRevD.46.4052.

Leutwyler, H. (1991). Chiral effective Lagrangians, *Lect.Notes Phys.* **396**, pp. 97–138, doi:10.1007/3-540-54978-1_8.

Luke, M. E. and Manohar, A. V. (1992). Reparametrization invariance constraints on heavy particle effective field theories, *Phys.Lett.* **B286**, pp. 348–354, doi:10.1016/0370-2693(92)91786-9.

Luke, M. E. and Manohar, A. V. (1997). Bound states and power counting in effective field theories, *Phys.Rev.* **D55**, pp. 4129–4140, doi:10.1103/PhysRevD.55.4129.

Luke, M. E., Manohar, A. V. and Rothstein, I. Z. (2000). Renormalization group scaling in nonrelativistic QCD, *Phys. Rev.* **D61**, p. 074025, doi:10.1103/PhysRevD.61.074025.

Mannel, T., Roberts, W. and Ryzak, Z. (1991). Baryons in the heavy quark effective theory, *Nucl. Phys.* **B355**, pp. 38–53, doi:10.1016/0550-3213(91)90301-D.

Mannel, T., Roberts, W. and Ryzak, Z. (1992). A Derivation of the heavy quark effective Lagrangian from QCD, *Nucl. Phys.* **B368**, pp. 204–220, doi:10.1016/0550-3213(92)90204-O.

Manohar, A. V. (1997). The HQET / NRQCD Lagrangian to order α/m^3, *Phys.Rev.* **D56**, pp. 230–237, doi:10.1103/PhysRevD.56.230.

Manohar, A. V. (2003). Deep inelastic scattering as $x \to 1$ using soft collinear effective theory, *Phys. Rev.* **D68**, p. 114019, doi:10.1103/PhysRevD.68.114019.

Manohar, A. V., Mehen, T., Pirjol, D. and Stewart, I. W. (2002). Reparameterization invariance for collinear operators, *Phys. Lett.* **B539**, pp. 59–66, doi:10.1016/S0370-2693(02)02029-4.

Manohar, A. V. and Stewart, I. W. (2007). The Zero-Bin and Mode Factorization in Quantum Field Theory, *Phys. Rev.* **D76**, p. 074002, doi:10.1103/PhysRevD.76.074002.

Manohar, A. V. and Wise, M. B. (2000). Heavy Quark Physics, *Camb.Monogr.Part.Phys.Nucl.Phys.Cosmol.* **10**, pp. 1–191.

Mantry, S., Pirjol, D. and Stewart, I. W. (2003). Strong phases and factorization for color suppressed decays, *Phys. Rev.* **D68**, p. 114009, doi:10.1103/PhysRevD.68.114009.

Neubert, M. (1994a). Heavy quark symmetry, *Phys.Rept.* **245**, pp. 259–396, doi:10.1016/0370-1573(94)90091-4.

Neubert, M. (1994b). Short distance expansion of heavy - light currents at order $1/m_Q$, *Phys.Rev.* **D49**, pp. 1542–1550, doi:10.1103/PhysRevD.49.1542.

Neubert, M. and Sachrajda, C. T. (1997). Spectator effects in inclusive decays of beauty hadrons, *Nucl.Phys.* **B483**, pp. 339–370, doi:10.1016/S0550-3213(96)00559-7.

Niedermaier, M. and Reuter, M. (2006). The Asymptotic Safety Scenario in Quantum Gravity, *Living Rev. Rel.* **9**, pp. 5–173, doi:10.12942/lrr-2006-5.

Nielsen, H. B. and Chadha, S. (1976). On How to Count Goldstone Bosons, *Nucl. Phys.* **B105**, p. 445, doi:10.1016/0550-3213(76)90025-0.

Olive, K. *et al.* (2014). Review of Particle Physics, *Chin.Phys.* **C38**, p. 090001, doi:10.1088/1674-1137/38/9/090001.

Parisi, G. (1988). *Statistical Field Theory* (Addison Wesley), ISBN-10: 0201059851, 352 pages.

Paz, G. (2009). Subleading Jet Functions in Inclusive B Decays, *JHEP* **06**, p. 083, doi:10.1088/1126-6708/2009/06/083.

Paz, G. (2010). An Effective Field Theory Look at Deep Inelastic Scattering, *Mod. Phys. Lett.* **A25**, pp. 2039–2049, doi:10.1142/S0217732310033803.

Peskin, M. E. and Schroeder, D. V. (1995). *An Introduction to Quantum Field Theory* (Westview Press), ISBN-10: 0201503972, 864 pages.

Petrov, A. A. and Shepherd, W. (2014). Searching for dark matter at LHC with Mono-Higgs production, *Phys. Lett.* **B730**, pp. 178–183, doi:10.1016/j.physletb.2014.01.051.

Pich, A., "Effective field theory: Course," hep-ph/9806303.

Pineda, A. and Soto, J. (1998). Effective field theory for ultrasoft momenta in NRQCD and NRQED, *Nucl.Phys.Proc.Suppl.* **64**, pp. 428–432, doi:10.1016/S0920-5632(97)01102-X.

Pirjol, D. and Stewart, I. W. (2003). A Complete basis for power suppressed collinear ultrasoft operators, *Phys. Rev.* **D67**, p. 094005, doi:10.1103/PhysRevD.69.019903,10.1103/PhysRevD.67.094005, [Erratum: Phys. Rev.D69,019903(2004)].

Polchinski, J. (1998a). *String theory. Vol. 1: An introduction to the bosonic string* (Cambridge University Press).

Polchinski, J. (1998b). *String theory. Vol. 2: Superstring theory and beyond* (Cambridge University Press).

Porto, R. A. (2006). Post-Newtonian corrections to the motion of spinning bodies in NRGR, *Phys. Rev.* **D73**, p. 104031, doi:10.1103/PhysRevD.73.104031.

Ramond, P. (1981). Field Theory: A Modern Primer, *Front.Phys.* **51**, pp. 1–397.

Rothstein, I. Z. (2014). Progress in effective field theory approach to the binary inspiral problem, *Gen.Rel.Grav.* **46**, p. 1726, doi:10.1007/s10714-014-1726-y.

Rovelli, C. (2011). Zakopane lectures on loop gravity, *PoS* **QGQGS2011**, p. 003.

Ryder, L. H. (1996). *Quantum Field Theory* (Cambridge University Press), ISBN 9780521478144, 9781139632393.

Scherer, S. and Schindler, M. R., "A Chiral perturbation theory primer," hep-ph/0505265.

Shifman, M. A. (1995). Lectures on heavy quarks in quantum chromodynamics. An extended version of the lectures given at Theoretical Advanced Study Institute *QCD and Beyond*, University of Colorado, Boulder, Colorado, June 1995, hep-ph/9510377.

Shifman, M. A. (1999). ITEP lectures on particle physics and field theory. Vol. 1, 2, *World Sci.Lect.Notes Phys.* **62**, pp. 1–875.

Shifman, M. A. and Voloshin, M. (1988). On Production of D and D^* Mesons in B Meson Decays, *Sov.J.Nucl.Phys.* **47**, p. 511.

Slansky, R. (1981). Group Theory for Unified Model Building, *Phys.Rept.* **79**, pp. 1–128, doi:10.1016/0370-1573(81)90092-2.

Stelle, K. S. (1978). Classical Gravity with Higher Derivatives, *Gen. Rel. Grav.* **9**, pp. 353–371, doi:10.1007/BF00760427.

't Hooft, G. (1980). Naturalness, chiral symmetry, and spontaneous chiral symmetry breaking, *NATO Sci. Ser. B* **59**, p. 135.

't Hooft, G. and Veltman, M. J. G. (1974). One loop divergencies in the theory of gravitation, *Annales Poincare Phys. Theor.* **A20**, pp. 69–94.

Trott, M. (2007). Jets in Effective Theory: Summing Phase Space Logs, *Phys. Rev.* **D75**, p. 054011, doi:10.1103/PhysRevD.75.054011.

Visser, M. (2002). Sakharov's induced gravity: A Modern perspective, *Mod.Phys.Lett.* **A17**, pp. 977–992, doi:10.1142/S0217732302006886.

Weinberg, S. (1972). *Gravitation and Cosmology: Principles and Applications of the General Theory of Relativity* (Cambridge University Press), ISBN-10: 0471925675, 657 pages.

Weinberg, S. (1979a). Baryon and Lepton Nonconserving Processes, *Phys. Rev. Lett.* **43**, pp. 1566–1570, doi:10.1103/PhysRevLett.43.1566.

Weinberg, S. (1979b). Phenomenological Lagrangians, *Physica* **A96**, p. 327.

Weinberg, S. (1980). Ultraviolet divergences in quantum theories of gravitation, in *General Relativity: An Einstein Centenary Survey*, pp. 790–831.

Weinberg, S. (1986). Superconductivity for Particular Theorists, *Prog. Theor. Phys. Suppl.* **86**, p. 43, doi:10.1143/PTPS.86.43.

Weinberg, S. (1990). Nuclear forces from chiral Lagrangians, *Phys.Lett.* **B251**, pp. 288–292, doi:10.1016/0370-2693(90)90938-3.

Weinberg, S. (1996). *The Quantum Theory of Fields. Vol. 2: Modern Applications* (Cambridge University Press), ISBN-10: 0521670543, 489 pages.

Wise, M. B. (1992). Chiral perturbation theory for hadrons containing a heavy quark, *Phys.Rev.* **D45**, pp. 2188–2191, doi:10.1103/PhysRevD.45.R2188.

Zhang, C. and Willenbrock, S. (2011). Effective-Field-Theory Approach to Top-Quark Production and Decay, *Phys. Rev.* **D83**, p. 034006, doi:10.1103/PhysRevD.83.034006.

Zinn-Justin, J. (2002). *Quantum field theory and critical phenomena* (International Series of Monographs on Physics, Vol. 113), ISBN-10: 0198509235, 1054 pages.

Zwiebach, B. (2004). *A first course in string theory* (Cambridge University Press), ISBN 0521880327, 558 pages.

Index

anomalies, 32–38
  't Hooft anomaly matching, 37, 38
  chiral anomaly, 35
  gauge anomalies, 36
  path integral derivation, 32, 34
anomalous dimension
  heavy-to-light current in HQET, 128
  QCD corrections to weak interactions, 84
Appelquist-Carrazonne decoupling theorem, 60–63

Callan-Coleman-Wess-Zumino construction, 27–30, 39
Callan-Symanzik equation, 49–52, 54, 55
  anomalous dimension
    critical scaling exponent η, 55
    definition, 49
  beta function
    ϕ^4 theory, 47
    fixed points, 52
    running couplings, 48, 52
chiral perturbation theory, 88
  baryons, 103
    heavy particle formalism, 139
  chiral symmetry breaking scale, 101
  Goldberger-Treiman relation, 103
  heavy meson molecules, 164
  heavy mesons, 133
  light mesons, 99
  power counting, 99
  sources, 93
chiral symmetry, 89, 277

dark matter, 212
dimensional transmutation, 48

Euler-Heisenberg Lagrangian, 70
examples
  blue sky, 64
  electromagnetic and hadronic decays of quarkonia, 155
  Gell-Mann-Okubo relation, 97
  gravity near Earth, 4, 5
  heavy quarkonium potential, 161
  hydrogen spectroscopy, 44–46, 64
  Newtonian potential, 233
extra dimensions, 250

Feynman rules
  heavy meson chiral PT, 135
  heavy quark effective theory, 114
  QCD, 272
  QED, 270
  quantum gravity, 229

gauge symmetries, 13–15
  QCD, 271
  standard model, 205
general relativity
  background field method, 224, 225

Christoffel symbols, 222
covariant derivative, 221
effective action, 227, 228
Einstein-Hilbert action, 224
energy-momentum tensor, 226, 227
  scalar fields, 226
general coordinate transformations, 223
geodesic, 221, 222
harmonic gauge, 229
Riemann-Christoffel curvature tensor, 223, 224
Weyl curvature tensor, 224
GIM mechanism, 75
Goldstone bosons, see Nambu-Goldstone bosons
Goldstone's theorem, 17, 18
group theory
  cosets, 28, 254
  defintion, 253
  group action, 259
  Lie group
    anomaly coefficient, 266, 267
    Cartan generators, 258
    definition, 255, 256
    Dynkin index, 257, 266, 267
    families, 258
    Lie algebra, 257, 258
    quadratic Casimir, 266, 267
    rank of a Lie group, 258
    representations, 258–262
    structure constants, 257
    Young tableaux, 262–265
  normal subgroups, 254, 255

heavy quark effective theory, 108
  Lagrangian, 113, 114
  projection operators, 111
  RPI, 117
heavy quark form factor, 189
Heisenberg ferromagnet, 19

irrelevant operator, 43
Isgur-Wise function, 143

jet physics, 197, 198

event shapes, 198
Monte Carlo simulation, 198, 201

Lagrangian
  QCD, 272
  QED, 270
  standard model, 205
Landau pole, 48, 57
lepton number violation, 206

Majorana mass, 207
marginal operator, 43
matching
  at one loop in QCD, 79
  general procedure, 57
  heavy-to-light current in HQET, 123
  NRQCD, 148
mean field theory
  comparison to EFT, 3
  order parameter action, 16
minimal subtraction, 47, 59, 60, 122, 232
  decoupling theorem, 62, 63
models
  Fermi of weak interactions, 71, 73
  Yukawa, 65

naive dimensional analysis, 101
Nambu-Goldstone bosons, 90
  definition, 17
  interactions with heavy mesons, 133
  magnons, 19, 20
  Nielsen-Chadha sum rule, 19, 20, 38
  QCD, 90
Noether current
  $U(1)$ scalar field, 12
  derivation, 10
  gauge symmetry, 14
  surface terms, 11
Noether's procedure, 14, 15
Noether's theorem, 9–11
non-relativistic QCD
  factorization theorem, 152

matching, 148
potential NRQCD, 160
power counting, 152
nonrelativistic general relativity, 235
  action, 237, 241, 243
  counterterms, 245–247
  deriving Newton, 242
  EIH Potential, 242, 248
  gravitational radiation, 242–245
  gravito-electric fields, 238
  gravito-magnetic fields, 238
  LIGO, 235
  multipole expansion, 240, 241
  post-Newtonian approximation, 235
  potential gravitons, 238, 239
  radiation gravitons, 238, 239
  spin degree of freedom, 237
  velocity power counting, 235, 239

parton distribution functions, 197
penguin operators, 86

quantum gravity
  deriving Newton, 233, 234
  Feynman rules, 229–231
  graviton propagator, 230
  one-loop corrections, 232

relevant operator, 43
renormalization group improvement
  HQET, 127
  weak interactions, 83

scaling dimensions, 41–43, 53, 54
small velocity (Shifman-Voloshin) limit, 144
soft collinear effective theory
  $B \to D\pi$, 188, 189
  $B \to X_s\gamma$, 171–174, 185–187
  collinear Lagrangian, 181
  collinear scaling, 176
  collinear Wilson lines, 180
  collinear-ultrasoft couplings, 181, 182
  deep inelastic scattering, 191–197
  factorized gauge invariance, 182–184
  jet anomalous dimension, 195
  jet function, 187
  jets, see jet physics
  offshellness, 176
  projectors, 179
  role of spectator, 188
  RPI, 184, 185
  $SCET_{II}$, 200
  ultrasoft scaling, 177
  ultrasoft Wilson lines, 181
spontaneous symmetry breaking
  BEHGHK, see Higgs mechanism
  definition, 16, 17
  Higgs mechanism, 20–25
    history, 20, 22, 23
    standard model, 206
spurion analysis
  definition, 30–32
superconductivity, 25–27, 252
supersymmetry, 249

technicolor, 251
Type-I EFT
  chiral perturbation theory, 88
  definition, 7
  Euler-Heisenberg Lagrangian, 70
  Fermi model, 71
  standard model, 205
Type-II EFT
  definition, 7
  heavy quark effective theory, 108
  non-relativistic QCD, 147
Type-III EFT
  definition, 8
  soft-collinear effective theory, 181

UV completion, 67

Veltman's formula, 284

Weinberg operator, 207
Weinberg's theorem, 6, 7